Springer Texts in Statistics

Advisors:
George Casella   Stephen Fienberg   Ingram Olkin

Springer

New York
Berlin
Heidelberg
Barcelona
Hong Kong
London
Milan
Paris
Singapore
Tokyo

Springer Texts in Statistics

(continued after index)

David Edwards

Introduction to
Graphical Modelling

Second Edition

With 83 Illustrations

 Springer

David Edwards
Statistics Department
Novo Nordisk A/S
DK-2880 Bagsvaerd
Denmark
ded@novo.dk

Library of Congress Cataloging-in-Publication Data

Edwards, David, 1949-
    Introduction to graphical modelling / David Edwards.-- 2nd ed.
      p. cm. -- (Springer texts in statistics)
    Includes bibliographical references and index.
    ISBN 0-387-95054-0 (alk. paper)
     1. Graphical modeling (Statistics) I. Title. II. Series.

  QA279 .E34 2000
  519.5'38--dc21

00-030760

Printed on acid-free paper.

Production managed by A. Orrantia; manufacturing supervised by Jeff Taub.
Photocomposed copy prepared from the authors' LaTeX files and formatted by The Bartlett Press, Marietta, GA.
Printed and bound by Hamilton Printing Co., Rensselaer, NY.
Printed in the United States of America.

9 8 7 6 5 4 3 2 1

ISBN 0-387-95054-0 Springer-Verlag New York Berlin Heidelberg   SPIN 10769313

I would rather discover a single causal relationship than be king of Persia.

Democritus.

Preface to the Second Edition

In the five years since the first edition of this book was published, the study of graphical models and their application has picked up momentum. New types of graphs have been introduced, so as to capture different types of dependence structure. Application of the methodology to what used to be called expert systems, but now are often called probabilistic networks, has grown explosively. Another active area of study has been in the ways directed acyclic graphs may contribute to causal inference.

To address some of these new developments, two topics have been extended in this edition, each now being given a whole chapter (arguably, each deserves a whole book). Chapter 7 describes the use of directed graphs of various types, and Chapter 8 surveys some work on causal inference, with particular reference to graphical modelling. I have not attempted a description of probabilistic networks, for which many excellent texts are available, for example Cowell et al. (1999).

In addition to the new chapters, there are some lesser additions and revisions: the treatment of mean linearity and CG-regression models has been expanded, the description of MIM has been updated, and an appendix describing various estimation algorithms has been added. A diskette with the program is not included with the book, as it was with the previous edition: instead it can be downloaded from the internet.

I am grateful to Don Rubin, Judea Pearl, Mervi Eerola, Thomas Richardson, and Vanessa Didelez for constructive comments to Chapter 8, and to Jan Koster and Elena Stanghellini for helpful advice.

March 20, 2000   David Edwards

Preface to the First Edition

Graphical modelling is a form of multivariate analysis that uses graphs to represent models. Although its roots can be traced back to path analysis (Wright, 1921) and statistical physics (Gibbs, 1902), its modern form is of recent origin. Key papers in the modern development include Darroch, Lauritzen, and Speed (1980), and Lauritzen and Wermuth (1989).

The purpose of this book is to provide a concise, application-oriented introduction to graphical modelling. The theoretical coverage is informal, and should be supplemented by other sources: the book by Whittaker (1990) would be a natural choice. Readers primarily interested in discrete data should consult the introductory-level book by Christensen (1990). Lauritzen (1992) provides a mathematically rigorous treatment: this is the source to consult about results stated here without proof.

Applications of graphical modelling in a wide variety of areas are shown. These analyses make use of MIM, a command-driven PC-program designed for graphical modelling. A student version of MIM is included with the book, and a reference guide is included as an appendix.

My interest in graphical modelling started in 1978–1980 under the influence of Terry Speed, who held a seminal lecture course on the topic in Copenhagen in 1978. Subsequent participation in a study group on graphical modelling, together with Steffen Lauritzen, Svend Kreiner, Morten Frydenberg, Jens Henrik Badsberg, and Poul Svante Eriksen, has served to stimulate and broaden my interest in the topic.

The first version of MIM was written in 1986–1987 at the Statistical Research Unit, Copenhagen University, when I was supported by a Danish

Social Science Research Council grant. I wish to thank all the people who helped in the development of MIM, including: Brian Murphy, for assistance and inspiration, and for kindly supplying the program LOLITA, which served as a starting point for MIM; Morten Frydenberg, for crucial assistance with the modified iterative proportional scaling (MIPS) algorithm; Steffen Lauritzen, Nanny Wermuth, Hanns-Georg Leimer, Svend Kreiner, Jens Henrik Badsberg, and Joe Whittaker for encouraging help; Brian Francis and Joe Whittaker for contributing the SUSPEND and RETURN commands; Egon Hansen for help with the interactive graphics; Tue Tjur for generously letting me use his module for distribution functions; Svend Kreiner for helping me code Patefield's (1981) algorithm; and Marta Horáková for programming the explicit estimation and EH-selection procedures.

Finally, thanks are also due to Peter Smith, Philip Hougaard, and Helle Lynggaard for helpful suggestions.

January 28, 1995   David Edwards
                                                   Roskilde

Contents

1

Preliminaries

This chapter introduces some of the theory behind graphical modelling. The basic concepts of independence and conditional independence are reviewed, and an explanation of how conditional independence structures can be represented graphically is given. A brief discussion of Simpson's paradox is given to further motivate the graphical modelling approach. The final section gives an overview of the book.

1.1 Independence and Conditional Independence

The concept of independence is fundamental to probability and statistics theory. Two events A and B are said to be independent if

$$\Pr(A \cap B) = \Pr(A) \Pr(B)$$

or, equivalently, if

$$\Pr(A|B) = \Pr(A).$$

In this book we distinguish between two types of variables: *continuous* variables, whose values lie in the real line $\Re$, and *discrete* variables (often called *factors*), which can take values from a finite set. For convenience, we label the values in this finite set as $\{1, 2, \ldots, \#X\}$, where $\#X$ is the number of *levels* of X.

When X is a random variable we write its density or mass function as $f_X(x)$. If X is discrete, we may also write this as $\Pr(X = j)$, for level $j \in \{1, 2, \ldots, \#X\}$.

Two random variables X and Y are said to be *independent* if their joint density factorizes into the product of their marginal densities:

$$f_{X,Y}(x,y) = f_X(x)f_Y(y),$$

or, equivalently, if the conditional density of, say, Y given $X = x$ is not a function of x, which we can write as

$$f_{Y|X}(y|x) = f_Y(y).$$

The advantage of this characterization is that it does not involve the density of X. For example, if I is a fixed grouping factor, and Y is a response, then it is natural to examine the densities $f_{Y|I}(y|i)$ for each level i of I. If these are constant over i, then we call this *homogeneity* rather than independence. Similarly, if X is a fixed continuous variable, then we will consider the conditional distributions $f_{Y|X}(y|x)$, for $x \in \Re$. Here, since x may take an infinity of values, it is necessary to adopt more specific parametric models— for example, the simple linear regression model

$$f_{Y|X}(y|x) \sim \mathcal{N}(a + bx, \sigma^2),$$

where a, b, and σ are unknown parameters. When $b = 0$, $f_{Y|X}(y|x)$ is not a function of x, and we have the same situation as before. However, we do not usually refer to this as homogeneity, but instead may use the expression *zero regression coefficients* or something similar.

We now turn to the concept of *conditional independence*, which is of central importance in graphical modelling. Consider now three random variables, X, Y, and Z. If, for each value z, X and Y are independent in the conditional distribution given $Z = z$, then we say that X and Y are conditionally independent given Z, and we write this as $X \perp\!\!\!\perp Y \mid Z$. This notation is due to Dawid (1979), who discusses alternative characterizations of the property. One of these is that

$$f_{X|Y,Z}(x|y,z)$$

does not depend on y. This is also appropriate when Y and/or Z are fixed.

As illustration, consider some data from a study of health and social charac-teristics of Danish 70-year-olds. Representative samples were taken in 1967 and again — on a new cohort of 70-year-olds — in 1984 (Schultz-Larsen et al., 1992). Body mass index (BMI) is a simple measure of obesity, de-fined as weight/height^2. It is of interest to compare the distribution between males and females, and between the two years of sampling.

Figures 1.1–1.4 show histograms of BMI in kg/m^2, broken down by gender and year. We write the true, unknown densities as

$$f_{B|G,Y}(b|G = i, Y = j) = f_{ij},$$

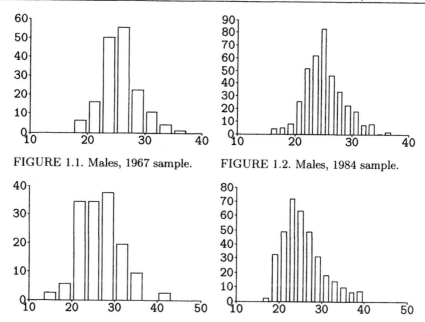

FIGURE 1.1. Males, 1967 sample.　　FIGURE 1.2. Males, 1984 sample.

FIGURE 1.3. Females, 1967 sample.　　FIGURE 1.4. Females, 1984 sample.

say, where $i = 1, 2$ (corresponding to male and female, respectively), and $j = 1, 2$ (corresponding to 1967 and 1984, respectively).

If the two sexes have differing distributions of BMI, but there has been no change in these from 1967 to 1984, so that $f_{11} = f_{12}$ and $f_{21} = f_{22}$, then this is equivalent to

BMI ⊥⊥ Year | Gender.

If the two sexes have the same distribution of BMI, but this changes from 1967 to 1984, so that $f_{11} = f_{21}$ and $f_{12} = f_{22}$, then this is equivalent to

BMI ⊥⊥ Gender | Year.

If the distribution of BMI is the same over year and sex, so that

$$f_{11} = f_{12} = f_{21} = f_{22},$$

then this is equivalent to

BMI ⊥⊥ (Gender, Year).

Another characterization of $X \perp\!\!\!\perp Y \mid Z$ discussed by Dawid (1979) is that the joint density $f_{X,Y,Z}(x, y, x)$ can be factorized into the product of two factors, one not involving x and the other not involving y, i.e., that

$$f_{X,Y,Z}(x, y, z) = h(x, z)k(y, z), \tag{1.1}$$

where h and k are some functions. We use this repeatedly below.

It is often helpful to regard conditional independence as expressing the notion of *irrelevance*, in the sense that we can interpret the statement $X \perp\!\!\!\perp Y | Z$ as saying something like:

If we know Z, information about Y is irrelevant for knowledge of X.

This formulation can be helpful, for example, when eliciting substantive assumptions from subject-matter specialists, or when using graphs to communicate the conclusions of an analysis.

1.2   Undirected Graphs

As we describe in the next section, the key tool in graphical modelling is the independence graph of a model. Before we do this, we briefly introduce some graph-theoretic terms that will be useful later. The definitions are collected here for convenience — they do not need to be absorbed at first reading.

A graph, $\mathcal{G} = (\mathcal{V}, \mathcal{E})$, is a structure consisting of a finite set $\mathcal{V}$ of *vertices* (also called nodes) and a finite set $\mathcal{E}$ of *edges* (also called arcs) between these vertices. We write vertices using roman letters— X, y, v and so on. In our context, they correspond to the variables in the model. We write an edge as $[XY]$ or equivalently as $[YX]$. In many of the graphs we consider, each pair of vertices can have either no or one edge between them, and the edges are undirected. We represent a graph in a diagram, such as that in Figure 1.6.

This graph is called *undirected*, since all the edges are undirected. We study other types of graphs in Chapter 7.

The vertices are drawn as dots or circles: *dots* represent *discrete* variables, and *circles* represent *continuous* variables. Edges are drawn as straight lines between the vertices. Clearly a given graph can be drawn in an infinite number of ways; this does not affect its essential nature, which is just defined through the vertex set $\mathcal{V}$ and the edge set $\mathcal{E}$.

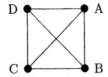

FIGURE 1.5. A complete graph.

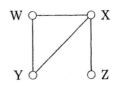

FIGURE 1.6. An incomplete graph.

We say that two vertices $X, Y \in \mathcal{V}$ are *adjacent*, written $X \sim Y$, if there is an edge between them, i.e., $[XY] \in \mathcal{E}$. For example, in Figure 1.6, X and Y are adjacent but Y and Z are not.

We call a graph *complete* if there is an edge between every pair of vertices. For example, the graph in Figure 1.5 is complete.

Any subset $u \subseteq \mathcal{V}$ induces a *subgraph* of $\mathcal{G}$. This is the graph $\mathcal{G}_u = (u, \mathcal{F})$ whose edge set $\mathcal{F}$ consists of those edges in $\mathcal{E}$ where both endpoints are in u. A subset $u \subseteq V$ is called *complete* if it induces a complete subgraph. In other words, if all the vertices in u are mutually adjacent, then it is complete. In Figure 1.6, the subset $\{X, Y, W\}$ is complete.

A subset $u \subseteq \mathcal{V}$ is called a *clique* if it is maximally complete, i.e., u is complete, and if $u \subset w$, then w is not complete. The concept of a clique is important in graphical modelling, and often one needs to identify the cliques of a given graph. For example, the cliques of the graph shown in Figure 1.6 are $\{X, Y, W\}$ and $\{X, Z\}$.

A sequence of vertices $X_0, \ldots X_n$ such that $X_{i-1} \sim X_i$ for $i = 1, \ldots n$ is called a *path* between X_0 and X_n of length n. For example, in Figure 1.6, Z, X, Y, W is a path of length 3 between Z and W. Similarly, Z, X, W is a path of length 2 between Z and W. A graph is said to be *connected* if there is a path between every pair of vertices.

A path $X_1, X_2, \ldots X_n, X_1$ is called an *n-cycle*, or a cycle of length n. For example, in Figure 1.5, A, B, D, C, A is a 4-cycle.

If the n vertices $X_1, X_2, \ldots X_n$ of an n-cycle $X_1, X_2, \ldots X_n, X_1$ are distinct, and if $X_j \sim X_k$ only if $|j - k| = 1$ or $n - 1$, then we call it a *chordless cycle*. Figure 1.7 shows two graphs, each of which contains a chordless 4-cycle. (The chordless 4-cycle in the first graph can be difficult to spot).

We call a graph *triangulated* if it has no chordless cycles of length greater than or equal to four. For example, if the edge $[AC]$ was included in Figure 1.7 (left), the resulting graph would be triangulated. The triangulated property turns out to be closely related to the existence of closed-form maximum likelihood estimates, as we see later.

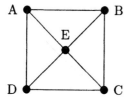
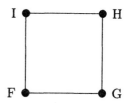

FIGURE 1.7. Two graphs with chordless 4-cycles.

For three subsets a, b, and s of $\mathcal{V}$, we say s *separates* a and b if all paths from a to b intersect s. For example, in Figure 1.6, $\{X\}$ separates $\{Y, W\}$ and $\{Z\}$.

Finally, we define the concept of a *boundary*. The boundary of a subset $u \subseteq \mathcal{V}$, written $\mathrm{bd}(u)$, is defined as those vertices in $\mathcal{V} \setminus u$ that are adjacent to a vertex in u, i.e., $\mathrm{bd}(u) = \{W \in \mathcal{V} \setminus u : W \sim X, X \in u\}$.

1.3   Data, Models, and Graphs

We are usually concerned with the analysis of datasets of the form

$$
\begin{array}{cccc}
v^{(1)} & w^{(1)} & \cdots & z^{(1)} \\
v^{(2)} & w^{(2)} & \cdots & z^{(2)} \\
\vdots & \vdots & \vdots & \vdots \\
v^{(N)} & w^{(N)} & \cdots & z^{(N)}
\end{array}
$$

that is, consisting of a set of measurements, say V, W, ..., Z, taken on N entities. The measurements are called *variables*, and the entities represent observations, for example, on different objects or at different times.

To model such data, we make a fundamental assumption, namely that the measurements V, W, ..., Z are random variables with a joint probability density function

$$f_\theta(v, w, \ldots, z),$$

say, where θ is some unknown parameter. We conceive of the data as resulting from an experiment in which the N observations are sampled independently from f_θ. We base our inference on statements about the unknown parameter θ. Usually this inference rests more or less explicitly on the idea of hypothetical replications of the experiment: this underlies the frequentist framework.

Sometimes we may regard the experiment slightly differently. We may fix some of the variables, so that their observed values become, as it were, a constant part of the experiment, and we regard the remaining variables as random variables drawn from a conditional density given the fixed variables. Thus, in any hypothetical replications, the fixed values are held constant.

In the statistical sense, a model is a family of possible densities $\{f_\theta : \theta \in \Theta\}$, say. Submodels correspond to parameter subsets; for example, if $\Theta_0 \subseteq \Theta$, then $\mathcal{M}_0 = \{f_\theta : \theta \in \Theta_0\}$ is a submodel of $\mathcal{M} = \{f_\theta : \theta \in \Theta\}$. Often we suppress the parameters and write simply $f \in \mathcal{M}$ for a density f in a model $\mathcal{M}$.

Although we base inference about the data on statements about parameters, it should be borne in mind that rarely is this the ultimate purpose of the analysis. Broadly speaking, statistical techniques have two main purposes: explanation or prediction. For example, studies may be set up with the broad goal to increase insight or understanding about a problem, without a more specific practical purpose. These stand in contrast to, for example, meteorological studies whose objective is to predict future weather, or economic studies for which the goal is to predict next year's gross national product.

How do models relate to graphs? In graphical modelling, the focus is on models under which some conditional independence relations of the form $X \perp\!\!\!\perp Y \mid$ (some other variables) hold for all densities in the model. In the first part of the book, we focus on models for which these relations take the form $X \perp\!\!\!\perp Y \mid$ (the rest), where by "the rest" is meant all the other variables in the model. For such a model, we can construct an undirected graph $(\mathcal{V}, \mathcal{E})$ where $\mathcal{V}$ is the set of variables in the model, and $\mathcal{E}$ consists of edges between variable pairs that are *not* conditionally independent given the rest. In other words, for all pairs $\{X, Y\}$ such that $X \perp\!\!\!\perp Y \mid$ (the rest), the edge between X and Y is omitted; for all other pairs, an edge is drawn between them. Thus, from the resulting graph, we can immediately see that if two variables are not adjacent, then they are conditionally independent given the rest. This is known as the *pairwise Markov property for undirected graphs*.

The key to interpreting such graphs is the *global Markov property for undirected graphs*, which states:

If two sets of variables u and v are separated by a third set of variables w, then $u \perp\!\!\!\perp v \mid w$.

For example, we may have a model for four variables, W, X, Y, and Z, for which we know that $W \perp\!\!\!\perp Z \mid (X, Y)$ and $Y \perp\!\!\!\perp Z \mid (W, X)$. So the edges $[WZ]$ and $[YZ]$ must be absent from the graph on $\{W, X, Y, Z\}$. We obtain the graph shown in Figure 1.6, from which we can infer that $W \perp\!\!\!\perp Z \mid X$ and $Y \perp\!\!\!\perp Z \mid X$. The global Markov property allows a simple translation from a graph-theoretic property, separation, to a statistical property, conditional independence.

Intermediate between the pairwise and global Markov properties is the *local Markov property for undirected graphs*, which states that each variable X is conditionally independent of its non-neighbours given its neighbours. More formally, it states that for all $X \in \mathcal{V}$,

$$X \perp\!\!\!\perp \mathcal{V} \setminus \{X \cup \mathrm{bd}(X)\} \mid \mathrm{bd}(X).$$

The pairwise, local, and global Markov properties are equivalent under quite general conditions. By this is meant that if all the pairwise conditional independences corresponding to a graph $\mathcal{G}$ hold for a given distribution, then all the global conditional independences corresponding to $\mathcal{G}$ also hold for that distribution (and vice versa). Specifically, Pearl and Paz (1986) showed that if the joint density of the variables with respect to a product measure is everywhere strictly positive, then the pairwise, local, and global Markov properties for undirected graphs are equivalent (see also Lauritzen, 1996).

The converse to the global Markov property is also true in the following sense: if for any sets $u, v, w \subseteq \mathcal{V}$, $u \perp\!\!\!\perp v \mid w$ holds under all densities in the model, then w separates u and v in the independence graph. (There may always be specific densities that obey additional conditional independences not implied by the graph). Thus, all marginal and conditional independences that hold for all densities in the model can be read directly off the graph. This property has been called *Markov perfection* (see Frydenberg, 1990, and Geiger and Pearl, 1993).

1.4 Simpson's Paradox

The inadequacy of studying only pairwise associations between variables is brought out by the phenomenon known as Simpson's paradox (Simpson, 1951). To illustrate this, suppose possible sexual discrimination in patterns of university admissions is under study. Table 1.1 shows some (fictive) data for a given university. There seems to be discrimination against female applicants, since only 41.6% of these are admitted as opposed to 58.3% males.

Suppose, however, that the figures are further broken down by university department, as shown in Table 1.2.

The departments differ greatly in their admission rates, and males tend to apply to the departments with high rates of admission, but there is no evidence of sexual discrimination with respect to intradepartmental admission rates. How do we express this in statistical terms? From Table 1.1, we see that Sex (S) and Admission (A) are marginally associated; in other words, $S \not\perp\!\!\!\perp A$. Table 1.2, on the other hand, indicates that Admission is

Sex	no. applying	no. admitted	percent admitted
Male	600	350	58.3
Female	600	250	41.6

TABLE 1.1. Applicants cross-classified by sex and admission.

Dept.	Sex	no. applying	no. admitted	percent admitted
I	Male	100	25	25
	Female	300	75	25
II	Male	200	100	50
	Female	200	100	50
III	Male	300	225	75
	Female	100	75	75

TABLE 1.2. Applicants cross-classified by sex, department, and admission.

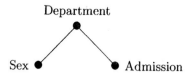

FIGURE 1.8. College admissions.

independent of Sex given Department (D), i.e., $S \perp\!\!\!\perp A \mid D$. We draw this graphically in Figure 1.8.

We call the association between Sex and Admission *spurious* because in some sense it is due to Department. We want to study whether there is a *direct* effect of Sex on Admission, after taking into account this *indirect* effect via Department. To do this, we must study the conditional association between Sex and Admission given Department, not their marginal association.

In this example, there is marginal association but no conditional association. Strictly speaking, Simpson's paradox refers to a reversal in the direction of association between the marginal and conditional distributions. To illustrate this, consider the Florida murder data (Radelet, 1981; Agresti, 1990). The numbers of murderers convicted and tried in Florida from 1976 to 1977, cross-classified by colour of murderer and sentence, are shown in Table 1.3.

The proportion of black murderers sentenced to death is about the same as that of white murderers: in fact, the latter is slightly higher. If the numbers are further broken down by colour of victim, however, Table 1.4 is obtained.

Colour of Murderer	Sentence		
	Death	Other	%
Black	17	149	10.2
White	19	141	11.8

TABLE 1.3. Murderers cross-classified by colour and sentence. Source: Radelet (1981).

Colour of	Colour of	Sentence		
Victim	Murderer	Death	Other	%
Black	Black	6	97	5.8
	White	0	9	0.0
White	Black	11	52	17.5
	White	19	132	12.6

TABLE 1.4. Murderers cross-classified by colour, victim's colour, and sentence. Source: Radelet (1981).

For black victims and white victims, the proportion of black murderers sentenced to death is higher than the proportion of white murderers: in other words, the interpretation is reversed. What is happening is that murderers tend to choose victims of their own colour, and that whether the death penalty is given or not depends on the colour of the victim: if the victim is white, the death penalty is given more often than if the victim is black. This results in a higher proportion of whites receiving the death penalty.

A similar phenomenon can occur with continuous variables. The data shown in Table 1.5 are based on an example in Wermuth (1990): ten observations are made on three variables — A binary, X and Y continuous.

If we regress Y on X for each value of A separately, in each case we obtain a positive coefficient of X. If we combine the groups, however, the coefficient becomes negative. This is illustrated in Figure 1.9.

As presented above, a reversal in association between the marginal and conditional distributions may be surprising, but is hardly paradoxical. Interpreted causally, however, it implies that a cause can have a positive effect in two subpopulations, but a negative effect when these are combined. This runs strongly against our intuition. Using methods (the so-called intervention calculus) described below in Section 8.3, Pearl (2000, section 6.1) shows that our intuition is correct: the apparent paradox is due to this mistaken causal interpretation.

For our present purposes, the lesson to be learnt is that reliance on pairwise marginal associations may be very misleading. It is necessary to take a multivariate approach by including all relevant variables in the analysis so

A	1	1	1	1	1
X	4.5	4.0	5.0	6.0	5.5
Y	19.3	18.7	20.3	20.9	20.4
A	2	2	2	2	2
X	10.5	12.0	12.0	11.0	11.5
Y	15.3	16.8	17.2	16.2	16.2

TABLE 1.5. Fictive data illustrating Simpson's paradox in regression.

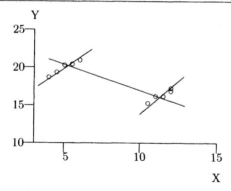

FIGURE 1.9. Simpson's paradox in regression. Within each group the regression coefficient is positive, but when the groups are combined it becomes negative.

as to study the conditional as well as the marginal associations. This is the basis of graphical modelling.

1.5 Overview of the Book

The emphasis in this book is on model structure, more precisely, conditional independence structure. In the following chapters, we examine various families of models that can be specified in terms of pairwise conditional independences: using the equivalence between pairwise and global Markov properties, the structure of such models is easily represented by independence graphs.

Chapter 2 describes one such family — loglinear models for discrete data. Here, pairwise conditional independences are equivalent to zero two-factor interactions. Loglinear models are well-known and widely used; the chapter provides a brief introduction to this extensive topic, with focus on aspects relevant to graphical modelling.

Chapter 3 treats another such family — graphical Gaussian models for continuous data. In this case, pairwise conditional independences correspond to zero partial correlations. These models are linked to other multivariate normal models, such as multivariate regression and recursive equation models.

Chapter 4 constitutes the core of the book in the sense that the models described there form the basis for the following chapters. These are models for mixed discrete and continuous data, constructed by combining the models of the two previous chapters. A central question treated here concerns when complex models with many variables can be broken down into a sequence of simpler models with fewer variables. This is important for model interpretation and other purposes. In addition, some closely related conditional models, the CG-regression models, are described here. Other topics covered in this chapter include use of the models in connection with latent variables (or missing data) and discriminant analysis.

Chapter 5 surveys hypothesis testing for the mixed models. Although asymptotic likelihood ratio tests are always available, they require large samples; various other tests, such as F-tests, are also valid in small samples. Exact conditional tests, which are valid under relaxed distributional assumptions, are also described.

Chapter 6 describes some methods for model selection and criticism. Explorative tools to identify simple models consistent with the data can be very useful in practice, although too much credence is sometimes given to the models they select. It is important to assess the selected models critically, and various techniques to do this are described.

Chapter 7 treats graphs with directed edges and associated models. The main emphasis in the previous chapters is on models for undirected graphs, in which no assumptions concerning directions of influence are made. However, when good prior knowledge of direction of influence is available, it should be taken into account in the modelling process. Particularly useful in many applications are the chain graphs. Brief mention is also made of other types of graphs, including local independence graphs, reciprocal graphs and covariance graphs.

It is natural to interpret graphical models in causal terms, and this is perhaps part of their attraction. Chapter 8 attempts to address the question of when such causal interpretations are justified. Two competing statistical accounts of causality are described and compared. Key concepts here are confounding and randomisation.

The worked examples in the book make use of the graphical modelling system MIM. A guide to this is found in Appendices A and B. Some results on symmetry tests are given in Appendix C, and Appendix D describes some estimation algorithms.

2

Discrete Models

This chapter gives a concise exposition of loglinear models for discrete data. Since their introduction in the 1960s and 1970s, loglinear models have been applied in a wide variety of fields, and the methodology is described in many textbooks on discrete data analysis. Notable among these are Bishop, Fienberg, and Holland (1975), Fienberg (1980), Agresti (1990), and Christensen (1990). The use of graphs and graphical models in connection with loglinear models is due to Darroch, Lauritzen, and Speed (1980); an application-oriented exposition is given in Edwards and Kreiner (1983). Christensen (1990) can be recommended as a modern introductory-level textbook on contingency table analysis with a good coverage of the graphical modelling approach. Whittaker's (1990) book on graphical modelling covers both discrete and continuous models. Lauritzen (1996) gives a mathematically rigorous exposition. This chapter gives only a very brief account of the topic, focusing on graphs and graphical models.

2.1 Three-Way Tables

Loglinear models are so called because they apply linear models to the logarithms of the expected cell counts. To motivate this, suppose we observe N observations of three discrete variables, A, B, and C. Let A have $\#A$ levels, B have $\#B$ levels, and C have $\#C$ levels. We form a three-way table of counts by cross-classifying A, B, and C, and write a typical count as n_{jkl} where j can take the values $1, \ldots, \#A$, k can be $1, \ldots, \#B$, and l can go from $1, \ldots, \#C$. Similarly, we write the cell probability— in other words, the probability that an observation falls in the given cell— as p_{jkl}

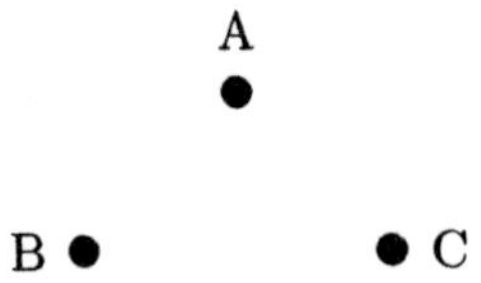

A
●

B ●   ● C

FIGURE 2.1. Graph of the independence model.

and the expected cell count as $m_{jkl} = N p_{jkl}$, where N is the total number of observations.

This sampling scheme, in which the number of observations is considered fixed, is called multinomial sampling. Other schemes can be used: for example, in row-multinomial sampling the number of observations with each level of, say, A is fixed in advance. As we will see later, much of the theory carries through regardless of which sampling scheme is adopted.

The simplest model for a three-way table writes the logarithm of the cell probabilities as

$$\ln(p_{jkl}) = u + u_j^A + u_k^B + u_l^C, \qquad (2.1)$$

where the u's are unknown parameters— usually called *interaction terms*.

Since additivity in the log scale is equivalent to multiplicity in the original scale, the model can be reformulated as $p_{jkl} = p_{j++} p_{+k+} p_{++l}$, where the $+$'s denote summation over the respective indices. In other words, the model (2.1) states that A, B, and C are completely independent.

To define the interaction terms uniquely, some ANOVA-like constraints would need to be applied to them, for example that $\sum_j u_j^A = 0$ or that $u_{\#A}^A = 0$, but we do not specify any particular set of constraints here. The point of view behind this is that the numerical values of the interaction terms are of little interest in themselves: primary interest is usually in the cell probabilities, or properties such as independence or conditional independence, or functions of the cell probabilities such as cross-product ratios or odds ratios. The constraints on the interaction terms we are here discussing do not limit the possible values of the cell probabilities, and hence do not affect the properties we are interested in.

To identify the model, a *model formula* can be used. This consists of a list of terms (*generators*) corresponding to the maximal interaction terms in the model. For the independence model, the formula is A, B, C, and the independence graph is shown in Figure 2.1.

A more complex loglinear model could be

$$\ln(p_{jkl}) = u + u_j^A + u_k^B + u_l^C + u_{jk}^{AB} + u_{jl}^{AC}.$$

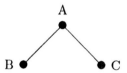

FIGURE 2.2. The graph of $B \perp\!\!\!\perp C \mid A$.

Since u_{jk}^{AB} and u_{jl}^{AC} are the maximal interaction terms, the model formula is AB, AC. The model can be reformulated in terms of cell probabilities as

$$p_{jkl} = \frac{p_{jk+}p_{j+l}}{p_{j++}} \tag{2.2}$$

or

$$\frac{p_{jkl}}{p_{j++}} = \left(\frac{p_{jk+}}{p_{j++}}\right)\left(\frac{p_{j+l}}{p_{j++}}\right),$$

which is equivalent to

$$\Pr(B = k, C = l|A = j) = \Pr(B = k|A = j)\Pr(C = l|A = j).$$

This last form reveals that in the conditional distribution given $A = j$, B and C are independent. In other words, $B \perp\!\!\!\perp C \mid A$. The graph of the model is shown in Figure 2.2.

The general loglinear model for three factors can be written as

$$\ln(p_{jkl}) = u + u_j^A + u_k^B + u_l^C + u_{jk}^{AB} + u_{jl}^{AC} + u_{kl}^{BC} + u_{jkl}^{ABC},$$

with corresponding model formula ABC. This imposes no restrictions on the $\{p_{jkl}\}$. The previous model AB, AC can be seen to be a special case of the general model defined by setting $u^{BC} = u^{ABC} = 0$, i.e., setting $u_{kl}^{BC} = 0$ for all k and l and $u_{jkl}^{ABC} = 0$ for all j, k, and l.

Similarly, the independence model is a special case of the general model defined by setting $u^{BC} = u^{AC} = u^{AB} = u^{ABC} = 0$.

Almost invariably, attention is restricted to *hierarchical* loglinear models. Here, the term hierarchical means that if a term is set to zero, then all its higher-order relatives are also set to zero. For example, if we set $u^{AB} = 0$, but allow nonzero three-factor interaction terms u_{jkl}^{ABC}, then this would define a nonhierarchical model. Such models are difficult to interpret— what does it mean to say that a two-factor interaction is zero, if a larger three-factor interaction is present?— and for this reason, they have not been used much in practice.

Under multinomial sampling with N observations, the likelihood of a given table $\{n_{jkl}\}$ is

$$\mathcal{L}(\{p_{jkl}\} \mid \{n_{jkl}\}) = \frac{N!}{\prod_{jkl} n_{jkl}!} \prod_{jkl} p_{jkl}^{n_{jkl}}.$$

The values of p_{jkl} that maximize this expression for a given model are called the maximum likelihood estimates (MLEs) and are written $\hat{p}_{jkl}$. Since the logarithmic function is monotonic, the MLEs also maximize the log likelihood

$$\ell(\{p_{jkl}\} \mid \{n_{jkl}\}) = \ln\left(\frac{N!}{\prod_{jkl} n_{jkl}!}\right) + \sum_{jkl} n_{jkl} \ln p_{jkl}.$$

The *deviance* of a model $\mathcal{M}_0$ is the likelihood ratio test of $\mathcal{M}_0$ versus the unrestricted (saturated) model $\mathcal{M}_f$, i.e.,

$$G^2 = 2(\hat{\ell}_f - \hat{\ell}_0),$$

where $\hat{\ell}_f$ and $\hat{\ell}_0$ are the maximized log likelihoods under $\mathcal{M}_f$ and $\mathcal{M}_0$, respectively. The deviance is a convenient measure of the goodness of fit of $\mathcal{M}_0$, due to three properties:

1. When $\mathcal{M}_0 = \mathcal{M}_f$, it is zero.
2. If $\mathcal{M}_0$ is true, then it is asymptotically $\chi^2_{(k)}$ distributed, where k (the degrees of freedom) is given as the difference in number of free parameters between $\mathcal{M}_0$ and $\mathcal{M}_f$. It follows that E(deviance) = k under $\mathcal{M}_0$.
3. When $\mathcal{M}_0$ is not true, the asymptotic distribution of the deviance is stochastically larger than $\chi^2_{(k)}$, so that, for example, E(deviance) > k.

Under $\mathcal{M}_f$, the MLEs of the p_{jkl} are n_{jkl}/N so that the deviance can be written as

$$G^2 = 2\left\{\sum_{jkl} n_{jkl} \ln(n_{jkl}/N) - \sum_{jkl} n_{jkl} \ln(\hat{p}_{jkl})\right\}$$

$$= 2\sum_{jkl} n_{jkl} \ln\left(\frac{n_{jkl}}{\hat{m}_{jkl}}\right),$$

where $\hat{p}_{jkl}$ are the MLEs of p_{jkl} under $\mathcal{M}_0$, and $\hat{m}_{jkl}$ are the corresponding fitted cell counts, i.e., $\hat{m}_{jkl} = N\hat{p}_{jkl}$. This expression for the deviance is also valid under other sampling schemes, for example, independent Poisson or row-multinomial sampling.

For two nested models, $\mathcal{M}_0 \subseteq \mathcal{M}_1$, the deviance difference is given as

$$d = 2\sum_{jkl} n_{jkl} \ln\left(\frac{\hat{m}^1_{jkl}}{\hat{m}^0_{jkl}}\right),$$

where $\hat{m}^0_{jkl}$ and $\hat{m}^1_{jkl}$ are the MLEs under $\mathcal{M}_0$ and $\mathcal{M}_1$, respectively. Under $\mathcal{M}_0$, d is asymptotically $\chi^2_{(k)}$ where k is the difference in the number of free parameters between $\mathcal{M}_0$ and $\mathcal{M}_1$.

As a rule, tests based on the deviance difference are more reliable than overall goodness-of-fit tests based on G^2, in the sense that their null distribution is closer to their asymptotic reference distribution.

Under the unrestricted model, we have

$$p_{jkl} = \exp(u + u_j^A + u_k^B + u_l^C + u_{jk}^{AB} + u_{jl}^{AC} + u_{kl}^{BC} + u_{jkl}^{ABC}).$$

For the model AB, AC, which is defined by setting $u^{BC} = 0$ (and hence $u^{ABC} = 0$), the probability can be factored into

$$p_{jkl} = \exp(u + u_j^A + u_k^B + u_{jk}^{AB}) \exp(u_l^C + u_{jl}^{AC}),$$

i.e., into two factors, the first not involving C and the second not involving B. It then follows from the factorization criterion (1.1) that $B \perp\!\!\!\perp C \mid A$. More generally, it is apparent that under any hierarchical model, two factors are conditionally independent given the rest if and only if the corresponding two-factor interaction term is set to zero.

This result forms the basis of a subclass of the hierarchical models, the so-called *graphical* models. Such models are defined by setting a set of two-factor interaction terms (and hence their higher-order relatives) to zero. In other words, the higher-order interactions to include in the model are specified by the two-factor interaction terms. The attractive feature of such models is that they can be interpreted solely in terms of conditional independence.

For example, the model AB, AC is graphical since it is formed by setting u^{BC} to zero. The simplest example of a hierarchical *nongraphical* model is AB, BC, AC, which sets the three-factor interaction u^{ABC} to zero. Its graph is as follows:

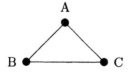

The graphical model corresponding to this graph is the saturated model ABC. See Table 2.2 for further examples of graphical and hierarchical models.

2.1.1 Example: Lizard Perching Behaviour

We here consider the analysis of a three-way contingency table, treated at length in Fienberg (1980). The table concerns data on the perching behaviour of two species of lizard, taken from Schoener (1968). For a sample of 164 lizards, the following variables were recorded: species (A); perch

| Species | Perch Diameter (inches) | Perch Height (feet) | |
		> 4.75	≤ 4.75
Anoli	≤ 4	32	86
	> 4	11	35
Distichus 2	≤ 4	61	73
	> 4	41	70

TABLE 2.1. Data on the perching behaviour of two species of lizards. Source: Schoener (1968).

diameter (B): 1 = narrow, 2 = wide; and perch height (C): 1 = high, 2 = low. The original data on perch diameter and height were continuous, but were later dichotomized. The data are shown in Table 2.1. We illustrate an analysis of these data, making use of MIM.

The data are defined as follows:

```
MIM->factor A2B2C2; statread ABC
DATA->32 86 11 35 61 73 41 70 !
Reading completed.
```

The Factor command defines the three binary factors, A, B, and C. The command StatRead reads the data in the form of cell counts.

A sensible analysis strategy is to examine which conditional independence relations, if any, hold. To do this, we examine which edges can be removed from the complete graph. We set the current model to ABC (the full model) and then use the Stepwise command with the O (one step only) option:

```
MIM->model ABC
MIM->stepwise o
Coherent Backward Selection
Decomposable models, Chi-squared tests.
DFs adjusted for sparsity.
Single step.
Critical value:   0.0500
Initial model: ABC
Model: ABC
Deviance:    0.0000 DF:   0 P:  1.0000
    Edge        Test
  Excluded   Statistic DF        P
     [AB]      14.0241  2      0.0009 +
     [AC]      11.8229  2      0.0027 +
     [BC]       2.0256  2      0.3632
No change.
Selected model: ABC
```

The output presents the deviances of the three models formed by removing an edge from the complete graph, together with the associated χ^2-tests.

Only one edge, $[BC]$, can be removed. We delete this edge and repeat the process:

```
MIM->delete BC
MIM->stepwise o
Coherent Backward Selection
Decomposable models, Chi-squared tests.
DFs adjusted for sparsity.
Single step.
Critical value:   0.0500
Initial model: AC,AB
Model: AC,AB
Deviance:    2.0256 DF:    2 P:  0.3632
     Edge         Test
 Excluded    Statistic DF          P
    [AB]        12.6062  1       0.0004 +
    [AC]        10.4049  1       0.0013 +
No change.
Deviance:        2.0256 DF: 2
Selected model: AC,AB
```

Deleting the edge $[BC]$ results in the model AC, AB. We then consider the models formed by removing an edge from AC, AB. The output presents the deviance differences, together with the associated χ^2-tests. Neither of the two edges can be removed if we test at, say, the 1% level. The model AC, AB is thus the simplest acceptable model. Its graph is

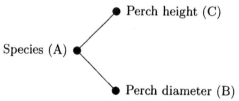

It states that for each species of lizard considered separately, perch height and perch diameter are unassociated.

2.2 Multi-Way Tables

In this section, we describe loglinear models for multi-way tables. Extending the results of the previous section to multi-way tables is essentially quite straightforward: the main difficulty is notational. Whereas for three-way tables the factor names and levels can be written explicitly, as for example in p_{jkl}^{ABC}, u_{jk}^{BC}, and n_{j+l}, this method becomes very cumbersome when the number of factors is arbitrary, and we need a more general notation.

Let Δ be a set of p discrete variables. We suppose that the data to be analyzed consist of N observations on these p variables, which can be summed up in a p-dimensional table of counts, formed by cross-classifying the p discrete variables.

We write a typical cell in the table as the p-tuple

$$i = (i_1, \ldots, i_p), \tag{2.3}$$

and denote the number of observations in cell i as n_i, the probability of cell i as p_i, and the expected cell count as $m_i = Np_i$. Let $\mathcal{I}$ be the set of all cells in the table. We only consider complete tables, so that the number of cells in $\mathcal{I}$ is the product of the number of levels of the factors in Δ.

Under multinomial sampling, the likelihood of a given table $\{n_i\}_{i \in \mathcal{I}}$ is

$$\frac{N!}{\prod_{i \in I} n_i!} \prod_{i \in I} p_i^{n_i}. \tag{2.4}$$

We also need a notation for marginal cells. For a cell $i \in \mathcal{I}$ and a subset $a \subseteq \Delta$, let i_a be the corresponding sub p-tuple of i, and let $\mathcal{I}_a$ be the set of all possible i_a.

We write a general interaction term as u_i^a, where it is understood that u_i^a depends on i only through i_a. Thus, the full (saturated) loglinear model for p factors can be written

$$\ln(p_i) = \sum_{a \subseteq \Delta} u_i^a.$$

It is easy to show that for this model, the MLEs are $\hat{p}_i = n_i/N$.

As before, models are formed by setting interactions and all their higher-order relatives to zero. A model can be specified through a model formula

$$d_1, \ldots, d_r,$$

where the sets $d_i \subseteq \Delta$ are called *generators*. These identify the maximal interactions that are not set to zero. Thus, we may write the model as

$$\ln(p_i) = \sum_{a \subseteq \Delta : a \subseteq d_j \text{ for some } j} u_i^a. \tag{2.5}$$

2.2.1 Likelihood Equations

Let $\{n_{i_a}\}_{i_a \in \mathcal{I}_a}$ be the marginal table of observed counts corresponding to a. Similarly, let $\{\hat{m}_{i_a}\}_{i_a \in \mathcal{I}_a}$ be the marginal table of fitted counts for some estimate $\{\hat{m}_i\}_{i \in \mathcal{I}}$ of $\{m_i\}_{i \in \mathcal{I}}$.

Consider a given model with formula $d_1, \ldots, d_r$. A set of minimal sufficient statistics is given by the set of marginal tables $\{n_{i_a}\}_{i_a \in \mathcal{I}_a}$ corresponding to the generators in the model, i.e., for $a = d_1, \ldots, d_r$. By equating the expected values of the minimal sufficient statistics to their observed values, we form the likelihood equations

$$\{\hat{m}_{i_a}\}_{i_a \in \mathcal{I}_a} = \{n_{i_a}\}_{i_a \in \mathcal{I}_a} \tag{2.6}$$

for $a = d_1, \ldots, d_r$. That is to say, the MLEs $\hat{m}_i$ under the model are given as solutions to (2.6) that also satisfy the model constraints as expressed in (2.5). Generally, the MLEs must be found by iterative methods, of which the iterative proportional scaling (IPS) algorithm, due to Deming and Stephan (1940), is the most widely used. Starting from some initial estimate, for example, $\hat{m}_i^0 = 1$ for all $i \in \mathcal{I}$, the kth iterative cycle performs the update

$$\hat{m}_i^k := \hat{m}_i^{k-1}(n_{i_a}/\hat{m}_{i_a}^{k-1})$$

for all $i \in \mathcal{I}$, for each $a = d_1, \ldots d_r$ in turn. It can be shown that these estimates converge to the MLEs (see, for example, Ireland and Kullback, 1968).

2.2.2 Deviance

The deviance under $\mathcal{M}_0$ is

$$G^2 = 2 \sum_{i \in I} n_i \ln \left(\frac{n_i}{\hat{m}_i^0} \right), \tag{2.7}$$

and the deviance difference for $\mathcal{M}_0 \subseteq \mathcal{M}_1$ is

$$\begin{aligned}
d &= G_0^2 - G_1^2 \\
&= 2 \sum_{i \in \mathcal{I}} n_i \ln \left(\frac{n_i}{\hat{m}_i^0} \right) - 2 \sum_{i \in \mathcal{I}} n_i \ln \left(\frac{n_i}{\hat{m}_i^1} \right) \\
&= 2 \sum_{i \in \mathcal{I}} n_i \ln \left(\frac{\hat{m}_i^1}{\hat{m}_i^0} \right),
\end{aligned}$$

where $\{\hat{m}_i^0\}_{i \in \mathcal{I}}$ and $\{\hat{m}_i^1\}_{i \in \mathcal{I}}$ are the fitted cell counts under $\mathcal{M}_0$ and $\mathcal{M}_1$, respectively. Under $\mathcal{M}_0$, d has an asymptotic χ^2-distribution with degrees of freedom given as the difference in number of free parameters between $\mathcal{M}_0$ and $\mathcal{M}_1$.

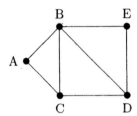

FIGURE 2.3. The graph of ABC, BCD, DE, BE.

2.2.3 Graphs and Formulae

Given a model formula, we form the independence graph by connecting all pairs of vertices that appear in the same generator. For example, the graph of ABC, BCD, DE, BE is shown in Figure 2.3.

Conversely, given a graph, we can construct the model formula of the graphical model corresponding to the graph by finding the *cliques* of the graph (see Section 1.2). For example, the cliques of the graph shown in Figure 2.3 are $\{A, B, C\}$, $\{B, C, D\}$, and $\{B, D, E\}$, so the corresponding graphical model has formula ABC, BCD, BDE.

Models for which explict formulae for the MLEs can be derived are called *decomposable*. Darroch, Lauritzen, and Speed (1980) characterized decomposable loglinear models as graphical models whose graphs are triangulated, i.e., contain no cycles of length greater or equal to four without a chord (see Section 1.2). In Table 2.2, a list of models is displayed, showing their model formula, graph, and whether they are graphical or decomposable. Decomposability is studied in greater depth in Section 4.4.

2.2.4 Example: Risk Factors for Coronary Heart Disease

In this section, we analyze a six-dimensional contingency table concerning risk factors for coronary heart disease. The data originated in a prospective study of coronary heart disease carried out in Czechoslovakia (Reinis et al., 1981), and were also analyzed by Edwards and Havránek (1985).

For a sample of 1841 car-workers, the following information was recorded: whether they smoked (A), whether their work was strenuous mentally (B), whether their work was strenuous physically (C), whether their systolic blood pressure was less than 140 mm (D), whether the ratio of beta to alpha lipoproteins was less than 3 (E), and whether there was a family history of coronary heart disease (F). In the program fragment below, the data are coded so that factor levels of 1 and 2 correspond to yes and no, respectively.

Model Formula	Graphical	Decomposable	Graph
B, AC	yes	yes	
AB, AC	yes	yes	
AB, AC, BC	no	no	
ABC	yes	yes	
ABC, BCD	yes	yes	
AB, AC, CD, BD	yes	no	
AB, BD, ACD	no	no	
ABE, ACE, BDE, CDE	yes	no	
AB, AC, BDE, CE	yes	no	

TABLE 2.2. Some hierarchical loglinear models and their graphs.

The fragment illustrates how a plausible model may be found using stepwise model selection. This is an explorative technique to identify a simple model consistent with the data, described in detail in Section 6.1. In the present example, the procedure starts out from the saturated model and proceeds by successively removing the least significant edge, that is to say, the edge for which the deviance difference test for edge removal results in the largest p-value greater than or equal to 0.05. Only decomposable models are considered. Moreover, the coherence principle is respected (see Section 6.1.4): in backward selection, this just means that if the removal of an edge is rejected at one step, then the edge is not subsequently eligible for removal. Such edges are marked with "+".

```
MIM->fact A2B2C2D2E2F2
MIM->label A "Smoker" B "Mental work" C "Phys. work"
MIM->label D "Syst. BP" E "Lipo ratio" F "Fam. history"
MIM->statread FEDCBA
DATA->44 40 112 67 129 145 12 23 35 12 80 33 109 67 7 9
DATA->23 32 70 66 50 80 7 13 24 25 73 57 51 63 7 16 5 7 21 9
DATA->9 17 1 4 4 3 11 8 14 17 5 2 7 3 14 14 9 16 2 3 4 0 13 11
DATA->5 14 4 4 !
Reading completed.
MIM->satmod
MIM->stepwise
Coherent Backward Selection
Decomposable models, Chi-squared tests.
DFs adjusted for sparsity.
Critical value:   0.0500
Initial model: ABCDEF
Model: ABCDEF
Deviance:   0.0000 DF:   0 P:  1.0000
     Edge          Test
  Excluded   Statistic DF          P
      [AB]     22.6518 16      0.1234
      [AC]     42.8039 16      0.0003 +
      [AD]     28.7241 16      0.0259 +
      [AE]     40.0240 16      0.0008 +
      [AF]     21.3052 16      0.1671
      [BC]    684.9893 16      0.0000 +
      [BD]     12.2256 16      0.7283
      [BE]     17.2263 16      0.3711
      [BF]     22.7875 16      0.1195
      [CD]     14.8084 16      0.5387
      [CE]     18.6293 16      0.2884
      [CF]     22.1529 16      0.1383
      [DE]     31.0594 16      0.0132 +
      [DF]     18.3454 16      0.3041
      [EF]     18.3160 16      0.3057
```

```
Removed edge [BD]
Model: ACDEF,ABCEF
Deviance:  12.2256 DF:  16 P:  0.7283
     Edge         Test
   Excluded    Statistic DF           P
     [AB]       15.5745  8         0.0489 +
     [BE]       11.3637  8         0.1819
     [BF]       15.7439  8         0.0462 +
     [CD]        7.1489  8         0.5207
     [DF]       11.3018  8         0.1852
Removed edge [CD]
Model: ADEF,ABCEF
Deviance:  19.3745 DF:  24 P:  0.7317
     Edge         Test
   Excluded    Statistic DF           P
     [BE]       11.3637  8         0.1819
     [CE]       12.7351  8         0.1213
     [CF]       12.2370  8         0.1409
     [DF]        6.3676  4         0.1733
Removed edge [BE]
Model: ADEF,ACEF,ABCF
Deviance:  30.7382 DF:  32 P:  0.5303
     Edge         Test
   Excluded    Statistic DF           P
     [CE]       22.5308  4         0.0002 +
     [DF]        6.3676  4         0.1733
Removed edge [DF]
Model: ADE,ACEF,ABCF
Deviance:  37.1058 DF:  36 P:  0.4178
     Edge         Test
   Excluded    Statistic DF           P
     [EF]        4.5767  4         0.3335
Removed edge [EF]
Model: ADE,ACE,ABCF
Deviance:  41.6825 DF:  40 P:  0.3975
     Edge         Test
   Excluded    Statistic DF           P
     [AF]        7.3858  4         0.1169
     [CF]        7.0181  4         0.1349
Removed edge [CF]
Model: ADE,ACE,ABF,ABC
Deviance:  48.7006 DF:  44 P:  0.2895
     Edge         Test
   Excluded    Statistic DF           P
     [AF]        2.6581  2         0.2647
Removed edge [AF]
Selected model: ADE,ACE,BF,ABC
```

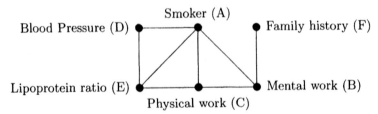

FIGURE 2.4. Risk factors for coronary heart disease: the selected model.

The graph of the selected model is shown in Figure 2.4. The association between blood pressure, smoking, and lipoproteins (cholesterol) is unsurprising, since it is well-known that smoking has significant cardiovascular effects. Similarly, the association between smoking, physical work, and mental work may be due to sociological factors. The association between physical and mental work is likely to be negative in the sense that work is rarely strenuous both physically and mentally. Physical work may affect cholesterol levels, both indirectly by influencing dietary habits, and directly by affecting metabolism. Perhaps the most surprising feature is the association between family history of coronary heart disease and strenuous mental work.

It should be noted that the strength of evidence for the different associations varies considerably. For example, removal of $[AE]$ was rejected with $p = 0.0008$, whereas the test for the removal of $[BF]$ was only marginally significant at the 5% level, with $p = 0.0462$.

2.2.5   Example: Chromosome Mapping

This example has its base in genetics, and it has the rather pleasing property that the independence graphs involved correspond, in a sense we describe below, to chromosome maps. It also illustrates a general approach to analyzing sparse contingency tables.

The data stem from a cross between two isolates of the barley powdery mildew fungus, described in Christiansen and Giese (1991). For each offspring, 38 binary characteristics, each corresponding to a single locus, were determined. The data analyzed summarize the results for six of these characteristics, selected from the 38 since they appeared to be closely linked. (The first two characteristics, A and B, are virulence genes, and the four others are genetic markers). The object of the analysis is to determine the order of the loci along the chromosome.

The total number of offspring is 70. The data form a 2^6 table so that there are 64 cells, many of which are empty. Note that the data are coded such

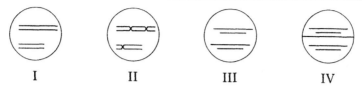

I   II   III   IV

FIGURE 2.5. Nuclear division. Stage I shows a nucleus after fusion, containing two pairs of homologous chromosomes. Each chromosome in a pair is derived from a different parent. For clarity, only two pairs, of different length, are shown. (Powdery mildew is thought to have seven chromosomes.) In stage II, the homologous chromosomes pair and a process known as crossing-over may occur at different points along the chromosomes. This results in exchange of sections between the two chromosomes. Stage III is characterized by the separation of the chromosome pairs, one of each type migrating to opposite poles of the nucleus. Eventually, in stage IV, the nucleus divides into two parts, each containing two chromosomes. Each of these subsequently develops into a spore, in time giving rise to separate mildew progeny. (See, for example, Suzuki et al. (1989), Ch. 5.)

that the two parental isolates have characteristics (1,1,...,1) and (2,2,...,2), respectively.

To understand these data, we must know more about the process of nuclear division (meiosis). The mildew fungus is a haploid organism, that is to say, each nucleus normally contains only one set of chromosomes. During the reproductive process, hyphae from two parents grow together, and the cells and nuclei fuse. Immediately afterwards, the nuclei undergo division, as shown in Figure 2.5.

Put briefly, the chromosome complements of the two parents mix and give rise to two progeny. If crossing-over did not occur, each chromosome of a progeny would simply be chosen at random from the two corresponding parental chromosomes.

Now consider the inheritance of individual loci. (We can think of these as blobs sitting on the chromosomes in Figure 2.5.) For a single locus, A, an offspring inherits either of the parental alleles, i.e., either $A = 1$ or $A = 2$. These are equiprobable, so that $\Pr(A = 1) = \Pr(A = 2) = \frac{1}{2}$. If two loci, A and B, are on different chromosomes, then they are inherited independently, so that

$$\Pr(A = i, B = j) = \Pr(A = i)\Pr(B = j) = \frac{1}{4}$$

for $i, j = 1, 2$. This is termed independent assortment. However, if they are on the same chromosome, then their inheritance is not independent. If crossing-over did not take place, then only parental combinations could occur, so that we would have

$$
\Pr(A = i, B = j) = \begin{cases} \dfrac{1}{2} & \text{if } i = j \\ 0 & \text{otherwise} \end{cases}
$$

for $i, j = 1, 2$. However, since crossing-over does take place, we have instead

$$
\Pr(A = i, B = j) = \begin{cases} (1 - p_r)/2 & \text{if } i = j \\ p_r/2 & \text{otherwise} \end{cases}
$$

for $i, j = 1, 2$, where p_r is the probability of recombination between A and B, i.e., the occurrence of nonparental combinations. Since loci that are close together have a low probability of recombination, p_r is a measure (albeit nonadditive) of the distance between the loci.

Independent assortment between two loci is often tested using the standard χ^2 goodness-of-fit test on the observed 2×2 contingency table (although a binomial test of $p_r = \frac{1}{2}$ against $p_r < \frac{1}{2}$ would have greater power). Similarly, if three loci are in the order A, B, C along a chromosome, then the occurrence of recombination between A and B will be independent of recombination between B and C, at least to a first approximation. If this is not the case, there is said to be *interference*. In other words, no interference is equivalent to $A \perp\!\!\!\perp C \mid B$. It follows that if we model the data set using loglinear models, we are interested in models whose graphs consist of variables linked in strings, which correspond to the chromosomes.

Again we proceed by selecting a parsimonious model using backwards selection, starting from the saturated model. Since the table is very sparse, asymptotic χ^2-tests would be quite inaccurate, and it is necessary to use exact tests— these are described later in Section 5.4:

```
MIM->fact A2B2C2D2E2F2
MIM->sread ABCDEF
DATA->0 0 0 0 3 0 1 0 0 1 0 0 0 1 0 0
DATA->1 0 1 0 7 1 4 0 0 0 2 1 3 0 11
DATA->16 1 4 0 1 0 0 0 1 4 1 4 0 0 0 1
DATA->0 0 0 0 0 0 0 0 0 0 0 0 0 0 0 !
Reading completed.
MIM->satmod
MIM->step e
Coherent Backward Selection
Decomposable models, Chi-squared tests.
Exact tests, exhaustive enumeration.
DFs adjusted for sparsity.
Critical value:   0.0500
Initial model: ABCDEF
Model: ABCDEF
```

```
Deviance:    0.0000 DF:   0 P:  1.0000
   Edge        Test
Excluded    Statistic DF        P
   [AB]      29.3360   5     0.0000 +
   [AC]       0.4027   1     1.0000
   [AD]      20.0479   3     0.0002 +
   [AE]       4.5529   3     1.0000
   [AF]       0.4027   1     1.0000
   [BC]       1.1790   2     1.0000
   [BD]       3.3078   3     0.6364
   [BE]       6.8444   3     0.1881
   [BF]       1.1790   2     1.0000
   [CD]       0.3684   2     1.0000
   [CE]       5.7735   4     0.3624
   [CF]      48.7976   6     0.0000 +
   [DE]       2.7335   4     1.0000
   [DF]       0.7711   3     1.0000
   [EF]       4.1392   4     0.6702
Removed edge [AE]
Model: BCDEF,ABCDF
Deviance:    4.5529 DF:  16 P:  0.9976
   Edge        Test
Excluded    Statistic DF        P
   [AC]       0.2227   1     1.0000
   [AF]       0.4230   1     1.0000
   [BE]       4.7105   4     0.5840
   [CE]       5.5934   4     0.3920
   [DE]       1.0446   4     1.0000
   [EF]       4.1595   4     0.6762
Removed edge [AC]
Model: BCDEF,ABDF
Deviance:    4.7756 DF:  24 P:  1.0000
   Edge        Test
Excluded    Statistic DF        P
   [AF]       2.9974   2     0.3452
   [BC]       1.8815   4     1.0000
   [BE]       4.7105   4     0.5840
   [CD]       1.4225   4     1.0000
   [CE]       5.5934   4     0.3920
   [DE]       1.0446   4     1.0000
   [EF]       4.1595   4     0.6762
Removed edge [BC]
Model: CDEF,BDEF,ABDF
Deviance:    6.6571 DF:  32 P:  1.0000
   Edge        Test
Excluded    Statistic DF        P
   [AF]       2.9974   2     0.3452
```

```
      [BE]      4.5564  4      0.6289
      [CD]      1.4032  3      1.0000
      [CE]      5.4393  4      0.5288
Removed edge [CD]
Model: CEF,BDEF,ABDF
Deviance:    8.0603 DF:  36 P:  1.0000
     Edge       Test
Excluded    Statistic DF          P
      [AF]      2.9974  2      0.3452
      [BE]      4.5564  4      0.6289
      [CE]      4.1912  2      0.2072
      [DE]      1.4387  4      0.9554
Removed edge [DE]
Model: CEF,BEF,ABDF
Deviance:    9.4990 DF:  40 P:  1.0000
     Edge       Test
Excluded    Statistic DF          P
      [AF]      2.9974  2      0.3452
      [BD]      2.5611  2      0.2230
      [BE]      4.4793  2      0.1323
      [CE]      4.1912  2      0.2072
      [DF]      3.0386  3      0.6152
Removed edge [DF]
Model: CEF,BEF,ABF,ABD
Deviance:   12.5375 DF:  44 P:  1.0000
     Edge       Test
Excluded    Statistic DF          P
      [AF]      0.0216  1      1.0000
      [BD]      0.0581  1      1.0000
      [BE]      4.4793  2      0.1323
      [CE]      4.1912  2      0.2072
Removed edge [AF]
Model: CEF,BEF,ABD
Deviance:   12.5592 DF:  46 P:  1.0000
     Edge       Test
Excluded    Statistic DF          P
      [BD]      0.0581  1      1.0000
      [BE]      4.4793  2      0.1323
      [BF]      2.2159  2      0.3396
      [CE]      4.1912  2      0.2072
Removed edge [BD]
Model: CEF,BEF,AD,AB
Deviance:   12.6172 DF:  48 P:  1.0000
     Edge       Test
Excluded    Statistic DF          P
      [BE]      4.4793  2      0.1323
      [BF]      2.2159  2      0.3396
```

```
      [CE]      4.1912  2        0.2072
Removed edge [BF]
Model: CEF,BE,AD,AB
Deviance:  14.8331 DF:  50 P:  1.0000
     Edge        Test
Excluded    Statistic DF           P
     [BE]      6.4075  1        0.0154 +
     [CE]      4.1912  2        0.2072
     [EF]      2.4240  2        0.4644
Removed edge [EF]
Model: CF,CE,BE,AD,AB
Deviance:  17.2571 DF:  52 P:  1.0000
     Edge        Test
Excluded    Statistic DF           P
     [CE]     10.5570  1        0.0016 +
No change.
Selected model: CF,CE,BE,AD,AB
```

The graph of the selected model CF, CE, BE, AD, AB, is shown in Figure 2.6. We note that it is in the form of a string, consistent with our expectation that there is no interference.

Note also that the edges retained in the graph have associated p-values of less than 0.0016, except [BE], whose p-value is 0.0154, and the edges removed have p-values > 0.3396. This implies that the selection procedure would have given the same result had the critical value been anywhere between 0.0155 and 0.3396. The data thus strongly suggest that the order of the loci is as shown. (See Edwards (1992) for a more detailed discussion of this example.)

FIGURE 2.6. The selected model — a chromosome map.

2.2.6   Example: University Admissions

Our final example with discrete data concerns the analysis of a table showing admissions to Berkeley in autumn 1973, cross-classified by department
and sex of applicant (Freedman et al., 1978). These data, shown in Table 2.3, are also analyzed in Agresti (1990, p. 225–228). A hypothetical
example of this kind was described in the illustration of Simpson's paradox
in Section 1.4.

If we label the variables admission (A), department (D), and sex (S), then
as we saw in Section 1.4, the interesting question is whether $A \perp\!\!\!\perp S \mid D$.
We examine this in the following fragment:

```
MIM->fact S2A2D6
MIM->label A "Admission" S "Sex" D "Department"
MIM->sread DSA
DATA->512 313  89  19 353 207  17   8
DATA->120 205 202 391 138 279 131 244
DATA->53  138  94 299  22 351  24 317 !
Reading completed.
MIM->model ADS
MIM->testdel AS
Test of H0: DS,AD
against H:  ADS
LR:  21.7355    DF:   6    P: 0.0014
```

The hypothesis is strongly rejected; this seems to imply that there is evidence of sexual discrimination in the admission process. To examine this
more closely, recall that $A \perp\!\!\!\perp S \mid D$ means that for every department (D),

Department	Sex	Whether admitted	
		Yes	No
I	Male	512	313
	Female	89	19
II	Male	353	207
	Female	17	8
III	Male	120	205
	Female	202	391
IV	Male	138	279
	Female	131	244
V	Male	53	138
	Female	94	299
VI	Male	22	351
	Female	24	317

TABLE 2.3. Admissions to Berkeley in autumn 1973. Source: The Graduate
Division, University of California, Berkeley.

admission (A) is independent of sex (S). We can break down the likelihood ratio test for conditional independence into a sum of tests for independence in each department, using

$$G^2 = 2 \sum_{d=1}^{6} \sum_{s=1}^{2} \sum_{a=1}^{2} n_{asd} \ln \left(\frac{n_{asd} n_{++d}}{n_{a+d} n_{+sd}} \right)$$

$$= \sum_{d=1}^{6} \left\{ 2 \sum_{s=1}^{2} \sum_{a=1}^{2} n_{asd} \ln \left(\frac{n_{asd} n_{++d}}{n_{a+d} n_{+sd}} \right) \right\},$$

where the expression inside {} is the deviance test for independence for the dth department. To decompose the deviance in this fashion we employ a special option (Z) with the `TestDelete` command:

```
MIM->testdel AS z
Test of HO: DS,AD
against H: ADS
Deviance Decomposition into Strata
   D  LRTest     df      P
   1  19.054      1   0.000
   2   0.259      1   0.611
   3   0.751      1   0.386
   4   0.298      1   0.585
   5   0.990      1   0.320
   6   0.384      1   0.536
```

Departure from independence is only in evidence for Department I: for none of the other departments is there evidence of discrimination. Thus, the data are best described using *two* graphs, as shown in Figures 2.7 and 2.8.

This example illustrates a phenomenon easily overlooked in graphical modelling: namely, that the dependence structure may well differ for different subsets of the data. It is not difficult to imagine other examples. For instance, different disease subtypes may have different etiologies (causes). Højsgaard (1998) describes a method of generalization to multidimensional tables called *split graphical models*.

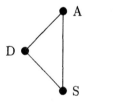

FIGURE 2.7. Department I.

FIGURE 2.8. Other departments.

Some problems exhibit a related type of hierarchical structure in which some variables are only defined for particular subsets of the data. For example, McCullagh and Nelder (1989, p. 160) describe a study of radiation mortality involving the following variables:

1. exposure (exposed/unexposed),
2. mortality (alive/dead),
3. death due to cancer or other causes,
4. leukemia or other cancers.

Clearly, (3) is only defined if (2) is dead, and (4) is only defined if (3) is cancer. Although the data could be summarized in a 2×4 table, with the row factor being exposure (yes/no) and the column factor being outcome (alive, leukemia death, other cancer death, noncancer death), it is more helpful to structure the problem as a sequence of conditional models (see Section 7.1).

Returning to the college admissions example, we may note that for Department I, 89 females (82%) as opposed to 512 males (62%) were admitted, i.e., the discrimination was in favour of females. For all the departments combined, however, the numbers were 557 females (30%) as opposed to 1198 males (44%). This is a clear example of Simpson's paradox.

3

Continuous Models

In this chapter, we describe models based on the multivariate normal distribution that are analogous to the loglinear models of the previous section. The best introduction to these models is given by Whittaker (1990), who aptly calls them *graphical Gaussian models*, although they are perhaps more widely known as *covariance selection* models, following Dempster (1972).

3.1   Graphical Gaussian Models

Suppose $Y = (Y_1, \ldots, Y_q)'$ is a q-dimensional random variable, with a multivariate normal distribution with mean

$$\mu = \begin{pmatrix} \mu^1 \\ \vdots \\ \mu^q \end{pmatrix} \tag{3.1}$$

and covariance matrix

$$\Sigma = \begin{pmatrix} \sigma^{11} & \cdots & \sigma^{1q} \\ \vdots & \ddots & \vdots \\ \sigma^{q1} & \cdots & \sigma^{qq} \end{pmatrix}. \tag{3.2}$$

(No apologies for the superscripts: these are used instead of the usual subscripts for notational reasons that will become apparent in the next chapter.)

We are very interested in the inverse covariance matrix, $\Omega = \Sigma^{-1}$, written as

$$\Omega = \begin{pmatrix} \omega^{11} & \cdots & \omega^{1q} \\ \vdots & \ddots & \vdots \\ \omega^{q1} & \cdots & \omega^{qq} \end{pmatrix}. \tag{3.3}$$

This is often called the *precision matrix*: some authors prefer the term *concentration matrix*.

It can be shown that the conditional distribution of (Y_1, Y_2) given $(Y_3, \ldots, Y_q)$ is a bivariate normal distribution with covariance

$$\begin{pmatrix} \omega^{11} & \omega^{12} \\ \omega^{21} & \omega^{22} \end{pmatrix}^{-1} = \frac{1}{\omega^{11}\omega^{22} - (\omega^{12})^2} \begin{pmatrix} \omega^{22} & -\omega^{21} \\ -\omega^{12} & \omega^{11} \end{pmatrix}. \tag{3.4}$$

The correlation coefficient in this bivariate distribution,

$$\rho^{12\cdot3\ldots q} = \frac{-\omega^{12}}{(\omega^{11}\omega^{22})^{\frac{1}{2}}}, \tag{3.5}$$

is called the partial correlation coefficient. We see that

$$\rho^{12\cdot3\ldots q} = 0 \iff \omega^{12} = 0. \tag{3.6}$$

In other words, two variables are independent given the remaining variables if and only if the corresponding element of the inverse covariance is zero. Thus, the elements of the inverse covariance matrix play the same role here as two-factor interaction terms in loglinear models.

It is instructive to derive this result in another way. The density of Y can be written

$$f(y) = |2\pi\Sigma|^{-\frac{1}{2}} \exp\{-\tfrac{1}{2}(y-\mu)'\Sigma^{-1}(y-\mu)\}. \tag{3.7}$$

Collecting terms inside the exponential brackets, we can rewrite this as

$$f(y) = \exp(\alpha + \beta'y - \tfrac{1}{2}y'\Omega y), \tag{3.8}$$

where $\Omega = \Sigma^{-1}$ as before, $\beta = \Sigma^{-1}\mu$, and α is the normalizing constant, given by

$$\alpha = -\tfrac{1}{2}\ln|\Sigma| - \tfrac{1}{2}\mu'\Sigma^{-1}\mu - \tfrac{q}{2}\ln(2\pi). \tag{3.9}$$

In exponential family terminology, β and Ω are called *canonical* parameters. Equation (3.8) can be rewritten as

$$f(y) = \exp(\alpha + \sum_{j=1}^{q} \beta^j y_j - \tfrac{1}{2}\sum_{j=1}^{q}\sum_{k=1}^{q} \omega^{jk} y_j y_k). \tag{3.10}$$

From this we can see, using the factorization criterion (1.1), that

$$Y_j \perp\!\!\!\perp Y_k \mid (\text{the rest}) \iff \omega^{jk} = 0. \tag{3.11}$$

Graphical Gaussian models are defined by setting specified elements of the inverse covariance matrix, and hence partial correlation coefficients, to zero. For example, if $q = 4$, then we could consider a model setting $\omega^{13} = \omega^{24} = 0$. Thus, the inverse covariance would look like

$$\Omega = \begin{pmatrix} \omega^{11} & \omega^{12} & 0 & \omega^{14} \\ \omega^{21} & \omega^{22} & \omega^{23} & 0 \\ 0 & \omega^{32} & \omega^{33} & \omega^{34} \\ \omega^{41} & 0 & \omega^{43} & \omega^{44} \end{pmatrix}. \tag{3.12}$$

The graph of this model is formed by connecting two nodes with an edge if the corresponding partial correlations are *not* set to zero:

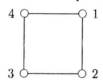

Notice that we use *circles* for the *continuous* variables, while in the previous chapter we used *dots* for the *discrete* variables. We now introduce a formula convention for graphical Gaussian models. For this purpose, we now label variables with letters rather than numbers. We call the set of variables Γ.

Just as with discrete graphical models, a model formula consists of a list of variables sets (*generators*) that are given as the cliques of the graph. For example, consider the following graph:

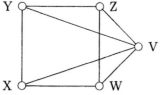

The missing edges are WY and XZ, so the model sets ω^{WY} and ω^{XZ} to zero. The cliques of the graph are $\{V, W, X\}$, $\{V, W, Z\}$, $\{V, Z, Y\}$, and $\{V, X, Y\}$, so the model formula is

$$//VWX, VWZ, VZY, VXY.$$

The double slashes are just a convention, necessary in connection with mixed models, as we shall see in the next chapter.

Note that all the models are graphical (hence the name: graphical Gaussian models): in contrast to the models for discrete data described in the

Model Formula	Decomposable	Graph
$//X,YZ$	yes	$\circ\,X$; $Y \circ\!\!-\!\!-\!\!\circ Z$
$//XY,YZ$	yes	X ; $Y \circ\!\!-\!\!\circ Z$
$//XYZ$	yes	X ; $Y \circ\!\!-\!\!\circ Z$ (triangle)
$//WX,XY,YZ,WZ$	no	W–X / Z–Y (square)
$//WXZ,XYZ$	yes	W–X / Z–Y with diagonal
$//WX,XYZ$	yes	W–X / Z–Y
$//VWX,VWZ,VXY,VYZ$	no	W–X / Z–Y with V

TABLE 3.1. Some graphical Gaussian models and their graphs.

previous chapter, there are no nongraphical models, and there is a one-to-one correspondence between models and graphs. Models with closed-form maximum likelihood estimates are called— as before— decomposable, and again a model is decomposable if and only if its graph is triangulated. Table 3.1 shows some models and their graphs, indicating whether they are decomposable.

3.1.1 Likelihood

Now suppose a sample of N observations $y^{(1)}, y^{(2)}, \ldots, y^{(N)}$ is taken; let $\bar{y} = \sum_{k=1}^{N} y^{(k)}/N$ be the sample mean vector, and let

$$S = \sum_{k=1}^{N} (y^{(k)} - \bar{y})(y^{(k)} - \bar{y})'/N$$

be the sample covariance matrix. The log density can be written as

$$\ln f(y) = -q \ln(2\pi)/2 - \ln |\Sigma|/2 - (y - \mu)'\Sigma^{-1}(y - \mu)/2,$$

so the log likelihood of the sample is

$$\ell(\mu, \Omega) = -Nq\ln(2\pi)/2 - N\ln|\Sigma|/2 - \sum_{k=1}^{N}(y^{(k)} - \mu)'\Omega(y^{(k)} - \mu)/2.$$

We can simplify the last term by writing

$$\sum_{k=1}^{N}(y^{(k)} - \mu)'\Omega(y^{(k)} - \mu) = \sum_{k=1}^{N}(y^{(k)} - \bar{y})'\Omega(y^{(k)} - \bar{y})$$
$$+ N(\bar{y} - \mu)'\Omega(\bar{y} - \mu),$$

and this can be further simplified using the trace function (sum of the diagonal elements of a square matrix). It can be shown (for example, by a tedious expansion of terms) that

$$\sum_{k=1}^{N}(y^{(k)} - \bar{y})'\Omega(y^{(k)} - \bar{y}) = N\mathrm{tr}(\Omega S).$$

Thus, we can write the log likelihood as

$$\ell(\mu, \Omega) = -Nq\ln(2\pi)/2 - N\ln|\Sigma|/2$$
$$- N\mathrm{tr}\,(\Omega S)/2 - N(\bar{y} - \mu)'\Omega(\bar{y} - \mu)/2. \tag{3.13}$$

3.1.2 Maximum Likelihood Estimation

For a variable subset $a \subseteq \Gamma$, let Σ^{aa} and S^{aa} denote the submatrices of Σ and S corresponding to a. For the model with generators $q_1, \ldots, q_t$, it can be shown that a set of minimal sufficient statistics is the sample mean $\bar{y}$ together with the set of marginal submatrices of the sample covariance matrix corresponding to the generators, i.e., $\{S^{aa}\}$, for $a = q_1, \ldots, q_t$. The likelihood equations are constructed by equating these with their expected values under the model so that we obtain $\hat{\mu} = \bar{y}$, and $\hat{\Sigma}^{aa} = S^{aa}$ for $a = q_1, \ldots, q_t$. For example, if $\Gamma = \{X, Y, Z\}$ and

$$S = \begin{pmatrix} 3.023 & 1.258 & 1.004 \\ 1.258 & 1.709 & 0.842 \\ 1.004 & 0.842 & 1.116 \end{pmatrix},$$

then under the model $//XY, YZ$, the maximum likelihood estimate of Σ is

$$\hat{\Sigma} = \begin{pmatrix} 3.023 & 1.258 & 0.620 \\ 1.258 & 1.709 & 0.842 \\ 0.620 & 0.842 & 1.116 \end{pmatrix},$$

whose inverse is

$$\hat{\Omega} = \begin{pmatrix} 0.477 & -0.351 & 0.000 \\ -0.351 & 1.190 & -0.703 \\ 0.000 & -0.703 & 1.426 \end{pmatrix}.$$

The likelihood equations constrain the elements of $\hat{\Sigma}$ to be equal to the corresponding elements of S, except for $\hat{\sigma}^{xz}$. The model constrains $\hat{\omega}^{xz}$ to be zero. Only $\hat{\sigma}^{xz}$ differs from its counterpart in S, and its value is such that $\hat{\omega}^{xz} = 0$.

An iterative algorithm for computing maximum likelihood estimates for graphical Gaussian models was described by Speed and Kiiveri (1986). This proceeds by incrementing submatrices of the precision matrix Ω so as to satisfy the likelihood equations. The precision matrix is initially set to the identity matrix. A step in the algorithm corresponding to a generator a adds an increment E to the current Ω so as to satisfy $\hat{\Sigma}^{aa} = S^{aa}$. A cycle in the algorithm repeats this for each generator.

The required increment matrix E is easily found. We require that

$$\begin{pmatrix} S^{aa} & * \\ * & * \end{pmatrix} = \begin{pmatrix} \Omega^{aa} + E & \Omega^{ab} \\ \Omega^{ba} & \Omega^{bb} \end{pmatrix}^{-1}, \qquad (3.14)$$

where $*$ denotes unspecified. Using standard results on inverses of partitioned matrices we obtain

$$S^{aa} = (\Omega^{aa} + E + \Omega^{ab}(\Omega^{bb})^{-1}\Omega^{ba})^{-1},$$

and so

$$E = (S^{aa})^{-1} - (\Omega^{aa} + \Omega^{ab}(\Omega^{bb})^{-1}\Omega^{ba}) \qquad (3.15)$$

$$= (S^{aa})^{-1} - (\Sigma^{aa})^{-1}. \qquad (3.16)$$

Convergence is generally rapid provided S is well-conditioned. Buhl (1993) studied conditions for the existence of the maximum likelihood estimates.

3.1.3 Deviance

Using these results, the expression for the maximized likelihood under a model can be simplified. Since $\hat{\mu} = \bar{y}$, the last term in (3.13) vanishes. And since $\hat{\Sigma}$ and S differ precisely for those elements for which $w^{ij} = 0$, it follows that $\text{tr}(\hat{\Omega}S) = \text{tr}(\hat{\Omega}\hat{\Sigma}) = q$. Thus, the maximized log likelihood under a model simplifies to

$$\hat{\ell}_m = -Nq\ln(2\pi)/2 - N\ln|\hat{\Sigma}|/2 - Nq/2.$$

Under the full model $\mathcal{M}_f$, $\hat{\Sigma} = S$, so that the maximized log likelihood for this model is

$$\hat{\ell}_f = -Nq\ln(2\pi)/2 - N\ln|\hat{S}|/2 - Nq/2.$$

The deviance of a model is thus given by

$$\begin{aligned} G^2 &= 2(\hat{\ell}_f - \hat{\ell}_m) \\ &= N\ln(|\hat{\Sigma}|/|S|), \end{aligned} \tag{3.17}$$

and the deviance difference for $\mathcal{M}_0 \subseteq \mathcal{M}_1$ by

$$d = N\ln(|\hat{\Sigma}_0|/|\hat{\Sigma}_1|),$$

where $\hat{\Sigma}_0$ and $\hat{\Sigma}_1$ are the estimates of Σ under $\mathcal{M}_0$ and $\mathcal{M}_1$, respectively. Under $\mathcal{M}_0$, d has an asymptotic χ^2 distribution with degrees of freedom given as the difference in number of free parameters ($\equiv$ number of edges) between $\mathcal{M}_0$ and $\mathcal{M}_1$. This test is satisfactory in large samples, but in small samples can be quite inaccurate.

Small samples tests (F-tests) are available in some circumstances; Eriksen (1996) provides a thorough treatment of this topic. A general result is the following: if $\mathcal{M}_0$ and $\mathcal{M}_1$ are decomposable models differing by one edge only, then an F-test of $\mathcal{M}_0 \subseteq \mathcal{M}_1$ is available. The statistic is

$$F = (N - k)(e^{d/N} - 1),$$

where k is the number of vertices in the clique in $\mathcal{M}_1$ containing the edge, and N is the number of observations. Under H_0, this has an F-distribution with 1 and $N - k$ degrees of freedom. (See Section 5.3.)

As shown by Whittaker (1990), there is a close connection between the deviance difference d associated with removing an edge $[VW]$, say, from the full model, and the sample partial correlation coefficient $\hat{\rho}^{VW.b}$, where $b = \Gamma \setminus \{V, W\}$. Namely, that

$$d = -N\ln\{(1 - (\hat{\rho}^{VW.b})^2)\}.$$

Finally, we note that expressions for asymptotic standard errors of moments and canonical parameters under the saturated model are given in Cox and Wermuth (1990), and generalized to arbitrary models in Roverato and Whittaker (1993).

3.1.4   Example: Digoxin Clearance

This is a very simple example of the use of graphical Gaussian models. The study of the passage of drugs through the body is important in medical science. Most drugs are metabolized (broken down) or eliminated primarily via the liver and kidneys. Table 3.2 shows data on 35 consecutive patients

Creatinine Clearance	Digoxin Clearance	Urine Flow	Creatinine Clearance	Digoxin Clearance	Urine Flow
19.5	17.5	0.74	66.8	37.5	0.50
24.7	34.8	0.43	72.4	50.1	0.97
26.5	11.4	0.11	80.9	50.2	1.02
31.1	29.3	1.48	82.0	50.0	0.95
31.3	13.9	0.97	82.7	31.8	0.76
31.8	31.6	1.12	87.9	55.4	1.06
34.1	20.7	1.77	101.5	110.6	1.38
36.6	34.1	0.70	105.0	114.4	1.85
42.4	25.0	0.93	110.5	69.3	2.25
42.8	47.4	2.50	114.2	84.8	1.76
44.2	31.8	0.89	117.8	63.9	1.60
49.7	36.1	0.52	122.6	76.1	0.88
51.3	22.7	0.33	127.9	112.8	1.70
55.0	30.7	0.80	135.6	82.2	0.98
55.9	42.5	1.02	136.0	46.8	0.94
61.2	42.4	0.56	153.5	137.7	1.76
63.1	61.1	0.93	201.1	76.1	0.87
63.7	38.2	0.44			

TABLE 3.2. Digoxin clearance data. Clearances are given in $ml/min/1.73m^2$, urine flow in ml/min. Source: Halkin et al. (1975).

under treatment for heart failure with the drug digoxin. The data are from Halkin et al. (1975) (see also Altman, 1991, p. 323). The quantities shown are digoxin clearance (i.e., the amount of blood that in a given interval is cleared of digoxin), creatinine clearance (defined similarly and used as a measure of kidney function), and urine flow.

Since both creatinine and digoxin are mainly eliminated by the kidneys, creatinine clearance and digoxin clearance can be expected to be correlated. There is no obvious reason for correlation with urine flow, which depends on factors such as fluid intake and temperature. Halkin et al. suspected that the elimination of digoxin might be subject to reabsorption, which might give rise to a correlation with urine flow. We study this more closely by fitting certain graphical Gaussian models. After defining the data

```
MIM->cont XYZ; label X "Cl_cre" Y "Cl_dig" Z "Urine flow"
MIM->read XYZ
19.5 17.5 0.74
    ...
```

we calculate the empirical correlations and partial correlations:

```
MIM->print uv
Empirical counts, means and correlations.
     X    1.000
     Y    0.775   1.000
     Z    0.309   0.531   1.000
 Means  76.094  52.026   1.071   35.000
              X       Y       Z    Count
Empirical discrete, linear and partial correlation parameters.
     X    1.000
     Y    0.758   1.000
     Z   -0.191   0.485   1.000
Linear   0.038  -0.015   3.199  -11.426
              X       Y       Z Discrete
```

The discrete and linear parameters refer to the parameters α and β in (3.8). We see that the partial correlation between X and Z is rather small. To test whether each partial correlation is zero, we use the Stepwise command.

```
MIM->model //XYZ
MIM->stepwise o
Coherent Backward Selection
Decomposable models, Chi-squared tests.
Single step.
Critical value:   0.0500
Initial model: //XYZ
Model: //XYZ
Deviance:   0.0000 DF:   0 P:   1.0000
    Edge        Test
 Excluded   Statistic DF          P
     [XY]    29.9633  1      0.0000 +
     [XZ]     1.3057  1      0.2532
     [YZ]     9.3747  1      0.0022 +
No change.
Selected model: //XYZ
```

The deviances of the three models formed by removing each edge in turn are displayed, together with the corresponding asymptotic χ^2 tests. It is seen that the edge $[XZ]$ can be removed, i.e., the partial correlation between X and Z does not differ significantly from zero.

Since the sample is fairly small, it is of interest to compute the corresponding F-tests; these are obtained by using the S option:

```
MIM->stepwise os
Coherent Backward Selection
Decomposable models, F-tests where appropriate.
Single step.
Critical value:   0.0500
```

```
Initial model: //XYZ
Model: //XYZ
Deviance:   0.0000 DF:   0 P:  1.0000
      Edge       Test
Excluded   Statistic DF           P
    [XY]     43.3265 1,   32 0.0000 +
    [XZ]      1.2163 1,   32 0.2783
    [YZ]      9.8288 1,   32 0.0037 +
No change.
Selected model: //XYZ
```

The results are very similar. We conclude that there is good evidence that digoxin clearance and urine flow are correlated.

3.1.5   Example: Anxiety and Anger

Cox and Wermuth (1993) describe a set of psychological data obtained from C. Spielberger (see Spielberger et al., 1970, 1983). There are four variables measured on 684 female students. The variables are anxiety state (W), anger state (X), anxiety trait (Y), and anger trait (Z). The trait variables are viewed as stable personality characteristics and the state variables as pertaining to behaviour in specific situations. The example is also treated in Wermuth (1991). Psychological theory suggests that $W \perp\!\!\!\perp Z \mid (X, Y)$ and $X \perp\!\!\!\perp Y \mid (W, Z)$. We define the variables and read the data into the program as follows:

```
MIM->cont WXYZ
MIM->label W "Anxiety st" X "Anger st" Y "Anxiety tr" Z "Anger tr"
MIM->sread WXYZ
DATA->684
DATA->18.8744 15.2265 21.2019 23.4217
DATA->37.1926
DATA->24.9311 44.8472
DATA->21.6056 17.8072 32.2462
DATA->15.6907 21.8565 18.3523 43.1191 !
```

We examine the independence structure using the Stepwise command:

```
MIM->model //WXYZ
MIM->stepwise o
Coherent Backward Selection
Decomposable models, chi-squared tests.
Single step.
Critical value:   0.0500
Initial model: //WXYZ
Model: //WXYZ
Deviance:   0.0000 DF:   0 P:  1.0000
```

```
      Edge        Test
   Excluded   Statistic DF        P
      [WX]     153.8998  1    0.0000 +
      [WY]     171.5093  1    0.0000 +
      [WZ]       1.2212  1    0.2691
      [XY]       0.3313  1    0.5649
      [XZ]      78.0384  1    0.0000 +
      [YZ]      72.9794  1    0.0000 +
   No change.
   Selected model: //WXYZ
```

The tests for the removal of $[WZ]$ and $[XY]$ are not rejected. We delete these edges from the model and fit the resulting model.

```
MIM->delete WZ,XY
MIM->print
The current model is //WX,WY,XZ,YZ.
MIM->fit
Deviance:       2.1033 DF: 2
MIM->test
Test of H0: //WX,WY,XZ,YZ
against H:  //WXYZ
LR:   2.1033    DF:   2    P: 0.3494
```

The χ^2 test indicates that the model fits the data well; the graph is shown in Figure 3.1. We write out the parameter estimates:

```
MIM->print f
Fitted counts, means and covariances.
      W    37.193
      X    24.931  44.847
      Y    21.606  17.022  32.246
      Z    16.886  21.856  18.352  43.119
   Means  18.874  15.226  21.202  23.422  684.000
             W       X       Y       Z    Count
MIM->print i
Fitted discrete, linear and partial correlation parameters.
      W     1.000
      X     0.449   1.000
      Y     0.471   0.000   1.000
      Z     0.000   0.318   0.307   1.000
   Linear  0.179  -0.074   0.378   0.350  -19.537
             W       X       Y       Z Discrete
```

3.1.6 Example: Mathematics Marks

These data are taken from Mardia, Kent and Bibby (1979). The variables are examination marks for 88 students on five different subjects, namely

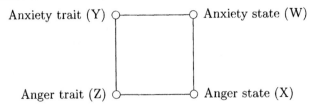

FIGURE 3.1. The anxiety and anger model. We see that like correlates with like: the pair of variables related to anxiety are connected, as are the pair related to anger, the two state variables, and the two trait variables.

mechanics (V), vectors (W), algebra (X), analysis (Y), and statistics (Z). All are measured on the same scale (0-100). Whittaker (1990) bases his exposition of graphical Gaussian models on this example.

The data set is moderately large, and is therefore stored on a file, say \data\mkb, to be read into MIM. The file has the following appearance:

```
cont VWXYZ
label V "mechanics" W "vectors" X "algebra" Y "analysis"
label Z "statistics"
read VWXYZ
77 82 67 67 81
63 78 80 70 81
75 73 71 66 81
    . . . .
    . . . .
5 26 15 20 20
0 40 21 9 14 !
```

The data are read into MIM, and the parameters for the saturated model are examined:

```
MIM->input \data\mkb
Reading completed.
MIM->print s
Calculating marginal statistics...
Empirical counts, means and covariances
      V   302.293
      W   125.777 170.878
      X   100.425  84.190 111.603
      Y   105.065  93.597 110.839 217.876
      Z   116.071  97.887 120.486 153.768 294.372
  Means  38.955  50.591  50.602  46.682  42.307   88.000
            V       W       X       Y       Z    Count
```

The correlation and partial correlation matrices are more informative:

```
MIM->print uv
Empirical counts, means and correlations
      V    1.000
      W    0.553  1.000
      X    0.547  0.610  1.000
      Y    0.409  0.485  0.711  1.000
      Z    0.389  0.436  0.665  0.607  1.000
  Means  38.955 50.591 50.602 46.682 42.307  88.000
           V      W      X      Y      Z    Count
Empirical discrete, linear and partial correlations
      V    1.000
      W    0.329  1.000
      X    0.230  0.281  1.000
      Y   -0.002  0.078  0.432  1.000
      Z    0.025  0.020  0.357  0.253  1.000
 Linear  -0.064  0.152  0.497 -0.021 -0.074 -29.882
           V      W      X      Y      Z Discrete
```

We note that all the correlations are positive, reflecting the fact that students that do well on one subject are apt to do well on the others. Apart from this, the correlation matrix does not appear to exhibit any particular structure. The partial correlation matrix, however, reveals a block of elements rather close to zero, namely between (Y, Z) and (V, W). To examine the edge removal deviances, we use the Stepwise command:

```
MIM->satmod; stepwise o
Coherent Backward Selection
Decomposable models, Chi-squared tests.
Single step.
Critical value:   0.0500
Initial model: //VWXYZ
Model: //VWXYZ
Deviance:   0.0000 DF:   0 P:  1.0000
    Edge       Test
 Excluded   Statistic DF         P
    [VW]      10.0999  1      0.0015 +
    [VX]       4.8003  1      0.0285 +
    [VY]       0.0002  1      0.9880
    [VZ]       0.0532  1      0.8176
    [WX]       7.2286  1      0.0072 +
    [WY]       0.5384  1      0.4631
    [WZ]       0.0361  1      0.8494
    [XY]      18.1640  1      0.0000 +
    [XZ]      11.9848  1      0.0005 +
    [YZ]       5.8118  1      0.0159 +
No change.
```

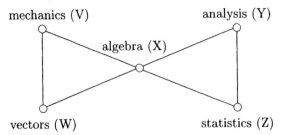

FIGURE 3.2. The graph of $//VWX, XYZ$, which resembles a butterfly.

The edge removal deviances for the four edges are very small. To fit the model without the edges, we specify:

```
MIM->delete VY,VZ,WY,WZ; fit; test
Deviance:      0.8957 DF: 4
Test of HO: //VWX,XYZ
against H:  //VWXYZ
LR:    0.8957   DF:    4    P: 0.9252
```

The model fits the data very well; its graph is shown in Figure 3.2. It states that the marks for analysis and statistics are conditionally independent of mechanics and vectors, given algebra. One implication of the model is that to predict the statistics marks, the marks for algebra and analysis are sufficient. Algebra is evidently of central importance.

The model parameters can be examined by specifying

```
MIM->print ihu
Empirical counts, means and correlations
     V    1.000
     W    0.553   1.000
     X    0.547   0.610   1.000
     Y    0.409   0.485   0.711   1.000
     Z    0.389   0.436   0.665   0.607   1.000
  Means  38.955  50.591  50.602  46.682  42.307  88.000
            V       W       X       Y       Z    Count
Fitted counts, means and correlations
     V    1.000
     W    0.553   1.000
     X    0.547   0.610   1.000
     Y    0.389   0.433   0.711   1.000
     Z    0.363   0.405   0.665   0.607   1.000
  Means  38.955  50.591  50.602  46.682  42.307  88.000
            V       W       X       Y       Z    Count
Fitted discrete, linear and partial correlation parameters
     V    1.000
     W    0.332   1.000
```

```
      X      0.235   0.327   1.000
      Y      0.000   0.000   0.451   1.000
      Z     -0.000   0.000   0.364   0.256   1.000
 Linear     -0.066   0.146   0.491  -0.010  -0.073  -29.830
              V       W       X       Y       Z Discrete
```

Observe that the fitted and empirical correlations agree on the submatrices corresponding to the cliques of the graph, i.e., $\{V, W, X\}$ and $\{X, Y, Z\}$, as required by the likelihood equations.

The following fragment calculates F-tests, first for removal of edges present in the selected model, and then for addition of edges not present.

```
MIM->stepwise os
Coherent Backward Selection
Decomposable models, F-tests where appropriate.
Single step.
Critical value:   0.0500
Initial model: //VWX,XYZ
Model: //VWX,XYZ
Deviance:    0.8957 DF:    4 P:   0.9252
      Edge        Test
   Excluded    Statistic DF          P
     [VW]       10.5009  1,  85 0.0017 +
     [VX]        9.5037  1,  85 0.0028 +
     [WX]       20.4425  1,  85 0.0000 +
     [XY]       31.0908  1,  85 0.0000 +
     [XZ]       17.9095  1,  85 0.0001 +
     [YZ]        5.9756  1,  85 0.0166 +
No change.
Selected model: //VWX,XYZ
MIM->stepwise ofs
Non-coherent Forward Selection
Decomposable models, F-tests where appropriate.
Single step.
Critical value:   0.0500
Initial model: //VWX,XYZ
Model: //VWX,XYZ
Deviance:    0.8957 DF:    4 P:   0.9252
      Edge        Test
    Added     Statistic DF          P
     [VY]        0.1057  1,  85 0.7459
     [VZ]        0.1432  1,  85 0.7061
     [WY]        0.7384  1,  85 0.3926
     [WZ]        0.2365  1,  85 0.6280
No change.
Selected model: //VWX,XYZ
```

There is no need to modify the model. We study these data further in Sections 4.6.4, 6.6, and 6.7.

3.2   Regression Models

Up to now in this chapter, we have assumed that the q continuous variables in Γ have a joint multivariate normal distribution. Within the same framework of graphical Gaussian models, we can work with *covariates*, i.e., we can consider some variables as fixed. For these variables, we need make no distributional assumptions.

Suppose that the q variables consist of q_1 response variables and $q_2 = q - q_1$ explanatory variables, so that we write y as (y_1, y_2) where y_1 is a q_1 vector of responses and y_2 is a q_2 vector of explanatory variables. We partition similarly $\Gamma = (\Gamma_1, \Gamma_2)$, $\mu = (\mu^1, \mu^2)$, etc. We assume the multivariate regression framework, which we can write as the model

$$Y_1 = A + B y_2 + V. \tag{3.18}$$

Here, A is a q_1 vector of intercepts, B is a $q_1 \times q_2$ matrix of regression coefficients, and V is a random q_1 vector with the $\mathcal{N}(0, \Psi)$ distribution.

To work with these conditional models in the joint framework, we restrict attention to graphical Gaussian models that contain all edges between the explanatory variables. As explained more carefully in Section 4.2 below, it then follows that likelihood ratio tests between nested conditional models are identical to tests between the corresponding joint models, i.e., the models in which the covariates are assumed to be random. Similarly, MLEs for the conditional models are directly obtained from MLEs for the joint models.

What do these conditional models look like? To see this, consider the conditional distribution of Y_1 given $Y_2 = y_2$. We know from standard results that this is multivariate normal with mean

$$\mu^{1\cdot 2} = \mu^1 + \Sigma^{12}(\Sigma^{22})^{-1}(y_2 - \mu^2) \tag{3.19}$$

and covariance

$$\Sigma^{11\cdot 2} = \Sigma^{11} - \Sigma^{12}(\Sigma^{22})^{-1}\Sigma^{21}. \tag{3.20}$$

So equating (3.18) with (3.19) and (3.20) we find that the matrix of regression coefficients $B = \Sigma^{12}(\Sigma^{22})^{-1}$, the intercept $A = \mu^1 - B\mu^2$, and the covariance matrix $\Psi = \Sigma^{11\cdot 2} = \Sigma^{11} - B\Sigma^{21}$.

Imposing a graphical model will constrain the A, B, and Ψ parameters in various ways. The most direct way to see this is to calculate the canonical parameters corresponding to $\mu^{1\cdot 2}$ and $\Sigma^{11\cdot 2}$, namely

$$\beta^{1\cdot 2} = (\Sigma^{11\cdot 2})^{-1}\mu^{1\cdot 2} = \beta^1 - \Omega^{12}y_2,$$

together with

$$\Omega^{11\cdot 2} = (\Sigma^{11\cdot 2})^{-1} = \Omega^{11}.$$

The expression $\beta^1 - \Omega^{12}y_2$ represents a model for the linear canonical parameters $\beta^{1\cdot 2}$. It includes covariates (y_2) and coefficients to these (the elements of Ω^{12}).

Note that the expressions for $\beta^{1\cdot 2}$ and $\Omega^{11\cdot 2}$ are considerably simpler than those for $\mu^{1\cdot 2}$ and $\Sigma^{11\cdot 2}$. It is interesting to observe that whereas the moments parameters are invariant under marginalization, as in (μ^1, Σ^{11}), the canonical parameters display similiar invariance under conditioning, as in $(\beta^1 - \Omega^{12}y_2, \Omega^{11})$.

For some models, the linear structure imposed on the $\beta^{1\cdot 2}$ induces the same linear structure on the $\mu^{1\cdot 2}$. This is the case when all response variables have identical formulae, i.e., the rows of Ω^{12} have zeros in the same places.

Conventional multivariate regression models posit the same model for the mean of each response variable and an unrestricted covariance matrix, so these can be handled as graphical Gaussian models in which the rows of Ω^{12} have zeros in the same places, and Ω^{11} has no zeros. (See Section 4.3 for further study of these models).

If some elements of Ω^{11} are set to zero, then this allows the covariance structure between the response variables to be modelled. For example, suppose that $\Gamma_1 = \{X, Y, Z\}$ and $\Gamma_2 = \{V, W\}$, and consider the model $//XYVW, YZVW$. This specifies that $Z \perp\!\!\!\perp X \mid (V, W, Y)$. The inverse covariance matrix has the form

$$\Omega = \begin{pmatrix} \omega^{XX} & \omega^{XY} & 0 & \omega^{XV} & \omega^{XW} \\ \omega^{YX} & \omega^{YY} & \omega^{YZ} & \omega^{YV} & \omega^{YW} \\ 0 & \omega^{ZY} & \omega^{ZZ} & \omega^{ZV} & \omega^{ZW} \\ \omega^{VX} & \omega^{VY} & \omega^{VZ} & \omega^{VV} & \omega^{VW} \\ \omega^{WX} & \omega^{WY} & \omega^{WZ} & \omega^{WV} & \omega^{WW} \end{pmatrix}.$$

The linear canonical parameter in the conditional distribution has the form

$$\beta^{1\cdot 2} = \beta^1 - \Omega^{12}\begin{pmatrix} v \\ w \end{pmatrix}$$

$$= \beta^1 - \begin{pmatrix} \omega^{XV} & \omega^{XW} \\ \omega^{YV} & \omega^{YW} \\ \omega^{ZV} & \omega^{ZW} \end{pmatrix}\begin{pmatrix} v \\ w \end{pmatrix},$$

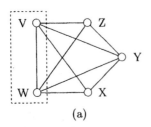
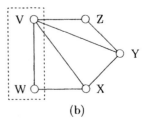

FIGURE 3.3. Graphs of two nonstandard multivariate regression models.

so using $\mu^{1 \cdot 2} = (\Omega^{11})^{-1} \beta^{1 \cdot 2}$, we see that the means of Y_1 conditional on $Y_2 = y_2$ are of the form

$$
\begin{aligned}
\mathrm{E}(X \mid V = v, W = w) &= a_1 + b_1 v + c_1 w \\
\mathrm{E}(Y \mid V = v, W = w) &= a_2 + b_2 v + c_2 w \\
\mathrm{E}(Z \mid V = v, W = w) &= a_3 + b_3 v + c_3 w,
\end{aligned}
\tag{3.21}
$$

where the parameters $\{a_i, b_i, c_i\}_{i=1\ldots 3}$ are unconstrained. Since the term ω^{XZ} is set to zero, the conditional covariance matrix is restricted, so the model is not a conventional multivariate regression model. Its independence graph is shown in Figure 3.3(a), where for clarity a box is drawn around the covariates.

If the response variables do not have identical linear structures, i.e., the rows of Ω^{12} do not have identical zero patterns, then this induces constraints on the parameters of the linear system (3.21). In some cases the model may correspond to a different linear system with unconstrained parameters; for example, if it is decomposable it will be expressible as a recursive linear system. An example is the model with graph shown in Figure 3.3(b) which is obtained from the previous graph by deleting edges $[ZW]$ and $[YW]$. It is expressible as the recursive system

$$
\begin{aligned}
X &= a_1 + b_1 v + c_1 w + \epsilon_1 \\
Y &= a_2 + b_2 v + c_2 x + \epsilon_2 \\
Z &= a_3 + b_3 v + c_3 y + \epsilon_3,
\end{aligned}
$$

where $\epsilon_i \sim \mathcal{N}(0, \sigma_i)$, for $i = 1 \ldots 3$, are independent error terms. Note that a given model may be expressible as recursive systems in many different ways (see Section 4.2 and Wermuth, 1990).

Finally, we consider a model for time series data. For example, a second order autoregressive, or AR(2), model for data measured at six timepoints can be represented as follows:

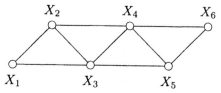

Here, we condition on $\{X_1, X_2\}$: the remaining variables are linearly dependent on the two previous variables.

We note in passing that graphical modelling for time series data is an active research topic. See Lynggaard and Walther (1993), Brillinger (1996), Dahlhaus et al. (1997), and Dahlhaus (2000).

3.2.1 Example: Determinants of Bone Mineral Content

As illustration of the application of graphical Gaussian models in the multivariate regression framework, we consider data arising in a clinical study of the effects of estrogen therapy on bone mineral content in 150 postmenopausal women (Munk-Jensen et al., 1994). There are three explanatory and three response variables. The three explanatory variables are menopausal age in weeks (U), body mass index, i.e., weight/height^2 (V), and alkaline phosphatase, an enzyme involved in bone metabolism (W). The three response variables correspond to three ways of estimating bone mineral content, namely bone mineral content determination in the upper arm (X), bone mineral density determination in the spine (Y), and bone mineral content determination in the spine (Z). All measurements were taken in a baseline period prior to active treatment. The object of the analysis is to use these cross-sectional data to see whether the explanatory variables are significant determinants of bone mineral content: this could be of use, for example, in identifying risk factors for osteoporosis.

The data are defined as follows:

```
MIM->cont UVWXYZ
MIM->label U "Men Age" V "BMI" W "Alk"
MIM->label X "BMC_arm" Y "BMD-spine" Z "BMC-spine"
MIM->read xyzuvw
.........
```

The sample count, mean, and covariance are displayed using the Print S command:

```
MIM->print s
Calculating marginal statistics...
Empirical counts, means and covariances.
     X    0.034
     Y    0.013   0.017
     Z    0.066   0.082   0.448
     U   -0.635  -0.465  -1.979 724.710
     V   -0.000   0.000   0.000   0.001   0.000
     W   -0.007  -0.008  -0.031   1.305   0.000   0.046
  Means   1.163   0.929   4.222  92.518   0.002   5.098 139.000
            X       Y       Z       U       V       W   Count
```

We see that 11 cases have missing values on one or more variables, so only 139 cases contribute to the analysis. We examine the correlation and partial correlation matrices:

```
MIM->print uv
Empirical counts, means and correlations.
     X    1.000
     Y    0.528   1.000
     Z    0.540   0.930   1.000
     U   -0.129  -0.131  -0.110   1.000
     V   -0.156   0.083   0.097   0.080   1.000
     W   -0.176  -0.265  -0.215   0.226   0.212   1.000
  Means   1.163   0.929   4.222  92.518   0.002   5.098 139.000
            X       Y       Z       U       V       W   Count
Empirical discrete, linear and partial correlation parameters.
     X    1.000
     Y    0.078   1.000
     Z    0.173   0.895   1.000
     U   -0.058  -0.039   0.023   1.000
     V   -0.239   0.044   0.073   0.029   1.000
     W    0.011  -0.172   0.072   0.183   0.223   1.000
 Linear  29.822 161.232 -15.381  -0.022-1899.335 133.305 -390.486
            X       Y       Z       U       V       W Discrete
```

We note that the partial correlations, except between the spinal measurements (Y and Z), are rather small. We can also examine the conditional distribution given the explanatory variables, by using the DisplayData command:

```
MIM->DisplayData XYZ,UVW
Empirical conditional means and covariances.
     X    1.900  -0.001  -0.112 -48.943   0.032
     Y    1.743  -0.000  -0.172  42.829   0.012   0.016
     Z    7.564  -0.002  -0.724 222.151   0.063   0.075   0.416
              U       W       V       X       Y       Z
```

The first four columns give the sample estimates of the coefficients of the conditional means of X, Y, and Z, and the last three columns show the sample estimate of the conditional covariance matrix (lower triangular part). We note that the coefficients of U are small, and that the conditional variances and covariances are only slightly smaller than the corresponding marginal quantities. Thus, only a small proportion in the variation of the response variables can be explained by the explanatory variables.

To find a simple model consistent with the data, we perform backward selection starting from the full model. We fix the edges between the covariates in the graph using the Fix command to prevent these edges from being removed. We restrict the selection process to decomposable models and use F-tests.

```
MIM->satmodel; fix UVW; stepwise s
Fixed variables: UVW
Coherent Backward Selection
Decomposable models, F-tests where appropriate.
Critical value:   0.0500
Initial model: //UVWXYZ
Model: //UVWXYZ
Deviance:  -0.0000 DF:   0 P:  1.0000
     Edge        Test
Excluded    Statistic DF         P
    [UX]       0.4430  1, 133 0.5068
    [UY]       0.1995  1, 133 0.6558
    [UZ]       0.0734  1, 133 0.7869
    [VX]       8.0852  1, 133 0.0052 +
    [VY]       0.2579  1, 133 0.6124
    [VZ]       0.7198  1, 133 0.3977
    [WX]       0.0166  1, 133 0.8977
    [WY]       4.0409  1, 133 0.0464 +
    [WZ]       0.6955  1, 133 0.4058
    [XY]       0.8239  1, 133 0.3657
    [XZ]       4.1250  1, 133 0.0442 +
    [YZ]     535.9620  1, 133 0.0000 +
Removed edge [WX]
Model: //UVWYZ,UVXYZ
Deviance:   0.0173 DF:   1 P:  0.8953
     Edge        Test
Excluded    Statistic DF         P
    [UX]       0.4297  1, 134 0.5133
    [WZ]       0.7622  1, 134 0.3842
    [XY]       0.8140  1, 134 0.3686
Removed edge [UX]
Model: //UVWYZ,VXYZ
Deviance:   0.4623 DF:   2 P:  0.7936
```

```
        Edge         Test
      Excluded    Statistic DF           P
        [UY]        0.2532   1, 134 0.6156
        [UZ]        0.0252   1, 134 0.8741
        [WZ]        0.7622   1, 134 0.3842
        [XY]        0.9172   1, 135 0.3399
    Removed edge [UZ]
    Model: //UVWY,VXYZ,VWYZ
    Deviance:   0.4885 DF:    3 P:   0.9214
        Edge         Test
      Excluded    Statistic DF           P
        [UY]        0.8954   1, 135 0.3457
        [WZ]        0.8491   1, 135 0.3584
        [XY]        0.9172   1, 135 0.3399
    Removed edge [WZ]
    Model: //UVWY,VXYZ
    Deviance:   1.3600 DF:    4 P:   0.8511
        Edge         Test
      Excluded    Statistic DF           P
        [UY]        0.8954   1, 135 0.3457
        [VZ]        1.2316   1, 135 0.2691
        [XY]        0.9172   1, 135 0.3399
    Removed edge [UY]
    Model: //UVW,VXYZ,VWY
    Deviance:   2.2789 DF:    5 P:   0.8094
        Edge         Test
      Excluded    Statistic DF           P
        [VZ]        1.2316   1, 135 0.2691
        [XY]        0.9172   1, 135 0.3399
    Removed edge [XY]
    Selected model: //UVW,VXZ,VWY,VYZ
```

The graph of the selected model is:

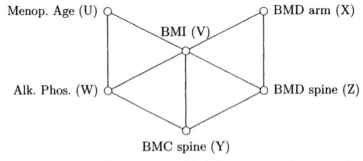

A surprising implication of the model selected is that given body mass index and the alkaline phosphatase level, menopausal age does not appear to influence the bone mineral content measurements. This would suggest that, in some sense, the influence of menopausal age on bone mineral content is mediated by the level of alkaline phosphatase and the body mass index. An alternative explanation is that these variables are confounded with menopausal age, and that the failure to find association is due to lack of power.

To see whether there are nondecomposable models that provide better fits to the data, we continue the stepwise selection procedure in unrestricted mode using χ^2-tests.

```
MIM->stepwise uo
Coherent Backward Selection
Unrestricted models, chi-squared tests.
Single step.
Critical value:    0.0500
Initial model: //UVW,VXZ,VWY,VYZ
Model: //UVW,VXZ,VWY,VYZ
Deviance:    3.2201 DF:    6 P:   0.7808
     Edge        Test
  Excluded  Statistic DF          P
     [VX]      8.8999  1       0.0029 +
     [VY]      0.0075  1       0.9311
     [VZ]      2.3733  1       0.1234
     [WY]     12.2502  1       0.0005 +
     [XZ]     53.3352  1       0.0000 +
     [YZ]    278.1634  1       0.0000 +
No change.
Selected model: //UVW,VXZ,VWY,VYZ
```

We see that the edges $[VY]$ and $[VZ]$, whose removal in each case leads to a nondecomposable model, yield nonsignificant p-values. To see whether they can both be removed, we first delete $[VY]$ and test whether $[VZ]$ can be removed:

```
MIM->delete VY; testdelete VZ
Test of H0: //UVW,VX,XZ,WY,YZ
against H:  //UVW,VXZ,WY,YZ
LR:   10.0447    DF:    1    P: 0.0015
```

This is strongly rejected. Similarly, we can replace $[VY]$, remove $[VZ]$, and then test for the removal of $[VY]$:

```
MIM->add VY; delete VZ; testdelete VY
Test of H0: //UVW,VX,XZ,WY,YZ
against H:  //UVW,VX,XZ,VWY,YZ
LR:    7.6789    DF:    1    P: 0.0056
```

This is also strongly rejected. Thus, either $[VZ]$ or $[VY]$ must be present, but it is not clear which. This is an example of nonorthogonality, due to multicollinearity between the two spinal measurements.

A reason for preferring the decomposable model first selected to the two nondecomposable submodels is that of interpretation. As we describe later in Section 4.4, a decomposable model is equivalent to a sequence of univariate regressions, whereas a nondecomposable model is not. So a decomposable model suggests a causal explanation for the data (see Cox, 1993). Of course, there is an important distinction between selecting a model that suggests a causal explanation, and claiming to have found evidence for causality. We discuss some related issues in Chapter 8.

4

Mixed Models

This chapter describes a family of models for mixed discrete and continuous variables that combine and generalize the models of the previous two chapters. Graphical models for mixed discrete and continuous variables were introduced by Lauritzen and Wermuth (1989), and both undirected and directed types of models were described. The undirected models (graphical interaction models) were extended to a broader class, the hierarchical interaction models, in Edwards (1990). The latter are constructed by combining loglinear models for discrete variables with graphical Gaussian models for continuous variables, as we now describe.

4.1 Hierarchical Interaction Models

Suppose we have p discrete variables and q continuous variables, and write the sets of variables as Δ and Γ, respectively. We write the corresponding random variables as (I, Y), and a typical observation as (i, y). Here, i is a p-tuple containing the values of the discrete variables, and y is a real vector of length q. We write $\mathcal{I}$ for the set of all possible i.

We suppose that the probability that $I = i$ is p_i, and that the distribution of Y given $I = i$ is multivariate normal $\mathcal{N}(\mu_i, \Sigma_i)$ so that both the conditional mean and covariance may depend on i. This is called the CG (conditional Gaussian) distribution. The density can be written as

$$f(i, y) = p_i |2\pi\Sigma_i|^{-\frac{1}{2}} \exp\{-\tfrac{1}{2}(y - \mu_i)'\Sigma_i^{-1}(y - \mu_i)\}. \tag{4.1}$$

The parameters $\{p_i, \mu_i, \Sigma_i\}_{i\in\mathcal{I}}$ are called the *moments* parameters.

We are often interested in models for which the covariance is constant over i, so that $\Sigma_i = \Sigma$. Such models are called homogeneous. As we shall see later, there are generally two graphical models corresponding to a given graph: a heterogeneous model and a homogeneous model.

We rewrite (4.1) in the more convenient form

$$f(i, y) = \exp\{\alpha_i + \beta_i' y - \tfrac{1}{2} y' \Omega_i y\}, \tag{4.2}$$

where α_i is a scalar, β_i is a $p \times 1$ vector, and Ω_i is a $p \times p$ symmetric positive definite matrix. These are called the *canonical* parameters. As in the previous chapter, we can transform between the moments and the canonical parameters using

$$\Omega_i = \Sigma_i^{-1}, \tag{4.3}$$
$$\beta_i = \Sigma_i^{-1} \mu_i, \tag{4.4}$$
$$\alpha_i = \ln(p_i) - \tfrac{1}{2} \ln |\Sigma_i| - \tfrac{1}{2} \mu_i' \Sigma_i^{-1} \mu_i - \tfrac{q}{2} \ln(2\pi), \tag{4.5}$$
and
$$\Sigma_i = \Omega_i^{-1}, \tag{4.6}$$
$$\mu_i = \Omega_i^{-1} \beta_i, \tag{4.7}$$
$$p_i = (2\pi)^{\frac{q}{2}} |\Omega_i|^{-\frac{1}{2}} \exp\{\alpha_i + \tfrac{1}{2} \beta_i' \Omega_i^{-1} \beta_i\}. \tag{4.8}$$

Hierarchical interaction models are constructed by restricting the canonical parameters in a similar fashion to loglinear models. That is to say, the canonical parameters are expanded as sums of interaction terms, and models are defined by setting higher-order interaction terms to zero. To introduce this, we examine some simple examples.

4.1.1   Models with One Discrete and One Continuous Variable

First, let us consider the case where $p = q = 1$, that is, where there is one discrete and one continuous variable. Let $\Delta = \{A\}$ and $\Gamma = \{Y\}$. The density can be written as

$$\begin{aligned} f(i, y) &= p_i \, (2\pi\sigma_i)^{-\frac{1}{2}} \exp\left\{-\tfrac{1}{2}(y - \mu_i)^2/\sigma_i\right\} \\ &= \exp\left\{\alpha_i + \beta_i y - \tfrac{1}{2}\omega_i y^2\right\}. \end{aligned} \tag{4.9}$$

Replacing the canonical parameters with interaction term expansions, we rewrite (4.9) as

$$f(i, y) = \exp\left\{(u + u_i^A) + (v + v_i^A)y - \tfrac{1}{2}(w + w_i^A)y^2\right\}.$$

The quadratic canonical parameter is $\omega_i = w + w_i^A$. Since $\sigma_i = \omega_i^{-1}$ we see that the cell variances are constant if $w_i^A = 0$. (We now see that we used superscripts in the previous chapter to reserve subscripts for cell indices).

The linear canonical parameter is $\beta_i = v + v_i^A$. Using $\mu_i = \omega_i^{-1}\beta_i$, we see that the cell means are constant if $w_i^A = v_i^A = 0$. Also, using the factorization criterion (1.1), we see that $A \perp\!\!\!\perp Y$ if and only if $w_i^A = v_i^A = 0$.

The discrete canonical parameter is $\alpha_i = u + u_i^A$; this corresponds to a main effects loglinear model and cannot be further simplified, i.e., we do not consider setting $u_i^A = 0$.

We are led to consider three possible models. The simplest is the model of marginal independence, formed by setting $v_i^A = w_i^A = 0$. The density is

$$f(i,y) = p_i \, (2\pi\sigma)^{-\frac{1}{2}} \exp\left\{-\tfrac{1}{2}(y-\mu)^2/\sigma^2\right\},$$

and the model formula is $A/Y/Y$.

The second model, formed by setting $w_i^A = 0$, allows the cell means to differ but constrains the variances to be homogeneous. The density is

$$f(i,y) = p_i \, (2\pi\sigma)^{-\frac{1}{2}} \exp\left\{-\tfrac{1}{2}(y-\mu_i)^2/\sigma^2\right\},$$

and the model formula is $A/AY/Y$.

The third model (the full model) has freely varying cell means and variances. The density is

$$f(i,y) = p_i \, (2\pi\sigma_i)^{-\frac{1}{2}} \exp\left\{-\tfrac{1}{2}(y-\mu_i)^2/\sigma_i^2\right\},$$

and the model formula is $A/AY/AY$.

As these examples show, model formulae for mixed models consist of three parts, separated by slashes (/). The three parts specify the interaction expansions of α_i, β_i, and Ω_i, respectively. Thus, in the second and third models above, the second part of the formulae was AY; this means that the element of β_i corresponding to Y has an expansion with formula A, i.e., has the form $\beta_i^Y = v + v_i^A$. In the model $A/AY/Y$, the term Y indicates that the quadratic canonical parameter ω_i^{YY} has a null formula; in other words, it is constant over the cells: $\omega_i^{YY} = w$.

These three models are, of course, closely related to familiar one-way ANOVA models: the only difference is that in the present setup, the cell counts are taken to be random. If we regard A as fixed, then we have precisely the one-way ANOVA setup, and the first model denotes homogeneity rather than independence (see the discussion in Section 1.1).

4.1.2 A Model with Two Discrete and Two Continuous Variables

As the next illustration, we consider a model with $p = q = 2$. Suppose that $\Delta = \{A, B\}$, $\Gamma = \{X, Y\}$, and that A and B are indexed by j and k, respectively. The canonical parameters are α_{jk},

$$\beta_{jk} = \begin{pmatrix} \beta_{jk}^X \\ \beta_{jk}^Y \end{pmatrix},$$

and

$$\Omega_{jk} = \begin{pmatrix} \omega_{jk}^{XX} & \omega_{jk}^{XY} \\ \omega_{jk}^{XY} & \omega_{jk}^{YY} \end{pmatrix}.$$

We can constrain the α_{jk} by requiring that they take the additive structure

$$\alpha_{jk} = u + u_j^A + u_k^B$$

for some parameters u, u_j^A, and u_k^B. We represent this expansion as A, B: this will be the first part of the model formula.

Similarly, we can constrain β_{jk}^X and β_{jk}^Y by requiring, for example,

$$\beta_{jk}^X = v^X + v_j^{X;A} + v_k^{X;B}, \text{ and}$$
$$\beta_{jk}^Y = v^Y + v_k^{Y;B}.$$

Thus, β_{jk}^X has additive A and B effects, and β_{jk}^Y depends on B only.

To form the second part of the formula, we combine the shorthand formula for the expansion for β_{jk}^X, A, B, with the formula for β_{jk}^Y, B, to obtain the expansion AX, BX, BY.

Finally, we model the elements of the inverse covariance Ω_{jk}. The simplest structure it can have is constant diagonal elements and zero off-diagonal elements: $\omega_{jk}^{XY} = 0$, $\omega_{jk}^{XX} = w^X$, and $\omega_{jk}^{YY} = w^Y$. The corresponding formula for Ω_{jk} is X, Y.

Now we can put together our shorthand formula for the whole model, namely,

$$A, B/AX, BX, BY/X, Y.$$

We can form the graph of this model by joining variables that occur in the same generator:

4.1.3 Model Formulae

We now continue with the general case. Suppose the model formula has the form

$$\underbrace{d_1,\ldots,d_r}_{\text{discrete}}\,/\,\underbrace{l_1,\ldots l_s}_{\text{linear}}\,/\,\underbrace{q_1,\ldots q_t}_{\text{quadratic}}\,. \tag{4.10}$$

The three parts have the following functions:

1. The discrete generators specify the expansion for α_i.

2. The linear generators specify the expansion for β_i. Each linear generator contains one continuous variable. The expansion for β_i^γ for some $\gamma \in \Gamma$ is given by the linear generators that contain γ.

3. The quadratic part gives the expansion for the inverse covariance matrix Ω_i. Each quadratic generator must contain at least one continuous variable. The expansion for $\omega_i^{\gamma\zeta}$ for $\gamma,\zeta \in \Gamma$ is given by the quadratic generators that contain γ,ζ.

Two syntax rules restrict the permissible formulae:

1. The linear generators must not be larger than the discrete generators, i.e., for each linear generator l_j there must correspond a discrete generator d_k such that $l_j \cap \Delta \subseteq d_k$. For example, $A, B/ABX/AX$ is not permitted since there is a linear generator ABX but no discrete generator containing AB.

2. The quadratic generators must not be larger than the corresponding linear generators, i.e., for each quadratic generator q_j and each continuous variable $\gamma \in q_j$, there must correspond a linear generator l_k such that $(q_j \cap \Delta) \cup \{\gamma\} \subseteq l_k$. For example, $ABC/AX, BY, CZ/AXY, CZ$ is not permitted since there is a quadratic generator AXY but no linear generator containing AY.

To motivate these rules, consider the requirement that a model be invariant under scale and location transformations of the continuous variables. Suppose $\bar{Y}$ is given by

$$\bar{Y} = A(Y + b), \tag{4.11}$$

where b is a q-vector and A is a diagonal matrix

$$A = \begin{pmatrix} a_1 & 0 & \cdots & 0 \\ 0 & a_2 & 0 & \vdots \\ \vdots & 0 & \ddots & 0 \\ 0 & \cdots & 0 & a_q \end{pmatrix},$$

with nonzero diagonal elements. Clearly, $(I, \bar{Y})$ is CG-distributed, and the moments parameters $\{\bar{p}_i, \bar{\mu}_i, \bar{\Sigma}_i\}_{i \in I}$ are given as

$$\bar{p}_i = p_i,$$
$$\bar{\mu}_i = A(\mu_i + b), \quad \text{and}$$
$$\bar{\Sigma}_i = A\Sigma_i A.$$

Using (4.3-4.5), we can derive the corresponding canonical parameters. In particular, we obtain

$$\bar{\beta}_i = \bar{\Sigma}_i^{-1}\bar{\mu}_i$$
$$= A^{-1}(\beta_i + \Omega_i b). \tag{4.12}$$

If the model is invariant under the transformation, the new linear canonical parameter $\bar{\beta}_i$ must be subject to the same constraints as the original β_i. In other words, it must have the same range as a function of i. From (4.12), we see that for each $\gamma \in \Gamma$, the range of $\bar{\beta}_i^\gamma$ as a function of i encompasses the range of $\omega_i^{\gamma\eta}$ terms for all $\eta \in \Gamma$. This is precisely the effect of the second syntax rule above.

Similarly, we obtain

$$\bar{\alpha}_i = \alpha_i - b'\beta_i - \tfrac{1}{2}b'\Omega_i b,$$

so that the range of $\bar{\alpha}_i$ as a function of i encompasses the ranges of β_i^γ and $\omega_i^{\gamma\eta}$ terms for $\gamma, \eta \in \Gamma$. This is ensured by the two syntax rules, so the rules ensure that the models are invariant under such transformations.

We note in passing that the model formula syntax implicitly introduces another constraint. For example, if an off-diagonal precision element $\omega^{\gamma\zeta}$ depends on i, so must both corresponding diagonal elements, $\omega^{\gamma\gamma}$ and $\omega^{\zeta\zeta}$. More generally, the range of $\omega_i^{\gamma\gamma}$ as a function of i must encompass the range of $\omega_i^{\gamma\zeta}$, for all $\gamma, \zeta \in \Gamma$. As pointed out by several authors (Edwards, 1990; Lauritzen, 1996, section 6.4.1) the model family could be usefully extended by removing this constraint (which would involve adopting a different model formula syntax). Some simple models fall outside the hierarchical interaction models as we have defined them above, but fall within this extended class: one such arises with $\Delta = \{I\}$ and $\Gamma = \{X, Y\}$. If we wish the regresssion of Y on X and I to involve nonparallel regression lines but homogeneous variances, then it turns out (see Section 4.3) that for the nonparallel lines we must introduce a term ω_i^{XY} by including a quadratic generator IXY in the formula. But this forces ω^{YY} also to depend on i, so the conditional variance becomes heterogeneous.

4.1.4 Formulae and Graphs

To study the correspondence between model formulae and graphs, we can expand (4.2) as

$$f(i,y) = \exp\{\alpha_i + \sum_{\gamma \in \Gamma} \beta_i^\gamma y_\gamma - \tfrac{1}{2} \sum_{\gamma \in \Gamma} \sum_{\eta \in \Gamma} \omega_i^{\gamma\eta} y_\gamma y_\eta\} \qquad (4.13)$$

and then apply the factorization criterion (1.1) to examine the pairwise Markov properties implied by a given model.

For two discrete variables in the model, say A and B, $A \perp\!\!\!\perp B \mid$ (the rest) holds whenever all of the interaction terms involving A and B are set to zero. That is to say, none of the expansions for α_i, β_i^γ, or $\omega_i^{\gamma\eta}$, for any $\gamma, \eta \in \Gamma$, may contain an AB interaction. In terms of the model formula, we just require that no discrete generator contains AB since the syntax rules then imply that no linear or quadratic generator may contain AB either.

If A is discrete and X is continuous, we see that $A \perp\!\!\!\perp X \mid$ (the rest) holds whenever all of the interaction terms involving A and X are set to zero. That is to say, none of the expansions for β_i^X or $\omega_i^{X\eta}$ for any $\eta \in \Gamma$ may contain an interaction term involving A. In terms of the model formula, we just require that no linear generator may contain AX since the syntax rules then imply that no quadratic generator will contain AX either.

For two continuous variables, say X and Y, $X \perp\!\!\!\perp Y \mid$ (the rest) holds whenever ω_i^{XY} is set to zero. In terms of the model formula, this means that no quadratic generator may include XY.

These results make it easy to derive the independence graph from a model formula. We simply connect vertices that appear in the same generator. For example, the graph of $AB/AX, BX, AY, BZ/XY, XZ$ is shown in Figure 4.1. Note that different models may have the same graph: for example, $AB/ABX, AY, BZ/AXY, BXZ$ and $AB/ABX, AY, BZ/XY, YZ$ both have the graph shown in Figure 4.1.

We now consider the reverse operation, that is, finding the formula of the graphical model corresponding to a given graph $\mathcal{G}$, by identifying the maximal interactions that are consistent with $\mathcal{G}$. To be more precise, we

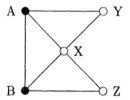

FIGURE 4.1. A graph on five vertices.

Model formula	in $\mathcal{H}_G^+$	in $\mathcal{H}_G$	in $\mathcal{H}_D$	Graph
$A/AX/X$	no	yes	yes	A •———○ X
$A/AX/AX$	yes	no	yes	A •- - -○ X
$A/AX/XY$	no	yes	yes	A connected to X, Y
$A/AX,Y/AX,XY$	yes	no	yes	A connected to X, Y
$A/AX,AY/X,Y$	no	yes	yes	A connected to X, Y
$A/AX,AY/AX,Y$	no	no	no	A connected to X, Y
$AB/AX,BX/X$	no	no	no	triangle A, B, X
$AB/ABX/X$	no	yes	yes	triangle A, B, X
$AB/AX,BY/AX,BY$	yes	no	yes	A—X, B—Y, A—B
$A,B/AX,AY,BX,BY/X,Y$	no	yes	no	A, X, Y, B square

TABLE 4.1. Some hierarchical interaction models. $\mathcal{H}_G^+$ is the class of heterogeneous graphical models, $\mathcal{H}_G$ is the class of homogeneous graphical models, and $\mathcal{H}_D$ is the class of decomposable models.

associate *two* graphical models with a given graph— a homogeneous and a heterogeneous one. Consider again Figure 4.1.

The discrete generators are given as the cliques of $\mathcal{G}_\Delta$, i.e., the subgraph of $\mathcal{G}$ on the discrete variables. In Figure 4.1, this is just AB, so the first part of the formula is AB.

For the linear part of the formula, we need to find the cliques of $\mathcal{G}_{\Delta \cup \{\gamma\}}$ that contain γ, for each $\gamma \in \Gamma$. In Figure 4.1, this will give generators ABX, AY and BZ, so the linear part is ABX, AY, BZ.

For the quadratic part, it depends on which graphical model we are interested in: the homogeneous or the heterogeneous one. For the homogeneous model, we need to identify the cliques of $\mathcal{G}_\Gamma$. In the present example, the cliques of $\mathcal{G}_\Gamma$ are $\{X, Y\}$ and $\{X, Z\}$, so we get the formula

$$AB/ABX, AY, BZ/XY, XZ.$$

For the heterogeneous model, we need to find the cliques of $\mathcal{G}$ that intersect Γ. In Figure 4.1, the cliques are $\{A, X, Y\}$, $\{A, B, X\}$, and $\{B, X, Z\}$, so that we obtain the formula

$$AB/ABX, AY, BZ/AXY, ABX, BXZ.$$

4.1.5 Maximum Likelihood Estimation

Models, by their nature, need data. Suppose we have a sample of N independent, identically distributed observations $(i^{(k)}, y^{(k)})$ for $k = 1 \ldots N$, where i is a p-tuple of levels of the discrete variables, and y is a q-vector. Let $(n_j, t_j, \overline{y}_j, SS_j, S_j)_{j \in \mathcal{I}}$ be the observed counts, variate totals, variate means, uncorrected sums of squares and products, and cell variances for cell j, i.e.,

$$n_i = \#\{k : i^{(k)} = i\},$$
$$t_i = \sum_{k:i^{(k)}=i} y^{(k)},$$
$$\overline{y}_i = t_i/n_i,$$
$$SS_i = \sum_{k:i^{(k)}=i} y^{(k)}(y^{(k)})',$$
$$S_i = \sum_{k:i^{(k)}=i} (y^{(k)} - \overline{y}_i)(y^{(k)} - \overline{y}_i)'/n_i$$
$$= SS_i/n_i - \overline{y}_i \overline{y}_i'.$$

We also need a notation for some corresponding marginal quantities. For $a \subseteq \Delta$, we write the marginal cell corresponding to i as i_a and likewise for $d \subseteq \Gamma$, we write the subvector of y as y^d. Similarly, we write the marginal cell counts as $\{n_{i_a}\}_{i_a \in \mathcal{I}_a}$, marginal variate totals as $\{t_{i_a}^d\}_{i_a \in \mathcal{I}_a}$, and marginal uncorrected sums of squares and products as $\{SS_{i_a}^d\}_{i_a \in \mathcal{I}_a}$.

Consider now a given model with formula $d_1, \ldots, d_r/l_1, \ldots, l_s/q_1, \ldots, q_t$. From (4.2), it is straightforward to show that a set of minimal sufficient statistics is given by

1. A set of marginal tables of cell counts $\{n_{i_a}\}_{i_a \in \mathcal{I}_a}$ corresponding to the discrete generators, i.e., for $a = d_1, \ldots, d_r$.

2. A set of marginal variate totals $\{t_{i_a}^\gamma\}_{i_a \in \mathcal{I}_a}$ corresponding to the linear generators, i.e., for $a = l_j \cap \Delta$, $\gamma = l_j \cap \Gamma$, for $j = 1, \ldots, s$.

3. A set of marginal tables of uncorrected sums and squares $\{SS_{i_a}^d\}_{i_a \in \mathcal{I}_a}$ corresponding to the quadratic generators, i.e., for $a = q_j \cap \Delta$, and $d = q_j \cap \Gamma$, for $j = 1, \ldots, t$.

As we have seen, models are constructed by constraining the canonical parameters through factorial interaction expansions. Given a set of data, we wish to estimate the model parameters subject to these constraints by maximum likelihood estimation. From exponential family theory, we know that the MLEs can be found by equating the expectations of the minimal sufficient statistics with their observed values. That is, for $a = d_1 \ldots d_r$,

$$\{m_{i_a}\}_{i_a \in \mathcal{I}_a} = \{n_{i_a}\}_{i_a \in \mathcal{I}_a}; \tag{4.14}$$

for $a \cup \gamma = l_1 \ldots l_s$,

$$\{\sum_{j:j_a=i_a} m_j \mu_j^\gamma\}_{i_a \in \mathcal{I}_a} = \{t_{i_a}^\gamma\}_{i_a \in \mathcal{I}_a}; \tag{4.15}$$

and for $a \cup d = q_1 \ldots q_t$,

$$\left\{\sum_{j:j_a=i_a} m_j \left[\Sigma_j^{dd} + \mu_j^d(\mu_j^d)'\right]\right\}_{i_a \in \mathcal{I}_a} = \{SS_{i_a}^d\}_{i_a \in \mathcal{I}_a}. \tag{4.16}$$

These are the likelihood equations. The MLE, when this exists, is the unique solution to these equations that also satisfies the model constraints. General conditions for existence are complex even in the discrete case (Glonek, Darroch, and Speed, 1988). For decomposable models, explicit expressions are available, as described in Section 4.4, but in general the equations must be solved iteratively (see Appendix D.1).

4.1.6 Deviance

In this section, we derive some expressions for the deviance. The log density can be written as

$$\ln f(i, y) = \ln p_i - q \ln(2\pi)/2 - \ln |\Sigma_i|/2 - (y - \mu_i)'\Sigma_i^{-1}(y - \mu_i)/2,$$

so the log likelihood of a sample $(i^{(k)}, y^{(k)})$, $k = 1, \ldots, N$, is

$$\begin{aligned}
\ell &= \sum_i n_i \ln p_i - N q \ln(2\pi)/2 - \sum_i n_i \ln |\Sigma_i|/2 \\
&\quad - \sum_{k=1}^N (y^{(k)} - \mu_{i^{(k)}})'\Sigma_{i^{(k)}}^{-1}(y^{(k)} - \mu_{i^{(k)}})/2.
\end{aligned}$$

We can simplify the last term using

$$\sum_i \sum_{k:i^{(k)}=i} (y^{(k)} - \mu_i)'\Sigma_i^{-1}(y^{(k)} - \mu_i) =$$

$$\sum_i \{n_i \text{tr}(S_i \Sigma_i^{-1}) + n_i(\bar{y}_i - \mu_i)' \Sigma_i^{-1}(\bar{y}_i - \mu_i)\}.$$

So an alternative expression for the log likelihood is

$$\ell = \sum_i n_i \ln p_i - Nq \ln(2\pi)/2 - \sum_i n_i \ln|\Sigma_i|/2$$
$$- \sum_i n_i \text{tr}(S_i \Sigma_i^{-1})/2 - \sum_i n_i(\bar{y}_i - \mu_i)' \Sigma_i^{-1}(\bar{y}_i - \mu_i)/2.$$

For the full heterogeneous model, from the likelihood equations (4.14–4.16) we obtain that $\hat{p}_i = n_i/N$, $(\hat{m}_i = N\hat{p}_i)$, $\hat{\mu}_i = \bar{y}_i$, and $\hat{\Sigma}_i = S_i$, so the maximized log likelihood for this model is

$$\hat{\ell}_f = \sum_i n_i \ln(n_i/N) - Nq \ln(2\pi)/2 - \sum_i n_i \ln|S_i|/2 - Nq/2. \qquad (4.17)$$

Thus, if a model has MLEs $\hat{p}_i$, $\hat{\mu}_i$, and $\hat{\Sigma}_i$, its deviance with respect to the full heterogeneous model is

$$2\sum_i n_i \ln(n_i/\hat{m}_i) - \sum_i n_i \ln|S_i \hat{\Sigma}_i^{-1}| + \sum_i n_i \{\text{tr}(S_i \hat{\Sigma}_i^{-1}) - q\}$$
$$+ \sum_i n_i(\bar{y}_i - \hat{\mu}_i)' \hat{\Sigma}_i^{-1}(\bar{y}_i - \hat{\mu}_i).$$

The full homogeneous model has MLEs $\hat{p}_i = n_i/N$, $(\hat{m}_i = N\hat{p}_i)$, $\hat{\mu}_i = \bar{y}_i$, and $\hat{\Sigma} = \hat{\Sigma}_i = S$, where $S = \sum_i n_i S_i/N$, so the maximized log likelihood for this model is

$$\hat{\ell}_f^h = \sum_i n_i \ln(n_i/N) - Nq \ln(2\pi)/2 - N \ln|S|/2 - Nq/2, \qquad (4.18)$$

and the deviance of a homogeneous model with MLEs $\hat{p}_i$, $\hat{\mu}_i$, and $\hat{\Sigma}$ with respect to the full homogeneous model simplifies to

$$2\sum_i n_i \ln(n_i/\hat{m}_i) - N \ln|S \hat{\Sigma}^{-1}| + N \{\text{tr}(S \hat{\Sigma}^{-1}) - q\}$$
$$+ \sum_i n_i(\bar{y}_i - \hat{\mu}_i)' \hat{\Sigma}^{-1}(\bar{y}_i - \hat{\mu}_i).$$

For two models $\mathcal{M}_0 \subseteq \mathcal{M}_1$, the deviance difference has under $\mathcal{M}_0$ an asymptotic χ^2 distribution with degrees of freedom given as the difference in the number of free parameters between the two models. For small samples, tests based on this asymptotic χ^2 distribution are unreliable. We describe a number of alternative tests in the next chapter.

Diet	Coagulation times						
1	62	60	63	59			
2	63	67	71	64	65	66	
3	68	66	71	67	68	68	
4	56	62	60	61	63	64	63	59

TABLE 4.2. Coagulation time (seconds) for blood drawn from 24 animals randomly allocated to four different diets.

4.1.7 A Simple Example

It is instructive to illustrate these results with a simple numerical example. We consider the simplest of the mixed models— those that involve one discrete and one continuous variable. Table 4.2 shows data from an experiment examining the effect of diet on coagulation time for a group of animals.

We can summarize the data in terms of cell counts, means, and sample variances defined through

$$n_j = \#\{k : i^{(k)} = j\},$$

$$\bar{x}_j = \sum_{k:i^{(k)}=j} x^{(k)}/n_j, \quad \text{and}$$

$$s_j = \sum_{k:i^{(k)}=j} (x^{(k)} - \bar{x}_j)^2/n_j.$$

In the example, these are

j	n_j	$\bar{x}_j$	s_j
1	4	61.0	2.50
2	6	66.0	6.67
3	6	68.0	2.33
4	8	61.0	6.00

As described in Section 4.1.1, there are three possible models we can apply. These are

$$\mathcal{M}_0 : A/X/X,$$
$$\mathcal{M}_1 : A/AX/X, \quad \text{and}$$
$$\mathcal{M}_2 : A/AX/AX.$$

Consider $\mathcal{M}_2$ first. This says that the cells have different means and variances. From the likelihood equations, we find, after some coaxing, that the MLEs are given by

$$\hat{p}_j = n_j/N,$$
$$\hat{\mu}_j = \bar{x}_j, \quad \text{and}$$
$$\hat{\sigma}_j = s_j.$$

In this example, the parameter p_j and its estimate $\hat{p}_j$ are without interest. We regard A as fixed, and use the joint framework to estimate the parameters μ_j and σ_j. Section 4.2 studies the relation between conditional and joint models in general terms.

The maximized log likelihood under $\mathcal{M}_2$ is

$$\hat{\ell}_2 = \sum_i n_i \ln(n_i/N) - N\ln(2\pi)/2 - \sum_i n_i \ln(s_i)/2 - N/2.$$

Under $\mathcal{M}_1$, the cell means differ but the variances are constant. The likelihood equations supply the MLEs as

$$\hat{p}_j = n_j/N,$$
$$\hat{\mu}_j = \bar{x}_j, \quad \text{and}$$
$$\hat{\sigma} = \sum_j \sum_{k:i^{(k)}=j} (x^{(k)} - \bar{x}_j)^2/N = s.$$

In the example $\hat{\sigma} = s = 4.67$. The maximized likelihood under $\mathcal{M}_1$ is

$$\hat{\ell}_1 = \sum_i n_i \ln(n_i/N) - N\ln(2\pi)/2 - N\ln(s)/2 - N/2.$$

A test of $\mathcal{M}_1 \subseteq \mathcal{M}_2$ is a test of variance homogeneity. The deviance difference is given by $d = N\ln(s) - \sum_i n_i \ln(s_i)$. Under $\mathcal{M}_1$ this has an asymptotic χ^2 distribution with $\#A - 1$ degrees of freedom, where $\#A$ is the number of levels of A. In the example, $d = 2.5049$, giving $P = 0.4744$, so there is no evidence of heterogeneity.

This test for homogeneity has been studied by various authors and it has been found that the asymptotic approximation is rather poor. Various modifications to improve the test have been proposed; in particular, one due to Box (1949) is described in Section 5.13. In the present example, Box's test has a value of 1.6539. If we compare this with the $\chi^2_{(3)}$ distribution, we obtain $p = 0.6472$, which confirms our previous conclusion that there is no evidence of variance heterogeneity.

The model $\mathcal{M}_0$, which states that A and X are independent, has MLEs given as

$$\hat{p}_j = n_j/N,$$
$$\hat{\mu}_j = \bar{x} = \sum_{k=1}^{N} x^{(k)}/N, \quad \text{and}$$

$$\hat{\sigma}_j = \sum_{k=1}^{N} (x^{(k)} - \bar{x})^2/N = s_0.$$

The maximized log likelihood under this model is

$$\hat{\ell}_0 = \sum_i n_i \ln(n_i/N) - N\ln(2\pi)/2 - N\ln(s_0)/2 - N/2.$$

A test of $\mathcal{M}_0 \subseteq \mathcal{M}_1$ is a test of equality of means assuming homogeneous variances. In contrast, a test of $\mathcal{M}_0 \subseteq \mathcal{M}_2$ will seek to detect inequality of both means and variances.

The deviance difference between $\mathcal{M}_0$ and $\mathcal{M}_1$ is $d = N\ln(s_0/s)$. Under $\mathcal{M}_0$, this has an asymptotic χ^2 distribution with $\#A - 1$ degrees of freedom. In the present example, we find $d = 26.6507$, corresponding to $P = 0.0000$, so it is very clear that the diets are associated with differing mean coagulation times.

There is, however, no good reason to rely on the asymptotic distribution of d under $\mathcal{M}_0$, since the small sample distribution of a simple function of d is known. If we write

$$d = N\ln\left(\frac{\sum_{k=1}^{N}(x^{(k)} - \bar{x})^2}{\sum_j \sum_{k:i^{(k)}=j}(x^{(k)} - \bar{x}_j)^2}\right),$$

then we see the close link with the ordinary F statistic, which we can here write as

$$F = \frac{\{\sum_{k=1}^{N}(x^{(k)} - \bar{x})^2 - \sum_j \sum_{k:i^{(k)}=j}(x^{(k)} - \bar{x}_j)^2\}/(\#A - 1)}{\sum_j \sum_{k:i^{(k)}=j}(x^{(k)} - \bar{x}_j)^2/(N - \#A)},$$

giving the simple relation

$$F = \frac{(\#A - 1)}{(N - \#A)}(e^{d/N} - 1).$$

In the current example, $F = 13.5714$, which under $\mathcal{M}_0$ should be F-distributed with 3 and 20 degrees of freedom, giving rise to $P = 0.0000$. The two tests thus lead to the same conclusion.

4.1.8   Example: A Drug Trial Using Mice

These data are taken from Morrison (1976) and concern a drug trial using mice. It is suspected that use of a drug may affect the level of three biochemical compounds in the brain. After randomisation, the drug was administered to 12 mice, with an additional 10 mice serving as controls. Assays of the brains of the mice were taken and the amounts of the three compounds were measured.

Group	X	Y	Z	Group	X	Y	Z
1	1.21	0.61	0.70	1	0.92	0.43	0.71
1	0.85	0.48	0.68	1	0.98	0.42	0.71
1	1.10	0.50	0.75	1	1.02	0.53	0.70
1	1.09	0.40	0.69	2	1.40	0.50	0.71
1	0.80	0.43	0.71	2	1.17	0.39	0.69
1	1.15	0.52	0.72	2	1.38	0.42	0.71
1	1.18	0.45	0.70	2	1.30	0.47	0.67
2	1.23	0.44	0.70	2	1.12	0.27	0.72
2	1.17	0.45	0.70	2	1.19	0.37	0.72
2	1.22	0.29	0.68	2	1.31	0.41	0.70
2	1.09	0.35	0.73	2	1.00	0.30	0.70

TABLE 4.3. Data from a drug trial on mice. Source: Morrison, *Multivariate Statistical Methods*, McGraw-Hill (1976). With permission.

We wish to examine how the treatment, A, affects the levels of the three compounds, which we denote simply as X, Y, and Z. The sample covariance, correlation, and partial correlation matrices are as follows:

```
MIM->print s
Calculating marginal statistics...
Empirical counts, means and covariances
A
1     X      0.018
      Y      0.004     0.004
      Z      0.000     0.000     0.000
      Means  1.030     0.477     0.707    10.000
             X         Y         Z        Count
2     X      0.013
      Y      0.006     0.005
      Z     -0.000    -0.000     0.000
      Means  1.215     0.388     0.703    12.000
             X         Y         Z        Count
MIM->print u
Empirical counts, means and correlations
A
1     X      1.000
      Y      0.489     1.000
      Z      0.209     0.111     1.000
      Means  1.030     0.477     0.707    10.000
             X         Y         Z        Count
2     X      1.000
      Y      0.704     1.000
      Z     -0.218    -0.211     1.000
      Means  1.215     0.388     0.703    12.000
             X         Y         Z        Count
```

```
MIM->print v
Empirical discrete, linear and partial correlation parameters
A
1     X      1.000
      Y      0.480     1.000
      Z      0.179     0.010     1.000
    Linear  -22.032   80.001   2206.654  -782.338
              X         Y          Z    Discrete
2     X      1.000
      Y      0.690     1.000
      Z     -0.100    -0.084     1.000
    Linear  178.426   18.476   2899.184 -1124.376
              X         Y          Z    Discrete
```

We observe that the partial correlations between Y and Z, and to a lesser extent between X and Z, are low.

To select a model, we can proceed as follows: first we test for variance homogeneity, using Box's test:

```
MIM->model A/AX,AY,AZ/AXYZ; fit; base
Deviance:        0.0000 DF: 0
MIM->model A/AX,AY,AZ/XYZ; fit; boxtest
Test of HO: A/AX,AY,AZ/XYZ
against H:  A/AX,AY,AZ/AXYZ
Box's test:  2.8688   DF:  6    P: 0.8251
```

There is no evidence of heterogeneity. We next test for zero partial correlation between Y and Z, and then between X and Z. Since the sample size is small, we use F-tests.

```
MIM->testdelete YZ s
Test of HO: A/AX,AY,AZ/XZ,XY
against H:  A/AX,AY,AZ/XYZ
F:    0.1440    DF:   1,  18    P: 0.7087
MIM->delete YZ
MIM->testdelete XZ s
Test of HO: A/AX,AY,AZ/Z,XY
against H:  A/AX,AY,AZ/XZ,XY
F:    0.0002    DF:   1,  19    P: 0.9888
MIM->delete XZ
```

These hypotheses can also be accepted. Next we can test for zero partial correlation between X and Y:

```
MIM->testdelete XY s
Test of HO: A/AX,AY,AZ/Z,Y,X
against H:  A/AX,AY,AZ/Z,XY
F:   10.6016    DF:   1,  19    P: 0.0042
```

This is rejected, so we proceed by attempting to remove the linear AZ term:

```
MIM->testdel AZ s
Test of H0: A/Z,AY,AX/Z,XY
against H: A/AX,AY,AZ/Z,XY
F:   0.3433   DF:   1, 20   P: 0.5645
MIM->delete AZ
```

We thus arrive at the model $A/AX, AY, Z/XY, Z$. It is not possible to simplify the model further (we omit the details). The independence graph is:

The interpretation is clear: the level of the compound Z is independent of the treatment and the compounds X and Y. These are both affected by the treatment, and are mutually correlated.

4.1.9 Example: Rats' Weights

We next consider a simple example studied in Morrison (1976). Mardia, Kent, and Bibby (1979) also use the example. The data stem from another drug trial, in which the weight losses of male and female rats under three drug treatments are studied. Four rats of each sex are assigned at random to each drug. Weight losses are observed after one and two weeks. There are thus 24 observations on four variables: sex (A), drug (B), and weight loss after one and two weeks (X and Y, respectively).

Again, we first examine whether the covariances are homogeneous using Box's test:

```
MIM->mod AB/ABX,ABY/ABXY; fit; base
Deviance:     0.0000 DF: 0
MIM->mod AB/ABX,ABY/XY; fit; boxtest
Deviance:    27.8073 DF: 15
Test of H0: AB/ABX,ABY/XY
against H:  AB/ABX,ABY/ABXY
Box's test: 14.9979   DF: 15   P: 0.4516
```

There is no evidence of heterogeneity. We adopt the homogeneous model, and attempt to simplify the covariance structure by removing the edge $[XY]$:

Sex	Drug	Wt 1	Wt 2	Sex	Drug	Wt 1	Wt 2
1	1	5	6	1	1	5	4
1	1	7	6	1	2	7	6
1	2	9	12	1	2	6	8
1	3	14	11	1	3	17	12
2	1	7	10	2	1	6	6
2	1	8	10	2	2	10	13
2	2	7	6	2	2	6	9
2	3	14	9	2	3	14	8
1	1	9	9	2	1	9	7
1	2	7	7	2	2	8	7
1	3	21	15	2	3	16	12
1	3	12	10	2	3	10	5

TABLE 4.4. Data from drug trial on rats. Source: Morrison, *Multivariate Statistical Methods*, McGraw-Hill (1976). With permission.

```
MIM->mod AB/ABX,ABY/XY; fit
Deviance:     27.8073 DF: 15
MIM->testdelete XY s
Test of H0: AB/ABX,ABY/Y,X
against H:  AB/ABX,ABY/XY
F:   20.2181    DF:   1, 17    P: 0.0003
```

This is strongly rejected. We can further simplify the mean structure:

```
MIM->testdelete AX s
Test of H0: AB/ABY,BX/XY
against H:  AB/ABX,ABY/XY
F:    0.0438    DF:   3, 17    P: 0.9874
MIM->delete AX
MIM->testdelete AY s
Test of H0: AB/BY,BX/XY
against H:  AB/ABY,BX/XY
F:    1.7368    DF:   3, 18    P: 0.1953
MIM->delete AY
MIM->testdelete BX s
Test of H0: AB/BY,X/XY
against H:  AB/BY,BX/XY
F:   36.1991    DF:   2, 20    P: 0.0000
MIM->testdelete BY s
Test of H0: AB/Y,BX/XY
against H:  AB/BY,BX/XY
F:    5.3695    DF:   2, 20    P: 0.0136
```

We arrive at the model $AB/BX, BY/XY$. It is graphical, with the following graph:

Sex (A) ● ○ Weight loss 2 (X)

Drug (B) ● ─── ○ Weight loss 1 (Y)

It has a simple interpretation. If we write the indices corresponding to factors A and B as j and k, respectively, we can write a cell in the two-way table as $i = (j, k)$. The distribution of X given A and B is clearly $\mathcal{N}(\mu_k, \sigma^x)$, i.e., depending on B only. The conditional distribution of Y given X, A, and B is normal with mean

$$E(Y|I = i, X = x) = (\beta_i^Y - \omega_i^{XY} x)/\omega_i^{YY},$$

and variance

$$\text{Var}(Y|I = i, X = x) = 1/\omega_i^{YY}.$$

For the present model, ω_i^{XY} and ω_i^{YY} do not depend on i, and β_i^Y is a function of k only.

We can re-express the model through the recursive equations

$$X = \mu_k + \varepsilon^x, \tag{4.19}$$
$$Y = \lambda_k + \eta x + \varepsilon^y, \tag{4.20}$$

where $\varepsilon^x \sim \mathcal{N}(0, \sigma^x)$ as above, $\varepsilon^y \sim \mathcal{N}(0, \tau^y)$, say, and ε^x and ε^y are independent. In other words, the expected weight loss at the second week is a constant proportion of the previous weight loss, plus a constant that depends on the treatment. Estimates of the regression coefficients in (4.19) and (4.20) can be obtained using the Display command:

```
MIM->Display Y,AXB
Parameters of the conditional distribution of Y given A,B,X.
A B
1 1    Y    0.953    0.900    2.433
                       X        Y

1 2    Y    1.753    0.900    2.433
                       X        Y

1 3    Y   -3.018    0.900    2.433
                       X        Y

2 1    Y    0.953    0.900    2.433
                       X        Y

2 2    Y    1.753    0.900    2.433
                       X        Y

2 3    Y   -3.018    0.900    2.433
```

We thus obtain the following estimated regression equation:

$$y = \left\{ \begin{array}{c} 0.953 \\ 1.753 \\ -3.018 \end{array} \right\} + 0.9x + \varepsilon^y, \quad \varepsilon^y \sim \mathcal{N}(0, 2.433)$$

for the three levels of B. The estimates for μ_k and σ^x in (4.19) can be obtained similarly.

Plots of the data (not shown here) suggest that the effects of the first two drugs may not differ widely from each other. This can be tested by omitting the third drug from the analysis:

```
MIM->fact C2; calc C=B; restrict B<3
MIM->mod AC/CX,CY/XY
MIM->fix CA
Fixed variables: AC
MIM->step u
Coherent Backward Selection
Unrestricted models, Chi-squared tests.
Critical value:   0.0500
Initial model: AC/CX,CY/XY
Model: AC/CX,CY/XY
Deviance:   7.7404 DF:  13 P:  0.8601
    Edge       Test
 Excluded   Statistic DF         P
    [CX]       0.0020  1       0.9646
    [CY]       0.6645  1       0.4150
    [XY]       9.8134  1       0.0017 +
 Removed edge [CX]
 Model: AC/CY,X/XY
 Deviance:   7.7423 DF:  14 P:  0.9023
    Edge       Test
 Excluded   Statistic DF         P
    [CY]       1.1549  1       0.2825
 Removed edge [CY]
 Selected model: AC/Y,X/XY
```

The results indicate that there is no difference between the first and second drugs with regard to weight loss.

The tests shown here are not identical to likelihood ratio tests under the model including all three treatment levels, since some drug-independent parameters are estimated from the complete data.

4.1.10   Example: Estrogen and Lipid Metabolism

This is a more extended example that comes from a clinical trial comparing various estrogen replacement therapies (Munk-Jensen et al., 1994). One hundred thirteen postmenopausal women were randomised to one of three treatment groups, corresponding to cyclic therapy (in which the doses of estrogen and progestin vary cyclically— mimicking the natural menstrual cycle), continuous therapy (fixed dose of estrogen and progestin daily), and placebo. Plasma samples were taken pretreatment and after 18 months' treatment, and the samples were assayed for lipoprotein fractions.

The effect of estrogen replacement therapies on lipid metabolism is of considerable interest, since lipoproteins are believed to represent risk factors for coronary heart disease. Here, we analyze the results for a high-density lipoprotein fraction (HDL), low-density lipoproteins (LDL), and very low-density lipoproteins (VLDL). Note that HDL is believed to be beneficial, whereas LDL and VLDL are believed to be deleterious.

The data consist of seven variables: treatment group (A), pretreatment VLDL (U), pretreatment LDL (V), pretreatment HDL (W), posttreatment VLDL (X), posttreatment LDL (Y), and posttreatment HDL (Z). The pretreatment values are treated as covariates in the analysis.

After reading the data, we first examine the desirability of transforming the variables using the techniques proposed by Box and Cox (1964), described in Section 6.5. We first look at X:

```
MIM->satmod
MIM->pr
The current model is A/AU,AV,AW,AX,AY,AZ/AUVWXYZ.
MIM->boxcox X -2 2 4
Box-Cox Transformation of X:
              -2*loglikelihood -2*loglikelihood
   Lambda     (full model)    (current model)    Deviance
   ------------------------------------------------------------
   -2.0000      6302.6822        6302.6822         0.0000
   -1.0000      5947.1467        5947.1467         0.0000
    0.0000      5739.6486        5739.6486         0.0000
    1.0000      5773.7387        5773.7387         0.0000
    2.0000      5953.1883        5953.1883         0.0000
   ------------------------------------------------------------
```

The log transformation is clearly indicated. The same applies to the other variables (the output is omitted). We therefore transform the variables and then perform stepwise selection.

```
MIM->calc X=ln(X); calc Y=ln(Y); calc Z=ln(Z)
MIM->calc U=ln(U); calc V=ln(V); calc W=ln(W)
```

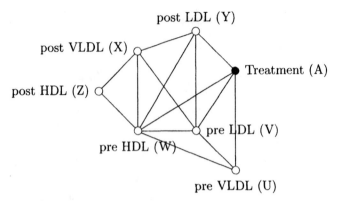

FIGURE 4.2. The model selected initially, using stepwise selection.

```
MIM->fix auvw
Fixed variables: AUVW
MIM->stepwise z
MIM->pr
The current model is: A/Z,X,AY,AW,AV,AU/WXZ,VWXY,AVWY,AUVW
```

The Fix command fixes edges between variables in $\{A, U, V, W\}$ in the model, i.e., they are not candidates for removal in the selection process. The Z option suppresses output. The graph of the selected model is shown in Figure 4.2.

The model has two striking implications: first, the treatment only has a direct effect on LDL (Y), since VLDL (X) and HDL (Z) are independent of treatment given LDL and the pretreatment variables. This suggests that the mechanism of action works primarily on LDL; any effect on VLDL and HDL is mediated by LDL. Secondly, the responses LDL, VLDL, and HDL are independent of pretreatment VLDL (U) given the remaining variables. In other words, pretreatment VLDL has no explanatory value and can be omitted from the analysis. We therefore remove it from the model, arriving at

$$A/AV, AW, X, AY, Z/WXZ, VWXY, AVWY,$$

whose graph is shown in Figure 4.3.

The deviance is 32.1893 on 24 degrees of freedom, corresponding to a p-value of 0.1224. It is a relatively simple and well-fitting model. It also has a straightforward interpretation, in the sense that it can be decomposed into three parts, describing

1. the conditional distribution of Y given A, V, and W,
2. the conditional distribution of X given V, W, and Y, and
3. the conditional distribution of Z given W, Y, and Z,

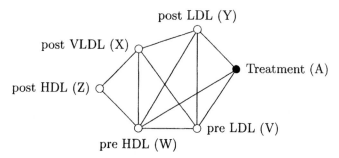

FIGURE 4.3. The model obtained by omitting pretreatment VLDL from the model shown in Figure 4.2.

corresponding respectively to a one-way ANOVA with two covariates (and variance heterogeneity), and two multiple regressions.

Our tentative conclusion is that it appears that the estrogen therapy has a direct effect on LDL only, suggesting a mechanism of action primarily through LDL. Any effect of therapy on HDL and VLDL appears to be mediated by LDL.

This style of analysis is essentially explanatory, seeking to identify possible causal relations. In contrast, if the purpose of the analysis was to address more pragmatic questions like, Does estrogen therapy increase HDL levels? a univariate analysis of each response variable would be more appropriate.

In the next three sections, we study more systematically the question of when models can be broken down into simpler models. This is one of the more mathematical parts of the book, and may be skipped at first reading. However, the concepts described are important not only for theoretical reasons, but also for practical ones; in particular, they are useful for interpreting complex models.

4.2   Breaking Models into Smaller Ones

Let a be a subset of the model variables, i.e., $a \subset V$, and let b be its complement, i.e., $b = V \setminus a$. We want to be able to decompose a model $\mathcal{M}$ for V into two models: one ($\mathcal{M}_a$) describing the marginal distribution of a, and the other ($\mathcal{M}_{b|a}$) describing the conditional distribution of b given a. Of course, given any density f on V, we can always decompose this into a marginal and conditional density $f = f_a f_{b|a}$, but it is unclear how these densities correspond to the models we are interested in. So first let us clarify what we mean by the marginal model ($\mathcal{M}_a$) and the conditional model ($\mathcal{M}_{b|a}$).

By $\mathcal{M}_a$ we mean simply the hierarchical interaction model for a that contains all the interactions between variables in a that were present in $\mathcal{M}$. If $\mathcal{M}$ is (homogeneous) graphical with graph $\mathcal{G}$, then $\mathcal{M}_a$ is also (homogeneous) graphical, with graph given as the subgraph $\mathcal{G}_a$. To obtain the model formula for $\mathcal{M}_a$, we delete all variables in b from the formula for $\mathcal{M}$.

For example, if $\mathcal{M} = AB/ABX, AY, BZ/AXY, ABX, BXZ$ (see Figure 4.1) and $a = \{A, B, X\}$, then $\mathcal{M}_a = AB/ABX/ABX$.

The conditional model $\mathcal{M}_{b|a}$ needs more explanation. For some $f \in \mathcal{M}$, consider $f_{b|a}$. We partition Δ into $\Delta_1 = \Delta \cap b$ and $\Delta_2 = \Delta \cap a$, indexed by $i = (j, k)$, and Γ into $\Gamma_1 = \Gamma \cap b$ and $\Gamma_2 = \Gamma \cap a$, writing the corresponding random variable as (Y, Z).

The joint density is given in (4.2). The conditional density is found by renormalizing the joint density:

$$f_{b|a}(j, y | k, z) = \kappa_{k,z} \exp(\alpha_i + \beta_i^{1\prime} y - \tfrac{1}{2} y' \Omega_i^{11} y$$
$$+ \beta_i^{2\prime} z - y' \Omega_i^{12} z - \tfrac{1}{2} z' \Omega_i^{22} z),$$

where β_i and Ω_i have been partitioned commensurately with Γ, and $\kappa_{k,z}$ is a normalizing constant. If we write

$$\alpha_{j|k,z} = \alpha_i + \beta_i^{2\prime} z - \tfrac{1}{2} z' \Omega_i^{22} z + \ln(\kappa_{k,z}) \tag{4.21}$$

$$\beta_{j|k,z} = \beta_i^1 - \Omega_i^{12} z \tag{4.22}$$

$$\Omega_{j|k} = \Omega_i^{11}, \tag{4.23}$$

then we can rewrite the conditional density as

$$f_{b|a}(j, y | k, z) = \exp(\alpha_{j|k,z} + \beta_{j|k,z}' y - \tfrac{1}{2} y' \Omega_{j|k} y).$$

So $f_{b|a}$ follows a CG-distribution whose canonical parameters are functions of k and z. In other words, for given $K = k$ and $Z = z$, (J, Y) is CG-distributed with parameters given in (4.21–4.23). This is called a CG-regression. The conditional model $\mathcal{M}_{b|a}$ consists of all the CG-regressions $f_{b|a}(j, y | k, z)$ that can be generated in this way; in other words, $\mathcal{M}_{b|a} = \{f_{b|a} : f \in \mathcal{M}\}$.

Example 1: $\mathcal{M} = //YZ$ and $a = \{Z\}$

Here $\mathcal{M}_{b|a}$ consists of a family of univariate normal distributions, one for each z. The canonical parameters of these are given by (4.21–4.23), i.e.,

$$\delta = \delta^{YY} \quad \text{and}$$
$$\beta = \beta^Y - \delta^{YZ} z.$$

Here, α is just a normalizing constant. So we obtain

$$E(Y|Z = z) \quad = \delta^{-1}\beta$$
$$= \delta^{-1}(\beta^Y - \delta^{YZ}z), \quad \text{and}$$
$$\text{Var}(Y|Z = z) = \delta^{-1}.$$

The conditional variance is constant over z, and the conditional mean is a linear function of z. If we write, say,

$$\gamma_0 = \delta^{-1}\beta^Y,$$
$$\gamma_1 = \delta^{-1}\delta^{YZ}, \quad \text{and}$$
$$\eta \quad = \delta^{-1},$$

then we obtain

$$E(Y|Z = z) \quad = \gamma_0 + \gamma_1 z,$$
$$\text{Var}(Y|Z = z) = \eta,$$

and we see that we are just dealing with an unfamiliar parametrization of a very familiar model, the linear regression of Y on Z. So $\mathcal{M}_{b|a}$ is just an ordinary linear regression model.

Example 2: $\mathcal{M} = A/AX/X$ and $a = \{X\}$

This model was discussed in Section 4.1.1. From (4.21), we know that the conditional distribution of A given X has canonical parameter

$$\alpha_{i|x} = \alpha_i + \beta_i x - \tfrac{1}{2}\omega x^2 + \ln(\kappa_x),$$

and so

$$p_{i|x} = \frac{\exp(\alpha_i + \beta_i x - \tfrac{1}{2}\omega x^2)}{\sum_i \exp(\alpha_i + \beta_i x - \tfrac{1}{2}\omega x^2)}$$
$$= \frac{\exp(\alpha_i + \beta_i x)}{\sum_i \exp(\alpha_i + \beta_i x).} \tag{4.24}$$

So $\mathcal{M}_{b|a}$ is just the familiar logistic regression model.

We see that these conditional models, the CG-regression models, include both ordinary linear regression and logistic regression models. They generalize these to incorporate multiple response variables, both discrete and continuous, and multiple explanatory variables, also both discrete and continuous.

Having (hopefully) clarified what we mean by $\mathcal{M}_a$ and $\mathcal{M}_{b|a}$, we return to the original question: When can we decompose $\mathcal{M}$ into $\mathcal{M}_a$ and $\mathcal{M}_{b|a}$? In other words, given a density $f \in \mathcal{M}$, when is $f_a \in \mathcal{M}_a$ and $f_{b|a} \in \mathcal{M}_{b|a}$? The last part of this question is unproblematic: by the way we constructed

$\mathcal{M}_{b|a}$, we know that $f_{b|a} \in \mathcal{M}_{b|a}$ is always true. The key question is when $f_a \in \mathcal{M}_a$. Accordingly, we define the property of collapsibility as follows:

A model $\mathcal{M}$ is said to be *collapsible* onto $a \subset V$ if for all $f \in \mathcal{M}$, $f_a \in \mathcal{M}_a$.

Collapsibility in this sense was first studied by Lauritzen (1982) in the contingency table context. Later references are Asmussen and Edwards (1983) and Frydenberg (1990). Other authors have used the term in different senses (for example, Whittemore, 1978, and Wermuth, 1987).

Before going any further, we consider some examples.

Example 3: $\mathcal{M} = AB, AC$ and $a = \{A, B\}$

The graph is

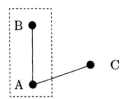

where we have drawn a box around a for emphasis. We have that $\mathcal{M}_a$ is AB, the unrestricted model on a. From

$$p_{ijk} = \exp(\alpha_{ijk})$$
$$= \exp(u + u_i^A + u_j^B + u_k^C + u_{ij}^{AB} + u_{ik}^{AC}),$$

we see that the marginal distribution on a is given by

$$p_{ij+} = \exp(u + u_i^A + u_j^B + u_{ij}^{AB}) \sum_k \exp(u_k^C + u_{ik}^{AC})$$
$$= \exp(\tilde{u} + \tilde{u}_i^A + u_j^B + u_{ij}^{AB}).$$

Clearly, $f_a \in \mathcal{M}_a$, so $\mathcal{M}$ is collapsible onto a. The conditional distribution $f_{b|a}$ is given by

$$p_{k|ij} = \kappa_{ij} \exp(\alpha_{ijk}),$$

where $\kappa_{ij}^{-1} = \sum_k \exp(\alpha_{ijk})$, so that we obtain

$$p_{k|ij} = \frac{\exp(u_k^C + u_{ik}^{AC})}{\sum_k \exp(u_k^C + u_{ik}^{AC})}.$$

The parameters of $\mathcal{M}_a$ and of $\mathcal{M}_{b|a}$ vary independently of each other so the MLE under $\mathcal{M}_a$ is

$$\hat{p}_{ij+} = \frac{n_{ij+}}{N},$$

and the MLE under $\mathcal{M}_{b|a}$ is

$$\hat{p}_{k|ij} = \frac{n_{i+k}}{n_{i++}}.$$

The MLE under $\mathcal{M}$ can be constructed as

$$\begin{aligned}
\hat{p}_{ijk} &= \hat{p}_{ij+}\hat{p}_{k|ij} \\
&= \frac{n_{ij+}}{N}\frac{n_{i+k}}{n_{i++}}.
\end{aligned}$$

Example 4: $\mathcal{M} = AC, BC$ and $a = \{A, B\}$

The graph is

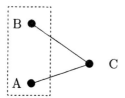

and we have $\mathcal{M}_a = A, B$, the model of independence. Since

$$p_{ij+} = \exp(u + u_i^A + u_j^B)\sum_k \exp(u_k^C + u_{jk}^{BC} + u_{ik}^{AC}), \tag{4.25}$$

in general, $f_a \notin \mathcal{M}_a$, so $\mathcal{M}$ is not collapsible onto a.

Example 5: $\mathcal{M} = A/AX/X$ and $a = \{X\}$

The marginal model $\mathcal{M}_a$ has formula $//X$ — in other words, under $\mathcal{M}_a$, X is normally distributed. But the marginal distribution of X is a mixture of normals, which is not normal, so $\mathcal{M}$ is not collapsible onto $\{X\}$.

In many respects, a good grasp of collapsibility is the key to understanding and interpreting more complex models, as we see in the remainder of this and in the following section. So it is important to characterize precisely when the property holds. We do this incrementally, starting with the purely discrete or continuous models, and then proceeding to the mixed models. Readers short of time or patience may note that the conditions given are implemented in the command Collapse, which tests whether the current model is collapsible onto a given variable set. See Section A.3.4 for further details.

For pure (that is, entirely discrete or entirely continuous) graphical models, collapsibility can be read off the graph, using the following result.

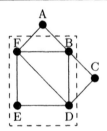
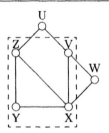

FIGURE 4.4. Collapsibility for pure graphical models. The first graph represents a discrete model with six variables. Here, $a = \{B, D, E, F\}$ and $b = \{A, C\}$. To find out whether the model is collapsible onto a, the subgraph $\mathcal{G}_b$ is examined. This consists of two disconnected vertices. The boundary of each of these must be complete in $\mathcal{G}$. The boundary of $\{A\}$ is $\{B, F\}$ and the boundary of $\{C\}$ is $\{B, D\}$: both are complete in $\mathcal{G}$, hence the model is collapsible onto a. The second graph represents a continuous model, also with six variables. To find one whether the model is collapsible onto $a = \{V, X, Y, Z\}$, the boundaries of $b = \{U, W\}$ are examined. One of the boundaries, $\{V, Z\}$, is not complete so the model is not collapsible onto a.

Let $a \subseteq V$, $b = V \setminus a$, and the connected components of $\mathcal{G}_b$ be $b_1, \ldots, b_k$ and their boundaries be $e_1, \ldots, e_k$. Then $\mathcal{M}$ is collapsible onto a if and only if for each $j = 1 \ldots k$, e_j is complete.

This is illustrated in Figure 4.4.

For pure, hierarchical models the criterion is almost unchanged: instead of requiring each boundary to be complete in the graph, we require a generator including the boundary to be present in the model formula.

For example, the model ACD, ABD, BCD is not collapsible onto the variable set $\{A, B, C\}$. This is because the boundary of $\{D\}$ is $\{A, B, C\}$, and there is no generator including ABC in the model formula.

Now for the mixed models. First we consider the mixed graphical models. There is an additional requirement: we can collapse over discrete variables only when the corresponding boundary is discrete. In other words, the condition for collapsibility is as follows.

Let $a \subseteq V$ and let the connected components of $\mathcal{G}_b$ be $b_1, \ldots b_k$, and their boundaries be $e_1, \ldots e_k$. Then a mixed graphical model is collapsible onto a if and only if for each $j = 1, \ldots k$, (i) e_j is complete, and (ii) either $b_j \subseteq \Gamma$ or $e_j \subseteq \Delta$.

For a mixed model that is hierarchical and homogeneous, we need to sharpen condition (ii). As before, we require that for each j, either (a) $b_j \subseteq \Gamma$ or (b) $e_j \subseteq \Delta$. In case (b), we also require that there is a discrete generator including e_j. In case (a), we also require, for each pair of linear generators l_s, l_t intersecting b_j, the existence of a discrete generator including $(l_s \cup l_t) \setminus b_j$.

For example, the model $AB, AC, BC/ABX, CX/X$ is not collapsible onto $\{A, B, C\}$ since this last condition is not satisfied. In contrast, the model $AB, AC, BC/AX, BX, CX/X$ fulfills the condition and is collapsible onto $\{A, B, C\}$.

We have now given conditions for collapsibility for all the different classes of hierarchical interaction models except one: the mixed, nongraphical, heterogeneous models. The appropriate conditions for these models, which are rather technical (Edwards, 1990), are omitted here.

We now describe some ways that the collapsibility property can be used. If $\mathcal{M}$ is collapsible onto a, then we know that any $f \in \mathcal{M}$ can be written as

$$f_V = f_a f_{b|a}, \tag{4.26}$$

where $f_a \in \mathcal{M}_a$ and $f_{b|a} \in \mathcal{M}_{b|a}$. Moreover, f_a and $f_{b|a}$ are variation independent, i.e., they can be parametrized separately, so the MLEs satisfy

$$\hat{f} = \hat{f}_a \hat{f}_{b|a}. \tag{4.27}$$

This relation is fundamental since it means that we can work with conditional distributions in the joint framework. If $\mathcal{M}$ is collapsible onto a, then we can use $\hat{f}$, the MLE under $\mathcal{M}$, both to calculate the MLE under $\mathcal{M}_a$ by marginalizing, and to calculate the MLE under $\mathcal{M}_{b|a}$ by renormalizing. We calculate the parameters of $\hat{f}_a$ by summing probabilities and extracting submatrices, and the parameters of $\hat{f}_{b|a}$ by renormalizing probabilities (as in $p_{i|j} = p_{ij}/p_{i+}$), and by calculating conditional means and covariances (as in Section 3.2).

From (4.27), it follows that

$$\hat{\ell} = \hat{\ell}_a + \hat{\ell}_{b|a}, \tag{4.28}$$

where $\hat{\ell}$, $\hat{\ell}_a$, and $\hat{\ell}_{b|a}$ are the maximized log likelihoods under $\mathcal{M}$, $\mathcal{M}_a$, and $\mathcal{M}_{b|a}$.

Similarly, for testing in the conditional framework, suppose we have two nested models, $\mathcal{M}_0 \subseteq \mathcal{M}_1$, both collapsible onto a with the same marginal model. Using (4.28), we see that

$$2(\hat{\ell}_{b|a}^1 - \hat{\ell}_{b|a}^0) = 2(\hat{\ell}^1 - \hat{\ell}^0),$$

so that the deviance difference for the conditional models is the same as the deviance difference for the joint models. We have seen a simple special case of this in connection with loglinear models, where it was mentioned that the expression for the deviance is the same under different sampling schemes.

Note that the second condition for collapsibility for mixed graphical models, namely that either $b_j \subseteq \Gamma$ or $e_j \subseteq \Delta$, restricts the conditional models that can be fitted in the joint framework. If we equate b_j with the response variables and e_j with the covariates, then the condition states that either all responses must be continuous or all covariates must be discrete. In other words, models with discrete responses and continuous covariates are excluded. This *discrete before continuous* rule means that the logistic regression model (4.24) cannot be fitted by maximum likelihood in the joint framework. Later, in Section 4.5, we study these conditional models explicitly, and show how they may be fitted.

Another type of application is best illustrated with an example. Consider the graph

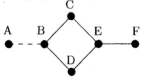

and the two corresponding models, i.e., with and without $[AB]$. Both models are collapsible onto $\{A, B\}$, with different marginal models but with identical conditional models. It follows using (4.28) that the likelihood ratio test

$$2(\hat{\ell}^1 - \hat{\ell}^0) = 2(\hat{\ell}_a^1 - \hat{\ell}_a^0),$$

i.e., the test can be performed in the two-way marginal $A \times B$ table as a test of $A \perp\!\!\!\perp B$.

A more subtle example of the same kind is the following. Consider the graph

$$\bullet\!\!-\!\!-\!\!-\!\!-\!\!\circ\!-\!-\!-\!-\circ$$
A   X   Y

The model is *not* collapsible onto $\{X, Y\}$, so at first sight it might seem that the test for removal of $[XY]$ cannot be performed in the marginal model $//XY$. However, the conditional model for Y given the remaining variables is the same in the model shown above and in the marginal model. So it follows that the test can be performed in the marginal model, contrary to expectations.

4.3   Mean Linearity

In this section we study some conditional models $\mathcal{M}_{b|a}$, for which all the response variables are continuous, that is, $b \subseteq \Gamma$. One difference between these models and conventional multivariate models, as used in, say, multivariate regression or the multivariate analysis of variance, is that the latter are defined by means of linear models for the means of the response variables, whereas the former are defined by means of factorial expansions of the elements of the canonical parameters β_i and Ω_i. Some of the models $\mathcal{M}_{b|a}$ have a property called *mean linearity*. For these models, the linear structure imposed on the canonical parameters induces a corresponding linear structure on the means of the response variables, so that we can write the conditional model in the familiar form,

$$\text{response} = \text{linear component} + \text{error.} \tag{4.29}$$

A simple example is the model $IJ/IY, JY/Y$. This corresponds to the two-way layout with homogeneous variances. Since β_{ij} is additive, clearly $\mu_{ij} = \sigma\beta_{ij}$ is additive too.

Similarly, the conditional model derived from $\mathcal{M} = IJ/IJY/IY$ is also mean linear, since β_{ij} is unrestricted and so $\mu_{ij} = \sigma_i\beta_{ij}$ is also unrestricted. On the other hand, the conditional model derived from $\mathcal{M} = IJ/IY, JY/IY$ is not mean linear, since the additivity of β_{ij} does not induce a simple additive structure for μ_{ij}.

Another example is $\mathcal{M} = IJ/IX, JX, IY, JY/XY$, with $b = \{X, Y\}$. Since β_{ij}^X and β_{ij}^Y are additive and $\mu_{ij} = \Sigma_{ij}\beta_{ij} = \Sigma\beta_{ij}$, we see that μ_{ij}^X and μ_{ij}^Y are also additive.

This property of mean linearity, which we now characterize, is primarily of interest when $\mathcal{M}$ is collapsible onto a, so that we can decompose $\mathcal{M}$ into a marginal model $\mathcal{M}_a$ and an easily interpretable conditional model $\mathcal{M}_{b|a}$.

Suppose that we partition Γ into responses (Γ_1) and covariates (Γ_2), and correspondingly write $Y = (Y_1, Y_2)'$. The conditional mean of Y_1 given (I, Y_2) is given by

$$\mu_i^{1\cdot2} = \Sigma_i^{11\cdot2}\beta_i^{1\cdot2},$$

where $\Sigma_i^{11\cdot2}$ is the conditional covariance given as $\Sigma_i^{11\cdot2} = (\Omega_i^{11})^{-1}$, and $\beta_i^{1\cdot2}$ is given as

$$\beta_i^{1\cdot2} = \beta_i^1 - \Omega_i^{12}y_2. \tag{4.30}$$

This last expression (4.30) is an additive linear function, including parameters (β_i^1 and Ω_i^{12}) and covariates (y_2). We want to know: When is the conditional mean $\mu_i^{1\cdot2}$ also linear?

Consider the connected components of $\mathcal{G}_{\Gamma_1}$, say, $b_1 \ldots b_t$. If $t > 1$, $\Sigma_i^{11\cdot2}$ is block-diagonal, and we can decompose the problem by considering the blocks $b_1 \ldots b_t$ separately.

Suppose first that $\Sigma_i^{11\cdot2}$ is homogeneous (i.e., constant over i). Then Ω_i^{12} is also homogeneous. We require that for each $\gamma \in b_s$, the formulae for $\beta_i^{\gamma\cdot2}$ represent identical linear functions. This is the case if

1. All β_i^γ with $\gamma \in b_s$ have identical factorial expansions.

2. For each $\eta \in \Gamma_2$, either $\omega^{\gamma\eta} = 0$ for all $\gamma \in b_s$ or $\omega^{\gamma\eta} \neq 0$ for all $\gamma \in b_s$. In other words, the rows of Ω^{12} have zeros in the same places (see Section 3.2). In terms of the independence graph, this means that either all edges $[\gamma, \eta]$ are present or all are absent.

When these conditions are fulfilled, the range of $\mu_i^{1\cdot2}$ as a function of i and y_2 is the same as that of $\beta_i^{1\cdot2}$, and so is also linear. Thus these conditions are sufficient for mean linearity. We examine some examples:

a: $\mathcal{M} = I/IX, IY, IZ/XYZ$ with $\Gamma_1 = \{X, Y, Z\}$. Here $\Gamma_2 = \emptyset$ so the second condition is empty, and the first condition clearly holds, so $\mathcal{M}_{b|a}$ is mean linear.

b: $\mathcal{M} = IJ/IX, JX, IY, JY/XY$ with $\Gamma_1 = \{X, Y\}$. Here again $\Gamma_2 = \emptyset$. Both β_{ij}^X and β_{ij}^Y are additive, so the first condition holds, and $\mathcal{M}_{b|a}$ is mean linear.

c: $\mathcal{M} = I/IX, IY, IZ/XY, YZ$ with $\Gamma_1 = \{X, Y, Z\}$. Here the conditional covariance is constrained, but otherwise the conditions still hold.

d: $\mathcal{M} = I/IX, IY, Z/XY, Z$ with $\Gamma_1 = \{X, Y, Z\}$. Here there are two blocks, $b_1 = \{X, Y\}$ and $b_2 = \{Z\}$. The conditions hold for each block, so $\mathcal{M}_{b|a}$ is mean linear.

e: $\mathcal{M} = I/IX, IY, IZ/XYZ$ with $\Gamma_1 = \{X, Y\}$. Both ω^{XZ} and ω^{YZ} are non-zero, so the second condition is fulfilled, and $\mathcal{M}_{b|a}$ is mean linear.

f: $\mathcal{M} = I/W, IX, IY, IZ/XYW, WYZ$ with $\Gamma_1 = \{X, Y, Z\}$. Similarly, all of ω^{XW}, ω^{YW} and ω^{ZW} are nonzero, so the second condition is fulfilled, and $\mathcal{M}_{b|a}$ is mean linear.

Clearly, if $\mathcal{M}$ is homogeneous and Γ_1 has only one variable, then $\mathcal{M}_{b|a}$ is mean linear.

Note that if $\mathcal{M}_{b|a}$ is mean linear and homogeneous (i.e., $\Sigma_i^{11\cdot2}$ is homogeneous), then it takes the form (4.30), where the error terms follow a graphical Gaussian model with zero mean, the latter model being given by the subgraph $\mathcal{G}_{\Gamma_1}$.

Homogeneous mean linear models generalize some well-known models used in multivariate analysis. Multivariate analysis of variance (MANOVA) models, for example, posit the same linear structure for the mean of each response variable and an unrestricted, homogeneous covariance matrix. The models **a** and **b** above are mean linear with homogeneous, unrestricted covariance matrices, and so are equivalent to standard MANOVA models. Models such as **c** extend this by allowing the errors to follow a graphical Gaussian model.

Similarly, standard MANCOVA (multivariate analysis of covariance) models posit the same linear structure for the mean of each response variable and an unrestricted, homogeneous conditional covariance matrix. The model **e** is of this kind. Again, models such as **f** extend this by allowing the errors to follow a graphical Gaussian model.

Some closely related models are described in Section 7.5.

We now turn to the case where $\Sigma_i^{11.2}$ is heterogeneous, and so Ω_i^{12} may also be heterogeneous. We noted above that $IJ/IY, JY/IY$ is not mean linear, but that $IJ/IJY/IY$ is: the heterogeneity needs to be "absorbed" by each term in (4.30). Accordingly we define the quadratic boundary of b_s as $\mathrm{bd_q}(b_s) = ((\cup_{q_k : q_k \cap b_s \neq \emptyset} q_k) \cap \Delta$. We require for $s = 1 \ldots t$,

1. All $\gamma \in b_s$ have identical formulae in the linear part.
2. For each $\eta \in \Gamma_2$, the formulae for $\omega_i^{\gamma\eta}$ are the same for all $\gamma \in b_s$. This ensures that the Ω_i^{12} contribute identical linear forms in (4.30).
3. Setting $d_s = \mathrm{bd_q}(b_s)$, we require that d_s is contained in all the terms just mentioned in 1 and 2: that is, we require that $d_s \subset l_k$ for all k such that $l_k \cap b_s \neq \emptyset$ and that $d_s \subseteq q_k$ for all k such that $q_k \cap b_s \neq \emptyset$ and $q_k \cap \Gamma_2 \neq \emptyset$.

We need to explain what is meant here by *formulae for* $\omega_i^{\gamma\eta}$ in the second condition. Recall that a precision element can in general be zero, be constant, or vary over i subject to a factorial expansion involving the discrete variables. Due to the third condition above, an element $\omega_i^{\gamma\eta}$ with $\gamma \in b_s$ and $\eta \in \Gamma_2$ can either be zero, or vary according to a factorial expansion, each of whose terms must contain d_s. The second condition states for each $\eta \in \Gamma_2$, either all $\omega_i^{\gamma\eta}$ with $\gamma \in b_s$ are set to zero, or they must all vary according to the same factorial expansion (each of whose terms must contain d_s).

Some examples of heterogenous mean linear models:

g: $\mathcal{M} = IJ/IJX/IX, JX$ with $\Gamma_1 = \{X\}$. Here $\Gamma_2 = \emptyset$ so the second condition is empty, and the third condition implies that $\{I, J\}$ should be contained in each linear generator, which it is, so $\mathcal{M}_{b|a}$ is heterogeneous mean linear.

h: $\mathcal{M} = I/IW, IX, IY, IZ/IXYW, IWYZ$ with $\Gamma_1 = \{X, Y, Z\}$. This extends model **f** above.

Clearly, if $\mathcal{M}$ is heterogenous graphical and Γ_1 has only one variable, then $\mathcal{M}_{b|a}$ is heterogeneous mean linear.

Some models can be recursively decomposed into a sequence of mean linear submodels. For example, consider a hypothetical repeated measures setup in which a treatment A is applied to some subjects, and measurements X, Y, and Z are subsequently taken at three consecutive points of time. The model $A/AX, Y, Z/XY, YZ$, whose graph is

has a simple interpretation. From the graph, we see that X is directly dependent on the treatment A, i.e., is normally distributed about a different mean value for each treatment group. Since X and Y are correlated, the mean value of Y may differ between groups, but there is no *direct* dependence of Y on A, since $Y \perp\!\!\!\perp A|X$. Similarly, $Z \perp\!\!\!\perp A|Y$.

More formally, if we condition on A, we can break the model down into three components:

1. The conditional model for X given A.

2. The conditional model for Y given X.

3. The conditional model for Z given Y.

All components are mean linear, and so we can write the model as

$$X = \mu_i^X + \varepsilon^X$$
$$Y = b_1 + b_2 x + \varepsilon^Y$$
$$Z = b_3 + b_4 y + \varepsilon^Z$$

for some constants $b_1 \ldots b_4$, and independent error terms $\varepsilon^X \sim \mathcal{N}(0, \sigma^x)$, $\varepsilon^Y \sim \mathcal{N}(0, \sigma^y)$, and $\varepsilon^Z \sim \mathcal{N}(0, \sigma^z)$.

We study this type of model more systematically in the next section.

4.4 Decomposable Models

In this section, we study some particularly simple models. These are the models that can be successively broken down into a sequence of smaller models, where each of these corresponds to an unrestricted marginal model. They are termed decomposable models. Frydenberg and Lauritzen (1989), Lauritzen (1992) study these models in detail.

Decomposable models are important in applied graphical modelling for various reasons. One reason is that closed form expressions for the MLEs can be found so parameter estimation can be performed more efficiently. A second, arguably more important, reason is that numerous theoretical and computational aspects are more tractable with decomposable models than with the general models. For example, we describe below in Section 5.4

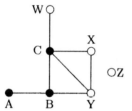

FIGURE 4.5. A strongly decomposable graph.

how exact conditional tests can be computed for edge deletion tests between decomposable models: in principle, these could also be computed in
more general circumstances, but this would appear to be very challenging
computationally. Finally, an attractive property of decomposable models is
ease of interpretation. As we see below, they are equivalent to a sequence
of univariate conditional models— regressions, if we take this in the wide
sense, including logistic regressions. So they suggest a stochastic mechanism
whereby the data may have been generated.

To introduce the topic, consider a graph such as the one shown in Figure 4.5.
Suppose the graph has k cliques $c_1, \ldots c_k$. Define the running intersection
sets $b_2, \ldots, b_k$ by $b_2 = c_2 \cap c_1$, $b_3 = c_3 \cap (c_1 \cup c_2)$, $\ldots, b_k = c_k \cap (\cup_{t=1}^{k-1} c_t)$.
We say that the clique ordering $c_1, \ldots, c_k$ is an SD-ordering if for each
$s = 2 \ldots k$,

1. for some $j < s$, b_s is contained in c_j, and
2. either $c_s \setminus b_s \subseteq \Gamma$ or $b_s \subseteq \Delta$.

For example, Figure 4.5 has five cliques, namely $c_1 = \{A, B\}$, $c_2 = \{B, C, Y\}$, $c_3 = \{C, W\}$, $c_4 = \{C, X, Y\}$, and $c_5 = \{X, Y, Z\}$. We find
that $b_2 = \{B\}$, $b_3 = \{C\}$, $b_4 = \{C, Y\}$, and $b_5 = \{X, Y\}$. It is easy to
verify that $c_1, \ldots, c_5$ is an SD-ordering.

We call a graph strongly decomposable if there is an SD-ordering of its
cliques, and we call a model decomposable if it is graphical and its graph
is strongly decomposable. These properties have been studied by various
authors (Lauritzen and Wermuth, 1989; Leimer, 1989; and Frydenberg
and Lauritzen, 1989). They are closely connected with the collapsibility
criteria for graphical models described in the previous section, since the
SD-ordering property implies that the model can be collapsed onto $\cup_{t=1}^{s} c_t$
for each $s = 1, \ldots k - 1$. By recursive use of the relation (4.27), we can
compute MLEs for a decomposable model. We illustrate this with the help
of three examples.

Example 1

The first example is the discrete model $\mathcal{M} = AB, BC$, with graph:

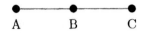

A   B   C

Let $a = \{A, B\}$ and $b = \{B, C\}$. We see that $\mathcal{M}$ is collapsible onto a, and the marginal model on a is AB, with estimates $\breve{p}_{ij} = n_{ij+}/N$. To estimate the conditional probabilities $\Pr(C = k | B = j)$, we estimate the joint probability $\tilde{p}_{jk} = n_{+jk}/N$, from which we see that $\tilde{p}_{j+} = n_{+j+}/N = \dot{p}_j$, say, giving $\Pr(C = k | B = j) = \tilde{p}_{jk}/\dot{p}_j$. Thus, using (4.27), we obtain the MLEs for $\mathcal{M}$ as

$$\hat{p}_{ijk} = (n_{ij+}/N)(n_{+jk}/n_{+j+}). \tag{4.31}$$

To tie this in with the next two examples, we rewrite this using the canonical parameters. Write $\hat{\alpha}_{ijk} = \ln \hat{p}_{ijk}$, $\breve{\alpha}_{ij} = \ln \breve{p}_{ij}$, $\tilde{\alpha}_{jk} = \ln \tilde{p}_{jk}$, and $\dot{\alpha}_j = \ln \dot{p}_j$. Then (4.31) can be written as

$$\hat{\alpha}_{ijk} = \breve{\alpha}_{ij} + \tilde{\alpha}_{jk} - \dot{\alpha}_j.$$

Example 2

Next we derive analogous calculations for a continuous model, with formula $\mathcal{M} = //XY, YZ$, and graph

X   Y   Z

Let $a = \{X, Y\}$ and $b = \{Y, Z\}$. We see that $\mathcal{M}$ is collapsible onto a, and the marginal model on a is $//XY$, with estimates

$$\breve{\mu}^a = \begin{pmatrix} \bar{x} \\ \bar{y} \end{pmatrix}$$

and

$$\breve{\Sigma}^a = \begin{pmatrix} s^{XX} & s^{XY} \\ s^{XY} & s^{YY} \end{pmatrix}.$$

The parameters of the conditional model $\mathcal{M}_{b|a}$ can be estimated by calculating $\hat{f}_b/\hat{f}_{b \cap a}$, i.e., using

$$\tilde{\mu}^b = \begin{pmatrix} \bar{y} \\ \bar{z} \end{pmatrix}$$

and

$$\tilde{\Sigma}^b = \begin{pmatrix} s^{YY} & s^{YZ} \\ s^{YZ} & s^{ZZ} \end{pmatrix},$$

together with $\dot{\mu}^{b\cap a} = (\bar{y})$ and $\dot{\Sigma}^{b\cap a} = (\dot{\sigma}^{YY}) = (s^{YY})$. Combining these parameter estimates to calculate the MLEs of $\mathcal{M}$ is done most conveniently in terms of the canonical parameters. We calculate

$$\check{\Omega}^a = \begin{pmatrix} \check{\omega}^{XX} & \check{\omega}^{XY} \\ \check{\omega}^{XY} & \check{\omega}^{YY} \end{pmatrix} = (\check{\Sigma}^a)^{-1},$$

$$\check{\beta}^a = \begin{pmatrix} \check{\beta}^X \\ \check{\beta}^Y \end{pmatrix} = \check{\Omega}^a \check{\mu}^a,$$

$$\tilde{\Omega}^b = \begin{pmatrix} \tilde{\omega}^{YY} & \tilde{\omega}^{YZ} \\ \tilde{\omega}^{YZ} & \tilde{\omega}^{ZZ} \end{pmatrix} = (\tilde{\Sigma}^b)^{-1},$$

$$\tilde{\beta}^b = \begin{pmatrix} \tilde{\beta}^Y \\ \tilde{\beta}^Z \end{pmatrix} = \tilde{\Omega}^b \tilde{\mu}^b,$$

and finally, $\dot{\omega}^{YY} = (\dot{\sigma}^{YY})^{-1}$ and $\dot{\beta}^y = \dot{\omega}^{yy} \dot{\mu}^y$. Then, using (4.27), we obtain the MLEs of the canonical parameters of $\mathcal{M}$ as

$$\hat{\Omega} = \begin{pmatrix} \check{\omega}^{XX} & \check{\omega}^{XY} & 0 \\ \check{\omega}^{YX} & \check{\omega}^{YY} & 0 \\ 0 & 0 & 0 \end{pmatrix} + \begin{pmatrix} 0 & 0 & 0 \\ 0 & \tilde{\omega}^{YY} & \tilde{\omega}^{YZ} \\ 0 & \tilde{\omega}^{ZY} & \tilde{\omega}^{ZZ} \end{pmatrix} - \begin{pmatrix} 0 & 0 & 0 \\ 0 & \dot{\omega}^{YY} & 0 \\ 0 & 0 & 0 \end{pmatrix}$$

and

$$\hat{\beta} = \begin{pmatrix} \check{\beta}^X \\ \check{\beta}^Y \\ 0 \end{pmatrix} + \begin{pmatrix} 0 \\ \tilde{\beta}^Y \\ \tilde{\beta}^Z \end{pmatrix} - \begin{pmatrix} 0 \\ \dot{\beta}^Y \\ 0 \end{pmatrix}.$$

Example 3

Finally, we sketch the corresponding calculations for a simple mixed model $\mathcal{M} = A/AY, X/XY$, whose graph is

We take $a = \{A, Y\}$ and $b = \{Y, Z\}$. The marginal model $\mathcal{M}_a = A/AY/Y$ has MLEs given through

$$\check{p}_i = n_i/N, \quad \check{\mu}_i^Y = t^x/n_i, \quad \check{\sigma}_i^{YY} = s^{xx},$$

with corresponding canonical parameters $\check{\omega}_i^{YY} = (\check{\sigma}_i^{YY})^{-1}$, $\check{\beta}_i^Y = \check{\omega}_i^{YY} \check{\mu}_i^Y$, and $\check{\alpha}_i = \ln(\check{p}_i) - \frac{1}{2}\ln(|\check{\sigma}_i^{YY}|) - \frac{1}{2}\check{\mu}_i^{Y'}\check{\sigma}_i^{YY}\check{\mu}_i^Y - \frac{q}{2}\ln(2\check{\pi})$. The conditional model is handled as in the previous example using $\hat{f}_b / \hat{f}_{b \cap a}$. We obtain in the same way

$$\hat{\Omega}_i = \begin{pmatrix} \breve{\omega}_i^{YY} & 0 \\ 0 & 0 \end{pmatrix} + \begin{pmatrix} \tilde{\omega}^{YY} & \tilde{\omega}^{YZ} \\ \tilde{\omega}^{ZY} & \tilde{\omega}^{ZZ} \end{pmatrix} - \begin{pmatrix} \dot{\omega}^{YY} & 0 \\ 0 & 0 \end{pmatrix},$$

where $\tilde{\omega}^{YY}, \tilde{\omega}^{YZ}, \tilde{\omega}^{ZZ}$, and $\dot{\omega}^{YY}$ are calculated as in Example 2. Similarly,

$$\hat{\beta}_i = \begin{pmatrix} \breve{\beta}_i^Y \\ 0 \end{pmatrix} + \begin{pmatrix} \tilde{\beta}_i^Y \\ \tilde{\beta}_i^Z \end{pmatrix} - \begin{pmatrix} \dot{\beta}_i^Y \\ 0 \end{pmatrix}.$$

We leave the remaining details to the reader. An interesting aspect of this example is that even though $\mathcal{M}$ is not collapsible onto $b = \{Y, Z\}$, we still fit the model $\mathcal{M}_b = //YZ$ in order to calculate the conditional density $f_{b|a}$.

It is straightforward to extend these results to the general case. To compute the MLEs of a decomposable model, we fit saturated models for margins corresponding to the cliques and the running intersection sets. The MLEs of the canonical parameters of the joint density are calculated by simply adding or subtracting as appropriate the MLEs of the corresponding canonical parameters of the marginal models (see Frydenberg and Lauritzen, 1989).

We now describe two useful characterizations of the strongly decomposable property. The simplest is the following, due to Leimer (1989). A necessary and sufficient condition for a graph to be strongly decomposable is that the graph is triangulated and does not contain any path between two non-adjacent discrete vertices passing through only continuous vertices. In other words, any path of the following type is forbidden.

Note that it is *not* sufficient that the graph is triangulated. For example, the graph shown in Figure 4.6 is triangulated but is not strongly decomposable.

Table 4.1 shows a list of hierarchical interaction models, together with their classification as decomposable or graphical. It is instructive to verify decomposability or otherwise of the models using the forbidden path property.

FIGURE 4.6. A triangulated graph that is not strongly decomposable.

Another characterization of strong decomposability given by the same authors is the following. Suppose the vertices are written $v_1, \ldots, v_{p+q}$. For each $t = 2, \ldots, p+q$, define $u_t = \cup_{s=1}^{t} v_s$, and let b_t be the boundary of $\{v_t\}$ in u_t. Then the condition is that for some ordering of the vertices, for each $t = 2, \ldots, p+q$, (i) b_t is complete, and (ii) either $v_t \in \Gamma$ or $b_t \subseteq \Delta$.

For example, consider the graph shown in Figure 4.5. One ordering satisfying the condition is $\{A, B, C, Y, X, Z, W\}$; another such ordering is $\{C, B, W, A, Y, X, Z\}$.

An interesting implication of this characterization is that it means that the model corresponding to the graph is collapsible onto $\{v_1, \ldots v_{p+q-1}\}$, the induced marginal model is collapsible onto $\{v_1, \ldots v_{p+q-2}\}$, and so on. In other words, we can recursively reduce the joint model into a succession of univariate conditional models.

This property of decomposable models— that they can be broken down into a recursive system of univariate models (regressions, if this is taken in the broad sense, including logistic regressions)— makes them relatively easy to interpret. They suggest a stochastic mechanism whereby the data have been generated: this may be interpreted as a causal ordering. Note that for a given model there may be many appropriate vertex orderings: in other words, a given model may be consistent with a large number of causal orderings.

We must emphasize that the definition of decomposability given above, in which we require decomposable models to be graphical, is somewhat arbitrary. There are nongraphical models with closed-form MLEs that are very similar to the models we have described. For example, as we see shortly, $A/AX, AY/AX, XY$ can be broken down into a sequence of univariate regression models, but one of these is heterogeneous and the other is homogeneous, and so $A/AX, AY/AX, XY$ is nongraphical and hence nondecomposable. There are also other types of nongraphical models with closed-form MLEs: for example, $AB/AX, BX/X$ and $AB/ABX/AX$. The definition used is chosen mainly out of convenience, to be consistent with the definition used by Frydenberg and Lauritzen (1989) and Lauritzen (1992), and also to be in line with the estimation algorithms used in MIM, for example, that use iterative estimation for nongraphical models, even when closed-form MLEs are available.

The decomposition of an undirected, decomposable model into a sequence of univariate conditional models is central to the interpretation of decomposable models, as we have seen above. When available, such decompositions are also important for nondecomposable models. Consider again the model $A/AX, AY/AX, XY$. Using criteria for collapsibility described in Section 4.2, we find that the model is collapsible onto $\{A, X\}$ but not onto $\{A, Y\}$. So the model is consistent with

but not with

Thus X may be prior to Y but not vice versa. The model can be broken down into two conditional models, say $f_{X|A}(x|i) = \mathcal{N}(\mu_i, \sigma_i)$, and $f_{Y|A,X}(y|i, x) = \mathcal{N}(\nu_i + \xi x, \tau)$. The marginal distribution of Y and A is heterogeneous, but controlling for X removes the heterogeneity.

4.5 CG-Regression Models

In this section we collect some results about the CG-regression models, and illustrate these with a simple application. These models are particularly useful in connection with analyses based on chain graphs: see Section 7.2.

As described in Section 4.2, a joint model $\mathcal{M}$ induces a conditional model $\mathcal{M}_{b|a}$ that describes the distribution of the response variables b given the covariates a. This is called a CG-regression model. If $\mathcal{M}$ is collapsible onto a, then it is straightforward to work with the conditional model in the joint framework. For example, we can obtain the maximum likelihood estimates for the parameters of the conditional model directly from the estimates in the joint model. Similarly, we can compute tests based on the deviance difference between two conditional models by using the deviance difference between the two corresponding joint models, as long as both are collapsible onto a. For pure models, we can ensure collapsibility onto a by just including all interactions between variables in a in all models considered. In other words, by restricting attention to models including all interations in a, in effect we condition on a.

When there are discrete response variables and continuous covariates, we cannot ensure collapsibility in this fashion, and so we cannot use the joint models to work implicitly in the conditional framework. We use the term CG-regression primarily for this situation, which is also where the novel models arise.

Rather than introduce a new model formula syntax for the CG-regression models, it is easier to continue to use the formulae for the joint models. So, for example, we talk about the CG-regression model induced from $A/AX/X$ by conditioning on X. But it should be noted that different joint models may induce the same conditional model. Consider, for example, the models

$AB/AX, BX/X$ and $AB/BX/X$. They differ in the marginal distribution of $\{A, X\}$ but not in the conditional distribution of B given $\{A, X\}$. It is not difficult to see that two joint models induce the same conditional model whenever the set of generators that contain one or more response variables is identical in both models. So, in effect, we can identify a CG-regression model $\mathcal{M}_{b|a}$ with the formula obtained by omitting all generators that are subsets of a from the formula for $\mathcal{M}$.

In MIM CG-regression models are specified indirectly via a joint model and the set of covariates a (the latter using the Fix command). They are fit using the CGFit command, which uses a recently developed method for maximum likelihood estimation, the ME-algorithm (see Appendix D). This algorithm resembles the EM-algorithm in many respects: in particular, convergence can be slow, and so at the time of writing it can be recommended only for models of small to moderate dimension.

For several reasons, when specifying a CG-regression model, it is best to use a full marginal model, that is to say, include all interactions between the covariates (a). Firstly, since the conditional model asserts no conditional independence relations between the covariates, it is appropriate that the corresponding independence graph should contain a complete subgraph on a. Secondly, the ME-algorithm is sensitive to the choice of joint model and generally is most efficient when the full marginal model is employed.

4.5.1 Example: Health Status Indicators

Consider models with two binary responses I and J, and one continuous covariate X. For example, suppose that I and J are two health status indicators that are both related to age (X). We have a sample of individuals, for whom we have registered these three variables, and we want to investigate whether, and how, I and J are related. We are not interested in the marginal association between I and J, since this only may reflect that both are age-dependent. We focus instead on their conditional association given X.

As in (4.24), we have that

$$p_{ij|x} = \frac{\exp(\alpha_{ij} + \beta_{ij}x - \frac{1}{2}\omega_{ij}x^2)}{\sum_{jk}\exp(\alpha_{jk} + \beta_{jk}x - \frac{1}{2}\omega_{jk}x^2)}. \tag{4.32}$$

A natural measure of the conditional association between I and J given X is the conditional log-odds ratio given by

$$\Psi(x) = \ln(\frac{p_{11|x}p_{22|x}}{p_{12|x}p_{21|x}}).$$

This is symmetric in I and J, and zero when these are conditionally independent. We see from (4.32) that

$$\Psi(x) = (\alpha_{11} + \alpha_{22} - \alpha_{12} - \alpha_{21}) + (\beta_{11} + \beta_{22} - \beta_{12} - \beta_{21})x$$
$$- \tfrac{1}{2}(\omega_{11} + \omega_{22} - \omega_{12} - \omega_{21})x^2.$$

If we restrict attention to homogeneous models, then we see that the ANOVA-model with interaction, $IJ/IJX/X$, corresponds to a CG-regression model in which $\Psi(x)$ is linear in x. Setting the interaction to zero in $IJ/IX, JX/X$ sets $\Psi(x)$ constant.

More generally, the joint model $IJ/IJX/IX, JX$ corresponds to a CG-regression model in which $\Psi(x)$ is linear, and $IJ/IX, JX/IX, JX$ to one in which $\Psi(x)$ is constant.

The following program fragments illustrate an analysis using this kind of model. The data are defined using the statements:

```
fact i2j2; cont x; read xij
69 2 1 59 2 1 69 2 1 68 2 2 66 2 1 63 2 1 62 2 1 62 2 1 62 2 1
54 1 2 62 2 2 64 2 2 59 1 2 61 2 1 60 2 2 57 1 2 65 2 1 59 1 2
60 2 1 64 2 2 58 1 2 58 2 1 50 1 2 58 2 2 54 1 2 55 1 1 50 2 2
51 1 1 68 2 1 63 1 2 51 2 2 67 2 1 53 2 2 66 2 1 64 2 1 58 1 2
59 2 2 51 2 2 58 1 2 53 2 2 54 1 1 65 2 1 58 2 2 60 2 2 50 1 2
53 1 2 60 2 2 59 1 2 60 2 1 69 2 1 54 1 1 63 2 1 68 1 1 64 2 1
62 2 1 54 1 2 55 1 2 60 2 1 59 2 2 50 2 2 64 2 1 54 1 2 59 2 2
59 1 2 50 1 2 60 1 2 57 2 2 52 2 2 52 2 2 54 2 2 58 2 2 57 1 2
61 2 1 56 1 2 59 1 2 55 1 2 53 1 2 65 2 1 58 1 2 61 2 1 64 2 1
53 2 1 69 2 1 53 1 2 64 2 1 68 2 1 65 2 1 63 2 1 56 1 2 56 1 2
54 1 2 50 1 2 67 2 1 55 1 2 52 1 2 65 1 2 55 2 2 66 1 1 55 2 1
54 1 1 !
```

To fit a simple linear conditional log-odds model, we define the corresponding joint model and use the CGFit command:

```
MIM->satmod; fix x; cgfit
Fixed variables: x
Convergence after 40 iterations.
-2*Conditional Loglikelihood:  199.046 DF:    0
MIM->base; homsat; cgfit; test
Convergence after 5 iterations.
-2*Conditional Loglikelihood:  205.284 DF:    3
Test of H0: ij/ijx/x
against H:  ij/ijx/ijx
LR:    6.2380    DF:   3    P: 0.1006
```

The homogeneous model (corresponding to linear $\Psi(x)$)fits the data reasonably well. To examine whether $\Psi(x)$ is constant we remove the interaction term:

```
MIM->base; model ij/ix,jx/x; cgfit; test
Convergence after 4 iterations.
-2*Conditional Loglikelihood:  209.788 DF:    4
Test of H0: ij/ix,jx/x
against H:  ij/ijx/x
LR:    4.5040    DF:    1    P: 0.0338
```

There is moderate evidence that $\Psi(x)$ is not constant over x. We print out the estimates of α_{ij} and β_{ij}:

```
MIM->backtobase
MIM->prf 12 6
Printing format: 12,6
MIM->disp ij,x
Linear predictors for i,j given x.
  i j        Constant          x
  1 1        0.000000    0.000000
  1 2        6.370484   -0.085128
  2 1      -16.809679    0.304779
  2 2        2.610482   -0.024100
```

We see that $\Psi(x)$ is estimated as $\Psi(x) = 13.05 - 0.24x$: that is, an increasingly negative association with age.

4.5.2 Example: Side Effects of an Antiepileptic Drug

Epilepsy is a common neurological disorder, characterized by unprovoked seizures. For some patients, medical treatment is only partially effective, so their seizures are inadequately controlled. This example uses data from a double-blind, parallel-group clinical trial studying an antiepileptic drug (Kalviainen et al., 1998). This included 154 patients with refractory epilepsy, randomised equally to receive in addition either tiagabine or placebo. Their previous drug regimen was held constant throughout the study. We here focus on the occurrence of three side effects of the drug: headache, tiredness, and dizziness. In the analysis we also include the following variables: age, sex, treatment group, the number of years with epilepsy, and the number of concomitant antiepileptic drugs. The last variable can be regarded as a proxy for how refractory the disorder is.

We illustrate an exploratory analysis of these data.

```
MIM->sh v
Var Label       Type Levels In   In    Fixed Block
                            Data Model
     ----------------------------------------------------
  s  Sex         disc   2    X    .      .       .
```

```
d  Dizziness    disc   2   X   .    .    .
t  Tiredness    disc   2   X   .    .    .
h  Headache     disc   2   X   .    .    .
g  Group        disc   2   X   .    .    .
a  Age          cont   .   X   .    .    .
y  Yrs epilepsy cont   .   X   .    .    .
n  No of AEDs   cont   .   X   .    .    .
-----------------------------------------------------
MIM->fix sgayn
Fixed variables: agnsy
MIM->homsat
MIM->step z
MIM->pr
The current model is: dt,ghs,dgs/ghsy,ghns,ags/any.
```

We have three binary responses and five covariates, two of which are discrete. As a first step, we choose a preliminary, undirected model using stepwise selection starting from the saturated homogeneous model: The Z option suppresses the output. The graph of the selected model is shown in Figure 4.7. It suggests that Dizziness and Tiredness are closely related, that the frequency of their occurrence differs between men and women, and that this is also affected by treatment. The same is true of Headache, but this is also related to the duration of the illness and how refractory it is. For given sex and treatment group, the occurrence of Headache is independent of the occurrence of Dizziness and Tiredness.

We now turn to CG-regression models and continue the selection process, starting out from this preliminary model. We do by using the Stepwise command again, this time with the G option, which searches by fitting and comparing CG-regression models.

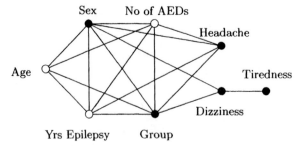

FIGURE 4.7. Side-effects of an anti-epileptic drug: the preliminary model.

```
MIM->step g
Coherent Backward Selection.
CG-regression models.
Unrestricted mode, Chi-squared tests.
Critical value:   0.0500
Initial model: dt,ghs,dgs/ghsy,ghns,ags/any
Model: dt,ghs,dgs/ghsy,ghns,ags/any
-2*LogLikelihood:      459.5590 DF: 262
     Edge        Test
  Excluded    Statistic DF          P
     [dg]       6.8130  2       0.0332 +
     [ds]      12.1744  2       0.0023 +
     [dt]       3.8869  1       0.0487 +
     [gh]      11.0734  6       0.0861
     [hn]      11.6532  4       0.0201 +
     [hs]      24.8656  6       0.0004 +
     [hy]      10.3847  4       0.0344 +
Removed edge [gh]
Selected model: hs,dt,dgs/hsy,gsy,hns,gns,ags/any
```

We see that the evidence that the treatment causes headache is weak, since the test for the removal of the edge $[gh]$ gives a p-value of 0.0861. As we see, the selection procedure removes this edge. Apart from this, the model is unchanged: the preliminary analysis using the undirected models was not too far from the mark.

In problems involving high-dimensional CG-regression models this may often be a sensible strategy: first select a preliminary undirected model, and then, as it were, fine-tune the analysis with CG-regression models, using the undirected model as a point of departure.

4.6 Incomplete Data

Up to now, we have assumed that all data are available. In practice, it often happens, for a variety of reasons, that some data are lost. Test tubes can be dropped, apples can be scrumped, and patients can withdraw from studies. Furthermore, in some applications it may be appropriate to include completely unobserved or *latent* variables in the analysis. In this section we show how the mixed models can be applied to incomplete data. A wide variety of latent variable problems, including mixture models, latent class models, factor analysis–type models, and other novel models can be handled in this manner.

To do this requires that the estimation algorithms previously described are augmented by means of the EM-algorithm (Dempster et al., 1977). (Appendix D.2 describes the computations in some detail.) The advantages of this algorithm are computational simplicity and stable convergence. There are, however, disadvantages. Although convergence is stable, it is often very slow, and moreover there is no guarantee that it arrives at the global maximum likelihood estimate; for many incomplete data problems, the likelihood has multiple local maxima. It is often difficult to know whether it is the global maximum that has been found. Furthermore, inference based on hypothesis testing for latent variable models is made difficult by the fact that the χ^2 approximation to the distribution of the likelihood ratio test statistic may be inaccurate or invalid: see Shapiro (1986) for some asymptotic results. In line with much literature on latent variable models, we emphasize estimation rather than hypothesis testing.

It is useful to distinguish between missing data problems and latent variable models. In the former, data are partially missing, i.e., values are observed for some cases. For such problems, the validity of the approach depends upon assumptions about the process whereby data become missing. As a rule, these assumptions are difficult to verify in practice. Nevertheless, their plausibility should be examined carefully since estimation may be strongly biased if they do not hold.

In the next section, we study these assumptions more closely. We then describe some latent variable models that fall within the current framework. The following section describes in detail the application of the EM-algorithm to the mixed models. Finally, we apply the methods to some examples.

4.6.1 Assumptions for Missing Data

To examine the assumptions required, suppose that Z, M, and Y are vector random variables such that Z is the (hypothetical) complete observation; M is the configuration of missing values, i.e., a vector of 1's and 0's indicating whether the corresponding element of Z is observed or missing; and Y is the (incomplete) observation. This is illustrated in Figure 4.8. The missing value process is encapsulated in the conditional probabilities $\Pr(m|z)$.

A given configuration of missing values m partitions z and y into two subvectors, $y_{\mathrm{obs}} = z_{\mathrm{obs}}$ and $y_{\mathrm{mis}} = *$, where z_{mis} are the values that are missing.

If we observed the complete data so that $Y = Z$, then we could work with the ordinary likelihood function for Y, i.e.,

$$\mathcal{L}(\theta; Y) = \prod_{k=1}^{N} f(y^{(k)}|\theta), \tag{4.33}$$

Y: the observed data

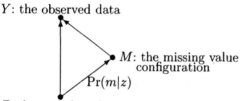

M: the missing value
configuration

$\Pr(m|z)$

Z: the complete data

FIGURE 4.8. The missing data framework.

but since the data are incomplete, this is not possible. If we ignore the process giving rise to the missing values, we can instead use the likelihood

$$\mathcal{L}^*(\theta; Y_{\text{obs}}) = \prod_{k=1}^{N} f^*(y_{\text{obs}}^{(k)}|\theta), \qquad (4.34)$$

where $f^*(y_{\text{obs}}^{(k)}|\theta)$ is shorthand for the marginal density for $y_{\text{obs}}^{(k)}$. This likelihood is the quantity maximized by the EM-algorithm.

It is not always valid to ignore the missing value process in this way. For example, if high values of a variable are apt to be missing, then clearly estimation of its mean using observed values only will be heavily biased. Rubin (1976) formulated a key condition for the validity of inference based on the likelihood (4.34), the so-called *missing at random* (MAR) property, which is defined as follows. For each configuration of missing values m, we require that the conditional probability $\Pr(m|z)$ is constant over all z' such that $z'_{\text{obs}} = z_{\text{obs}}$. For example, one outcome is "all data are missing"; here, z_{obs} is null so the probability of this outcome must be constant for all z. Another outcome is "no data are missing"; here, $z_{\text{obs}} = z$ so the probability of this pattern of missing values can vary freely for different z. An alternative way of formulating the MAR property is in terms of a binary random variable, say S_m, indicating whether the configuration is m. Then the MAR property is that for each m, $S_m \perp\!\!\!\perp z_{\text{mis}} \mid z_{\text{obs}}$.

For example, consider a survey in which height and gender are recorded. There may be a tendency for men's heights to be missing; as long as this increased probability does not depend on the heights of the men in question (only their gender), then the values are missing at random.

As a more realistic example, consider a clinical study comparing two anti-epileptic medicines in which the endpoint is weekly seizure rate. Suppose the patients in the study visit the clinician weekly and report any seizures they have experienced during the previous week. Patients may withdraw from the study for a variety of reasons, but suppose that the major reason is lack of treatment efficacy. During a visit, a patient may inform the clinician that he or she wishes to withdraw and report any seizures experienced during the previous week. Since the information is complete at the time of withdrawal, the decision cannot depend on the subsequent missing values. In this situation, it is reasonable to assume that the MAR property is satisfied. On the other hand, if the patient withdraws in such a fashion that any seizures occurring during the previous week remain unreported, then the MAR property is probably violated.

This example illustrates that it is possible to evaluate the plausibility of the MAR property, even without detailed knowledge of the missing data process. When there is a substantial proportion of missing data and these are clearly not missing at random, then an analysis based on the likelihood (4.34) will be heavily biased and should not be attempted. On the other hand, an analysis based on complete cases only will also be heavily biased. The only viable approach will be to model the missing data process explicitly, and this will often be very speculative.

We now examine some latent variable models. Note that for these, the MAR property is trivially fulfilled, provided no data are missing for the manifest (i.e., nonlatent) variables.

4.6.2   Some Latent Variable Models

The simplest latent variable model we can consider is

where L is a latent variable and X is a manifest (i.e., observed) variable. (Here and elsewhere, filled square vertices denote discrete latent variables and hollow squares denote continuous latent variables.) The model states that the distribution of X is a mixture of several Gaussian distributions. We use this model in Section 4.6.3.

Note that the following three bivariate models are *not* sensible:

Here, L is latent and X is manifest. Each model is collapsible onto the manifest variable so the latent variable has no explanatory power. If a model is collapsible onto a variable subset a, then the remaining variables contain no information about the marginal distribution of a (see, for example Asmussen and Edwards, 1983). In general, only latent variable models that are *not* collapsible onto the manifest variables are sensible.

Perhaps the most widely used type of latent variable model is exemplified by the graph

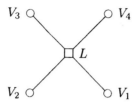

where L is latent and the remaining variables are manifest. This is a factor analysis model with one common factor (latent variable). The general factor analysis model for, say, manifest variables $V = (V_1, \ldots, V_q)'$ and latent factors $L = (L_1, \ldots, L_s)'$ can be written as

$$V = A + BL + E, \tag{4.35}$$

where A is a $q \times 1$ vector of intercepts, B is a $q \times s$ matrix of regression coefficients, L is normally distributed with zero mean and covariance Σ_L, and E is an error term. The covariance matrix of E is assumed to be diagonal so that the elements of V are conditionally independent given L.

For any invertible $s \times s$ matrix C, we can rewrite (4.35) as

$$V = A + (BC)(C^{-1}L) + E,$$

so B is not identifiable without further constraints. The usual approach is to assume that Σ_L is the identity matrix.

The conditional model (4.35) can be applied in the current framework by using models that are collapsible onto the latent variables. However, the marginal constraints on L cannot be incorporated; thus, there will be identifiability problems. If we use a model like

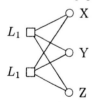

then we have that $L_1 \perp\!\!\!\perp L_2 \mid (X, Y, Z)$ rather than $L_1 \perp\!\!\!\perp L_2$, so the model is not a conventional factor analysis model.

Identifiability conditions for graphical Gaussian models with one latent variable have been studied by Stanghellini (1997) and Vicard (2000).

An apparently novel type of latent variable model is exemplified by the graph

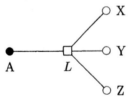

where again, L is latent and remaining variables manifest. Here, A could, for example, be gender, and X, Y, and Z could be manifest psychological traits. The model implies that the distribution of the latent trait differs for the two genders.

For purely discrete variables, latent class models have been widely applied, particularly in the social sciences. These posit that the manifest variables are conditionally independent given one or more latent variables. An example is

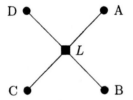

Such models also fit into the current framework.

Another broad class of problems for which latent variable models are useful concerns measurement error. In Section 4.3, we considered a model with the graph

and noted that it can be represented as a recursive equation system. If we suppose that the continuous variables are measured with error, and that the recursive equations apply to the true rather than the observed values, then we can write this as

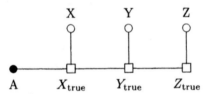

where X_{true}, Y_{true}, and Z_{true} are latent variables corresponding to the true values of the observed variables.

The discrete analogue of measurement error is misclassification. Although in some contexts discrete variables may be measured essentially without error, in others there may be errors of classification. This may be true, for example, when there is an element of subjectivity involved in the classification. A discrete analogue of the last model would be

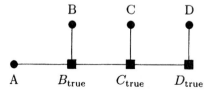

These last two models are small-scale examples of a class of models called *hidden Markov models*. As their name suggests, these assume that a system of variables obeys a Markov system of dependencies, when certain latent (hidden) variables are included. Often, the system is observed a number of times, and the latent variables represent the hidden state or memory of the system at the different times. For models with the above regular structure (and other, more complex, regular structures), particularly efficient estimation algorithms have been developed. Hidden Markov models have proven useful in a wide range of applications, including speech recognition, ion-channel kinetics and amino-acid sequence analysis. See Farewell (1998) for an overview and bibliography.

4.6.3 Example: The Components of a Normal Mixture

To illustrate some general features of the use of the EM-algorithm, we consider here the problem of separating a distribution into two components. We suppose that the following values

1.2, 1.3, 1.2, 1.5, 1.2, 2.3, 2.7, 1.2, 1.8, 3.2, 3.5, 3.7

have been drawn from a density of the form

$$f(y) = p_1(2\pi\sigma)^{-\frac{1}{2}} \exp\{-\frac{1}{2}(y - \mu_1)'\sigma^{-1}(y - \mu_1)\}$$
$$+ (1 - p_1)(2\pi\sigma)^{-\frac{1}{2}} \exp\{-\frac{1}{2}(y - \mu_2)'\sigma^{-1}(y - \mu_2)\},$$

where p_1 is the proportion of the first component, and the components are normal densities with means μ_1 and μ_2 and common variance σ. This is known as the two-component mixture model. We wish to estimate these parameters and, if possible, to test whether $\mu_1 = \mu_2$, i.e., whether the data are adequately described with one component. Day (1969) gives a careful treatment of this problem.

We define the data and fit the one-component model first:

```
MIM->cont X; read X
DATA->1.2 1.3 1.2 1.5 1.2 2.3 2.7 1.2 1.8 3.2 3.5 3.7 !
Reading completed.
MIM->mod //X; fit
Calculating marginal statistics...
Deviance:       0.0000 DF: 0
MIM->pr y
-2*LogLikelihood of //X:       32.3910       DF:    0
```

We see that minus twice the log likelihood is 32.3910. To fit the two-component model, we declare a discrete variable A, set its values to missing, and then fit the model $A/AX/X$:

```
MIM->fact A2; calc A=-1; model A/AX/X
MIM->emfit
EM algorithm: random start values.
  Cycle -2*Loglikelihood       Change
    1            32.3910
    2            32.3910       -0.000001
Successful convergence.
```

We see that the log likelihood has not changed and we therefore suspect that it was very flat at the starting point. We try some other starting points by repeating the command a few times:

```
MIM->emfit
EM algorithm: random start values.
  Cycle -2*Loglikelihood       Change
    1            32.3910
    2            32.3910       -0.000000
Successful convergence.
MIM->emfit
EM algorithm: random start values.
  Cycle -2*Loglikelihood       Change
    1            32.3912
    2            32.3912       -0.000002
Successful convergence.
MIM->emfit
EM algorithm: random start values.
  Cycle -2*Loglikelihood       Change
    1            32.3910
    2            32.3910       -0.000000
Successful convergence.
MIM->emfit
EM algorithm: random start values.
  Cycle -2*Loglikelihood       Change
    1            32.3891
    2            32.3889       -0.000157
```

3	32.3887	-0.000175
4	32.3885	-0.000195
5	32.3883	-0.000218
6	32.3881	-0.000246
7	32.3878	-0.000277
8	32.3875	-0.000314
9	32.3871	-0.000358
10	32.3867	-0.000410
11	32.3862	-0.000472
12	32.3857	-0.000547
13	32.3851	-0.000637
14	32.3843	-0.000748
15	32.3834	-0.000884
16	32.3824	-0.001054
17	32.3811	-0.001268
18	32.3796	-0.001542
19	32.3777	-0.001897
20	32.3753	-0.002364
21	32.3723	-0.002990
22	32.3685	-0.003846
23	32.3634	-0.005045
24	32.3567	-0.006769
25	32.3473	-0.009328
26	32.3341	-0.013271
27	32.3144	-0.019623
28	32.2840	-0.030418
29	32.2340	-0.050009
30	32.1455	-0.088541
31	31.9734	-0.172084
32	31.5993	-0.374114
33	30.6963	-0.903001
34	28.6080	-2.088266
35	25.9970	-2.610971
36	25.2835	-0.713516
37	25.2397	-0.043777
38	25.2356	-0.004159
39	25.2346	-0.001010
40	25.2343	-0.000275
41	25.2342	-0.000076
42	25.2342	-0.000021
43	25.2342	-0.000006

Successful convergence.

Success! We look at the maximum likelihood estimates:

```
MIM->pr f
Fitted counts, means and covariances.
 A
 1    X    0.147
      Means  3.208    4.285
             X        Count

 2    X    0.147
      Means  1.433    7.715
             X        Count
```

So $\hat{p}_1$, the estimated proportion of the first component, is $4.285/12 = 0.357$, and $\hat{\mu}_1$, $\hat{\mu}_2$, and $\hat{\sigma}$ are as shown. The estimated probabilities for each observation of deriving from the first or second component can be obtained using the Summary command (see Section A.6.2).

To test whether one component is adequate, we may be tempted to fit the model $A/X/X$:

```
MIM->base; delete AX; emfit
EM algorithm: random start values.
  Cycle -2*Loglikelihood      Change
    1            32.3910
    2            32.3910    0.000000
Successful convergence.
```

and to perform a likelihood ratio test:

```
MIM->test
Test of HO: A/X/X
against H:  A/AX/X
LR:   7.1568   DF:   1   P: 0.0075
```

However, this inference is wrong in two respects. First, there are two fewer parameters in the simpler model, not one: the parameter p_1 is inestimable, but this is not detected by MIM, so the degrees of freedom should be two, not one. Secondly, the likelihood ratio test does not have an asymptotic χ^2 distribution (Ghosh and Sen, 1985). McLachlan and Basford (1987) give a careful treatment of this and suggest use of bootstrapping methods.

The two-component normal mixture problem described here is the simplest case of a broad class of models called mixture models. McLachlan and Basford describe many applications of these, primarily in cluster analysis.

4.6.4 Example: Mathematics Marks, Revisited

Our second illustration of the use of the EM-algorithm is in an exploratory analysis of the mathematics marks data (Section 3.1.6). We suppose that there is a latent binary variable A and that the following model holds.

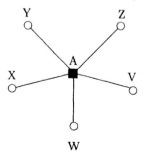

We fit this as follows:

```
MIM->fact A2; calc A=-1
MIM->model A/AV,AW,AX,AY,AZ/V,W,X,Y,Z
MIM->convcrit 0.001
Convergence Criterion:    0.00100000
MIM->emfit
EM algorithm: random start values.
  Cycle -2*Loglikelihood       Change
    1          3586.6322
    2          3544.4148   -42.217466
    3          3489.8524   -54.562365
    4          3475.6076   -14.244835
    5          3473.6885    -1.919064
    6          3473.1887    -0.499796
    7          3472.9763    -0.212403
    8          3472.8779    -0.098388
    9          3472.8326    -0.045326
   10          3472.8119    -0.020637
   11          3472.8026    -0.009319
   12          3472.7984    -0.004185
   13          3472.7966    -0.001874
   14          3472.7957    -0.000837
Successful convergence.
```

Convergence occurs after the 14th cycle. We save the predicted values of A and then plot these against the observation number, calculated as follows:

```
MIM->impute
MIM->calc O=obs
```

The plot is as follows:

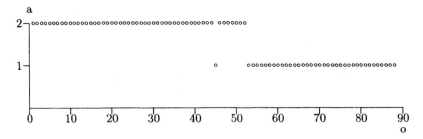

Curiously, the first 52 observations (except for number 45) all belong to the first group, and the remainder belong to the second group. This result is quite stable since repeated use of EMFit gives the same result. This casts doubt on the idea of the data as a pristine random sample from some ideal population. The data have been processed in some way prior to presentation. Are there confounding variables that explain the variation but are not reported, or have the data been subjected to a factor analysis and then sorted by factor score? Certainly they have been mistreated in some way, doubtless by a statistician.

To examine whether the data have been sorted by some criterion, for example, factor score, we try a factor analysis model.

```
MIM->cont S; calc S=ln(-1)
MIM->mod //SV,SW,SX,SY,SZ
MIM->convcrit 0.001
Convergence Criterion:    0.00100000
MIM->emfit
EM algorithm: random start values.
  Cycle -2*Loglikelihood       Change
      1         3592.5363
      2         3588.1598     -4.376522
      3         3556.1523    -32.007540
      4         3465.1367    -91.015609
      5         3414.1564    -50.980273
      6         3403.9184    -10.237988
      7         3401.3168     -2.601604
      8         3400.3137     -1.003101
      9         3399.8492     -0.464440
     10         3399.6055     -0.243702
     11         3399.4648     -0.140698
     12         3399.3783     -0.086549
     13         3399.3229     -0.055428
     14         3399.2864     -0.036469
     15         3399.2619     -0.024470
     16         3399.2453     -0.016669
     17         3399.2338     -0.011492
     18         3399.2258     -0.008001
```

```
  19            3399.2202    -0.005615
  20            3399.2162    -0.003966
  21            3399.2134    -0.002817
  22            3399.2114    -0.002010
  23            3399.2099    -0.001440
  24            3399.2089    -0.001035
  25            3399.2081    -0.000746
Successful convergence.
MIM->Disp VWXYZ,S
Fitted conditional means and covariances.
     V    38.955   -10.420  193.723
     W    50.591    -8.729   -0.000   94.684
     X    50.602    -9.679   -0.000    0.000   17.929
     Y    46.682   -11.409   -0.000   -0.000   -0.000  87.704
     Z    42.307   -12.425   -0.000   -0.000   -0.000  -0.000 140.001
                       S        V        W        X        Y       Z
MIM->Display S,VWXYZ
Fitted conditional means and covariances.
     S     4.314    -0.005   -0.009   -0.053   -0.013   -0.009   0.098
                       V        W        X        Y        Z       S
MIM->impute
MIM->calc O=obs; label S "Score" O "Obs"
```

The algorithm converges after 25 iterations. By fixing S, we examine the estimates of the conditional distributions of the manifest variables given the latent variable. For example, we see that

$$\mathrm{E}(X \mid S = s) = 50.602 - 9.679s.$$

We note that the conditional variance of X (Algebra) given $S = s$ is estimated to be 17.929: this is considerably smaller than that of the other manifest variables, and is consistent with the central role of Algebra noted in Section 3.1.6. Similarly, we can examine the estimates of the conditional distribution of S given the manifest variables. In factor analysis terminology, these are known as factor loadings. It is seen that the largest loading is that of Algebra (X). The index plot of the latent variable is shown in Figure 4.9. It appears to confirm that the data have been sorted by some criterion.

The discrete latent model we used above resembles the mixture model of Section 4.6.3 in that the joint density is posited to be a mixture of two distributions, say $\mathcal{N}(\mu_1, \Sigma)$ and $\mathcal{N}(\mu_2, \Sigma)$, where Σ is diagonal. Given the observed variables, each observation has a posterior probability of deriving from the first or second component. The parameters μ_1 and μ_2 are estimated as means of the observed values weighted by the respective posterior probabilities.

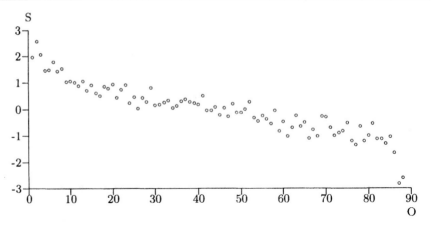

FIGURE 4.9. An index plot of the continuous latent variable.

If the primary purpose of the analysis is to classify the observations into two groups, another technique can be used. Suppose each observation is assigned to the group with the largest posterior probability, and that the parameters μ_1 and μ_2 are estimated on the basis of the observations assigned to the respective groups. This is a cluster-analytic approach, closely related to clustering by minimum Mahalanobis distance. The following example illustrates how this may be performed in MIM.

```
MIM->fact A2; calc A=1+uniform
MIM->mod A/AV,AW,AX,AY,AZ/VWXYZ
MIM->fit; classify AB; calc A=B
Deviance:      19.6597 DF: 15
MIM->fit; classify AB; calc A=B
Deviance:      18.2707 DF: 15
MIM->fit; classify AB; calc A=B
Deviance:      18.2707 DF: 15
MIM->fit
Deviance:      18.2707 DF: 15
MIM->pr f
Fitted counts, means and covariances.
 A
 1     V  145.593
       W  103.767 167.787
       X   83.589  81.825 109.794
       Y   75.499  89.444 107.663 212.298
       Z   94.143  94.807 118.130 149.631 291.303
   Means  46.182  51.606  51.379  48.045  43.318   66.000
               V       W       X       Y       Z   Count

 2     V  145.593
       W  103.767 167.787
```

```
X     83.589  81.825 109.794
Y     75.499  89.444 107.663 212.298
Z     94.143  94.807 118.130 149.631 291.303
Means 17.273  47.545  48.273  42.591  39.273  22.000
          V       W       X       Y       Z   Count
```

First a binary grouping factor A is declared and its values are set at random. The line

```
fit; classify AB; calc A=B
```

does three things: the model is fit, a factor B containing the group with the maximum posterior probability for each observation is calculated, and the grouping factor A is reset to the new grouping. These three steps are then repeated until convergence. In the example shown, convergence occurs rapidly, but whether the method will always lead to convergence is unclear.

4.7 Discriminant Analysis

In this section, we discuss application of the mixed models to discriminant analysis. The approach is closely related to work on mixed model discriminant analysis by Krzanowski (1975, 1980, 1988) and, in particular, Krusinka (1992).

Suppose that there are k groups, $\pi_1 \ldots \pi_k$, that we wish to discriminate between on the basis of $p-1$ discrete and q continuous variables. We write a typical cell as $i = (g, j)$ where g is a level of the grouping factor G, and j is a $(p-1)$-tuple of discrete measurement variables. The q-vector y contains the continuous measurement variables.

We wish to allocate an individual to a group on the basis of measurements (j, y). An optimum rule (see, for example, Krzanowski, 1988, p. 336) allocates the individual to the group π_g with maximum value of $f(j, y|G = g)q_g$, where q_g is the prior probability of an individual belonging to group π_g.

We assume the density (4.1) holds, i.e.,

$$f(i, y) = p_i |2\pi \Sigma_i|^{-\frac{1}{2}} \exp\{-\frac{1}{2}(y - \mu_i)' \Sigma_i^{-1}(y - \mu_i)\} \tag{4.36}$$

with $i = (g, j)$. Often we may use the sample proportions, n_g/N, as prior probabilities. In this case, we allocate the individual to the group with the maximum value of the joint density $f(g, j, y)$.

Following the graphical modelling approach, we select a parsimonious graphical model to the data and use this as a basis for allocation. Note that the allocation rule depends on the relative values of the $f(g, j, y)$ over g. This only involves variables adjacent to G in the independence graph, corresponding to, say, (J_1, Y_1). To see this, observe that these variables separate G from (J_2, Y_2), where $J = (J_1, J_2)$ and $Y = (Y_1, Y_2)'$. Thus, we obtain that

$$G \perp\!\!\!\perp (J_2, Y_2) \mid (J_1, Y_1).$$

So for some functions a and b, we can write $f(g, j, y) = a(g, j_1, y_1)b(j, y)$. It follows that

$$\frac{f(1, j, y)}{f(2, j, y)} = \frac{a(1, j_1, y_1)}{a(2, j_1, y_1)}.$$

The command `Classify` can be used to compute predicted classifications using the maximum likelihood discriminant analysis method. Each observation is assigned to the level g with the largest estimated density $\hat{f}(g, j, y)$. The density estimate can either use all available observations or use the leave-out-one method— that is, the density for each observation is estimated using all available observations except the one in question. This method is computationally intensive (see Section A.11 for details).

We now compare the graphical modelling approach to more conventional methods of discriminant analysis. Three widely used approaches are (i) classical linear discriminant analysis, (ii) quadratic discriminant analysis, and (iii) the independence model. These involve continuous measurement variables only. Method (i) assumes an unrestricted homogeneous covariance matrix, and method (ii) assumes unrestricted heterogeneous covariance matrices. Method (iii) assumes that the measurement variables are independent given the grouping variable— corresponding to a star graph as in Section 4.6.4. So (i) and (ii) can be thought of as unstructured approaches, while (iii) is highly structured. All three methods are special cases of hierarchical interaction models.

A fourth, related approach (iv) is the location model (Krzanowski, 1975, 1980, 1988). This involves both discrete and continuous measurement variables in a two-tiered (chain model) structure. First, the distribution of the discrete measurement variables given G is modelled using a loglinear model. Then the conditional distribution of the continuous measurement variables given the discrete is modelled using a $\mathcal{N}(\mu_i, \Sigma)$ distribution. A MANOVA-type model, with the same linear model for each continuous measurement variable and an unrestricted, homogeneous covariance matrix, is assumed. This approach might perhaps be described as being partially structured. Note that hierarchical interaction models that are collapsible onto Δ are mean linear and have unrestricted homogeneous covariance matrices are location models.

In practice, method (ii) often performs poorly due to overfitting. In other words, the model involves an excessive number of parameters that are estimated poorly, leading to poor prediction. At the other extreme, method (iii) is often patently unrealistic since measurement variables often show considerable correlation.

The rationale underlying the use of the hierarchical interaction models is to base discrimination on a small but accurate model that captures the important interrelationships between the variables. This would appear to be a promising approach, at least for those applications where the models describe the data well.

4.7.1 Example: Breast Cancer

As illustration of the approach to discriminant analysis sketched in the last section, we consider the fourth data set described in Krzanowski (1975). This summarizes the results of ablative surgery for advanced breast cancer. The grouping factor is treatment success, classified as successful or intermediate ($G = 1$), or failure ($G = 2$). The data set consists of 186 observations on ten variables, these being comprised of six continuous variables (U–Z) and four binary variables (A–C and G).

Initially, we examine the counts in the 16 cells:

```
MIM->mod ABCG; pr s
Calculating marginal statistics...
Empirical counts, means and covariances.
 A B C G    Count
 1 1 1 1    3.000
 1 1 1 2   10.000
 1 1 2 1    5.000
 1 1 2 2   17.000
 1 2 1 1   12.000
 1 2 1 2    6.000
 1 2 2 1    7.000
 1 2 2 2    4.000
 2 1 1 1   28.000
 2 1 1 2   17.000
 2 1 2 1   14.000
 2 1 2 2    7.000
 2 2 1 1   21.000
 2 2 1 2   18.000
 2 2 2 1    9.000
 2 2 2 2    8.000
```

Since there are six continuous variables, the empirical cell covariance matrix must be singular for those cells with $\leq$ six observations. It follows

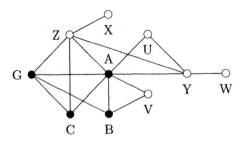

FIGURE 4.10. The model initially selected.

that the MLE of the heterogeneous saturated model cannot exist (see Section 5.2). We proceed therefore by performing backward selection from the homogeneous saturated model, suppressing the output by using the Z option:

```
MIM->mod ABCG/ABCGU,ABCGV,ABCGW,ABCGX,ABCGY,ABCGZ/UVWXYZ
MIM->fit
Likelihood:    11870.2091 DF: 315
MIM->maxmodel
MIM->stepwise z
```

The following model is selected:

$$ACG, ABG/X, W, AY, AU, ACGZ, ABV/YZ, XZ, WY, V, UY$$

whose graph is shown in Figure 4.10. This model is rather complex, but the important thing to note is that only the variables A, B, C, and Z are adjacent to G, so a substantial dimension reduction, from ten to five dimensions, is possible. We therefore consider the marginal distribution of these five variables. We are now able to test for variance homogeneity:

```
MIM->mod ABCG/ABCGZ/ABCGZ
MIM->fit
Deviance:      -0.0000 DF: 0
MIM->base
MIM->mod ABCG/ABCGZ/ABCZ; fit; test
Deviance:      16.7532 DF: 8
Test of H0: ABCG/ABCGZ/ABCZ
against H:   ABCG/ABCGZ/ABCGZ
LR:  16.7532    DF:  8    P: 0.0328
MIM->mod ABCG/ABCGZ/ABGZ; fit; test
Deviance:      16.2615 DF: 8
Test of H0: ABCG/ABCGZ/ABGZ
against H:   ABCG/ABCGZ/ABCGZ
LR:  16.2615    DF:  8    P: 0.0388
MIM->mod ABCG/ABCGZ/ACGZ; fit; test
Deviance:       8.9392 DF: 8
```

```
Test of H0: ABCG/ABCGZ/ACGZ
against H:  ABCG/ABCGZ/ABCGZ
LR:    8.9392    DF:   8    P: 0.3475
MIM->mod ABCG/ABCGZ/BCGZ; fit; test
Deviance:        14.2978 DF: 8
Test of H0: ABCG/ABCGZ/BCGZ
against H:  ABCG/ABCGZ/ABCGZ
LR:   14.2978    DF:   8    P: 0.0743
MIM->mod ABCG/ABCGZ/CGZ; fit; test
Deviance:        20.1567 DF: 12
Test of H0: ABCG/ABCGZ/CGZ
against H:  ABCG/ABCGZ/ABCGZ
LR:   20.1567    DF:  12    P: 0.0642
```

There is evidence of variance heterogeneity with respect to C and G. We now perform backwards selection starting from the last model tested:

```
MIM->stepwise u
Coherent Backward Selection
Unrestricted models, Chi-squared tests.
Critical value:   0.0500
Initial model: ABCG/ABCGZ/CGZ
Model: ABCG/ABCGZ/CGZ
Deviance:  20.1570 DF:  12 P:  0.0642
     Edge        Test
Excluded    Statistic DF        P
   [AB]      15.8959  8      0.0439 +
   [AC]      15.3270  8      0.0531
   [AG]      17.9454  8      0.0216 +
   [AZ]       9.2774  8      0.3194
   [BC]       5.2809  8      0.7272
   [BG]      11.6537  8      0.1673
   [BZ]       6.1698  8      0.6282
   [CG]      13.8980  9      0.1260
   [CZ]      29.1471 10      0.0012 +
   [GZ]      29.8069 10      0.0009 +
Removed edge [BC]
Model: ACG,ABG/ACGZ,ABGZ/CGZ
Deviance:  25.4379 DF:  20 P:  0.1852
     Edge        Test
Excluded    Statistic DF        P
   [AC]      13.1668  4      0.0105 +
   [AZ]       8.7295  6      0.1894
   [BG]      16.0391  4      0.0030 +
   [BZ]       4.0573  4      0.3983
   [CG]      20.9183  5      0.0008 +
Removed edge [BZ]
```

```
Model: ACG,ABG/ACGZ/CGZ
Deviance:  29.4952 DF:  24 P:  0.2021
    Edge         Test
 Excluded   Statistic DF        P
   [AZ]       6.7912  4      0.1473
 Removed edge [AZ]
 Selected model: ACG,ABG/CGZ/CGZ
```

The graph of the selected model is as follows:

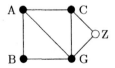

The allocation rule discussed in Section 4.7 has the form

$$lnf(1, j, y) - lnf(2, j, y) > 0.$$

It is convenient to use the canonical form (4.2) so that this becomes

$$\alpha_{1j} - \alpha_{2j} + (\beta_{1j} - \beta_{2j})y - \frac{1}{2}(\omega_{1j} - \omega_{2j})y^2 > 0.$$

The parameter estimates are as follows:

A	B	C	$\alpha_{1j} - \alpha_{2j}$	$\beta_{1j} - \beta_{2j}$	$\omega_{1j} - \omega_{2j}$
1	1	1	-1.51	-1.53×10^{-3}	-5.00×10^{-5}
1	1	2	-1.81	0.52×10^{-3}	0.22×10^{-5}
1	2	1	0.35	-1.53×10^{-3}	-5.00×10^{-5}
1	2	2	0.05	0.52×10^{-3}	0.22×10^{-5}
2	1	1	-0.01	-1.53×10^{-3}	-5.00×10^{-5}
2	1	2	0.27	0.52×10^{-3}	0.22×10^{-5}
2	2	1	-0.43	-1.53×10^{-3}	-5.00×10^{-5}
2	2	2	-0.15	0.52×10^{-3}	0.22×10^{-5}

To calculate the error rates, we use the Classify command:

```
MIM->class GM
MIM->class -GN
Leave-one-out method.
MIM->mod GMN
MIM->pr s
Calculating marginal statistics...
Empirical counts, means and covariances.
 G M N     Count
```

```
1 1 1    68.000
1 1 2     1.000
1 2 1     0.000
1 2 2    30.000
2 1 1    33.000
2 1 2     0.000
2 2 1     6.000
2 2 2    48.000
```

From the output, we obtain the following apparent and leave-one-out error rates:

Apparent error rates				Leave-one-out error rates		
	Predicted				Predicted	
Observed	1	2		Observed	1	2
1	69	30		1	68	31
2	33	54		2	39	48

As expected, the latter give a more conservative assessment of the performance of the allocation rule.

5

Hypothesis Testing

5.1 An Overview

This chapter describes and compares various significance tests that can be used in the framework of hierarchical interaction models. The overall structure is as follows.

First, the asymptotic χ^2 test is studied in more detail. This is a very useful test that forms, as it were, a backbone to inference in the model class. However, it does require that the sample is large. In small or moderate samples, the χ^2-test will often be inaccurate, and small-sample tests, when available, should be preferred.

The next test described, the F-test, is one such small-sample test. It is available in connection with models with continuous variables.

In the subsequent sections, a variety of exact conditional tests are described. In certain circumstances, tests can be constructed whose null distribution is completely known. These tests, known as permutation or randomisation tests, enjoy various advantages. First, the method of construction allows the exact conditional distribution of any test statistic to be calculated. This enables tests sensitive to specific alternatives to be constructed: for example, contingency table tests appropriate for ordinal discrete variables. Secondly, the tests are valid under weakened distributional assumptions— which is why they are widely known as distribution-free or nonparametric tests. Thirdly, large samples are not required. This is particularly useful in the analysis of high-dimensional contingency tables for which the asymptotic tests are often unreliable. A disadvantage of permutation tests is that they are computation-intensive.

The final three sections of the chapter describe some tests arising in classical multivariate analysis that can be formulated in the current framework.

5.2 χ^2-Tests

A general property of the likelihood ratio test is that under suitable regularity conditions, it has an asymptotic χ^2 distribution under the null hypothesis as the sample size tends to infinity. In the present context, this means that any pair of nested models can be compared by referring the deviance difference to a χ^2 distribution with the appropriate degrees of freedom.

The following example illustrates a test of

$$\mathcal{M}_0 : B, AC$$
$$\text{versus } \mathcal{M}_1 : AB, BC, AC$$

in a three-way contingency table.

```
MIM->fact a2b2c2; sread abc
DATA->2 5 4 7 6 8 9 5 !
Reading completed.
MIM->model ab,bc,ac; fit; base
Deviance:      0.1573 DF: 1
MIM->del ab,bc; fit
Deviance:      1.9689 DF: 3
MIM->test
Test of H0: b,ac
against H:  ab,bc,ac
LR:   1.8116   DF:   2   P: 0.4042
```

First, the model of no three-way interaction is fitted; it is then set as the base model (that is, the alternative hypothesis) using the **Base** command. A submodel is then fitted. The **Test** command tests this against the base model. The likelihood ratio test (deviance difference), degrees of freedom, and p-value are displayed.

The degrees of freedom are calculated as the difference in the number of free parameters in the two models, under the assumption that all parameters are estimable. If all cells are nonempty, and all cell SSP matrices are nonsingular, then all parameters are certainly estimable. For homogeneous models, it is sufficient that all cells are nonempty, and that the overall SSP matrix is nonsingular. For simple models, less stringent requirements will suffice, but precise conditions are difficult to determine. This is true even in the purely discrete case (Glonek et al., 1988). If there are not enough data for all parameters to be estimable, then the degrees of freedom as calculated by the `Test` command may be incorrect.

Problems of moderate to high dimension typically involve large numbers of cells with zero counts and/or singular SSP matrices. In such circumstances, it will generally be difficult to determine whether the standard degrees of freedom calculations are correct.

When one or more of the marginal tables of counts corresponding to the discrete generators contains empty cells, a question mark is printed after the degrees of freedom in the output from `Fit`. For example,

```
MIM->fact a3b3; sread ab
MIM->2 7 4
MIM->0 0 0
MIM->4 2 7 !
MIM->mod a,b; fit
Deviance:       4.4502 DF:       4 ?
```

In this example, it is clear that the correct number is two (using the formula $(R-1)(C-1)$ for an $R \times C$ table), but for more complex models, the correct number must be calculated from first principles— this calculation can be very difficult.

Fortunately, however, in some situations the calculations are straightforward. One of these is the test for variance homogeneity (see Section 5.13). Another is when both the alternative model $\mathcal{M}_1$ and the null model $\mathcal{M}_0$ are decomposable, and $\mathcal{M}_0$ is formed by removing an edge from $\mathcal{M}_1$. We call such a test a *decomposable edge deletion test*. Such tests have a particularly tractable structure, and we will encounter them repeatedly in this chapter.

It is useful to be able to characterize these tests precisely. In other words, given a decomposable model $\mathcal{M}_1$, which edges can be removed if the resulting model, $\mathcal{M}_0$, is to be decomposable? From Section 4.4, we know that $\mathcal{M}_1$ is decomposable if and only if its graph is triangulated and contains no forbidden paths. We require that the same be true of $\mathcal{M}_0$. It is not difficult to see that it is necessary and sufficient that (i) the edge is one clique only, and (ii) if both endpoints of the edge are discrete, then the whole clique is discrete. Condition (i) is necessary since its violation would give rise to a chordless 4-cycle, and condition (ii) because its violation would give rise to a forbidden path.

For example, the edge $[BC]$ cannot be removed from the model shown in Figure 5.1, ABC, BCD, but the others can. Similarly, the edges $[AX]$ and

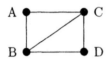

FIGURE 5.1. ABC,BCD.

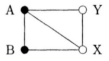

FIGURE 5.2. AB/ABX,AY/XY.

$[AB]$ cannot be removed from the model $AB/ABX, AY/XY$ (Figure 5.2), but the other edges can.

It can be shown that a decomposable edge deletion test is equivalent to a test for the removal of an edge from a saturated (marginal) model; see Frydenberg and Lauritzen (1989). For example, the test for the removal of $[AY]$ from the model $AB/ABX, AY/XY$ (see Figure 5.2) is equivalent to a test for the removal of $[AY]$ from the saturated marginal model $A/AX, AY/XY$.

Sometimes these tests are available for nondecomposable models. For example, the test for the removal of $[AE]$ from AB, BC, CD, ADE (see Figure A.1) can be performed in the marginal $\{A, D, E\}$ table as a decomposable edge deletion test.

Edge deletion tests are performed using the `TestDelete` command. For example,

```
MIM->mod ABCDEF
MIM->testdelete AB
Test of H0: BCDEF,ACDEF
against H:  ABCDEF
LR:  22.6518    DF:  16    P: 0.1234
```

Note that `TestDelete` calculates the correct degrees of freedom whenever the test corresponds to a decomposable edge deletion test. Further aspects of these tests are described later in this chapter.

5.3 F-Tests

To derive the F-test, suppose we have one continuous response variable X so that we are conditioning on $a = V \setminus \{X\}$. Consider first a model $\mathcal{M}$ that is collapsible onto a such that the conditional distribution of X given a is variance homogeneous. The conditional density is

$$(2\pi\sigma_{x|a})^{-\frac{1}{2}} \exp\left\{-\frac{1}{2}\left(x - \mu_{x|a}\right)^2 / \sigma_{x|a}\right\},$$

where $\mu_{x|a}$ and $\sigma_{x|a}$ are the conditional mean and variance. Thus, the conditional log likelihood is

$$-\left(\frac{N}{2}\right)\ln(2\pi) - \left(\frac{N}{2}\right)\ln(\sigma_{x|a}) - \frac{RSS}{2\sigma_{x|a}},$$

where RSS is the residual sum of squares. It can be shown that $\hat{\sigma}_{x|a} = RSS/N$, so the maximized log likelihood can be written as

$$-\left(\frac{N}{2}\right)\ln(2\pi) - \left(\frac{N}{2}\right)\ln(RSS/N) - \frac{N}{2}.$$

Consider now a test of two such models, that is, a test of $\mathcal{M}_0 \subseteq \mathcal{M}_1$ where both models are collapsible onto a, induce the same marginal model $\mathcal{M}_a$, and are such that the conditional distribution of X given a is variance homogeneous. Then the deviance difference between $\mathcal{M}_0$ and $\mathcal{M}_1$ is

$$\begin{aligned} d &= 2(\hat{\ell}_V^0 - \hat{\ell}_V^1) \\ &= 2(\hat{\ell}_{x|a}^0 - \hat{\ell}_{x|a}^1) \\ &= N\ln(RSS_0/RSS_1). \end{aligned} \qquad (5.1)$$

If we let r_0 be the difference in the number of free parameters between $\mathcal{M}_0$ and $\mathcal{M}_1$ (and hence between $\mathcal{M}_{x|a}^0$ and $\mathcal{M}_{x|a}^1$), and let r_1 be the number of free parameters in $\mathcal{M}_{x|a}^1$, then we can write the F-test for $\mathcal{M}_{x|a}^0$ versus $\mathcal{M}_{x|a}^1$ as

$$F = \frac{(RSS_0 - RSS_1)/r_0}{RSS_1/(N - r_1)}.$$

Using (5.1), we obtain

$$F = (e^{d/N} - 1)((N - r_1)/r_0),$$

which under $\mathcal{M}_0$ is F-distributed with r_0 and $N - r_1$ degrees of freedom.

The following fragment illustrates an F-test of no interaction in the two-way layout:

```
MIM->mod AB/ABX/X; fit; base
Deviance:       8.1337 DF: 5
MIM->mod AB/AX,BX/X; fix ab; fit
Fixed variables: AB
Deviance:      11.5229 DF: 7
MIM->ftest
Test of HO: AB/AX,BX/X
against H:  AB/ABX/X
F:    1.3651   DF:   2,  18   P: 0.2806
```

Like the **Test** command, **FTest** compares the current model to the base model. The **Fix** command is used to identify the continuous response variable (i.e., all the variables in the model except one continuous variable must be fixed). The output shows the test statistic, the degrees of freedom, and the p-value.

As with the **Test** command, the calculated degrees of freedom may be incorrect if the data are not sufficient to estimate all the parameters. In the same way as before, a way to get around this problem when it arises is to restrict attention to decomposable models and use the `TestDelete` command instead.

We can use the following result (Frydenberg and Lauritzen, 1989). Suppose that a test of $\mathcal{M}_0$ versus $\mathcal{M}_1$ is a decomposable edge deletion test between homogeneous models such that one or both of the endpoints of the edge are continuous. Then the test can be performed as an F-test.

This is illustrated in the following example:

```
MIM->model AB/AX,ABY/XY
MIM->testdelete AX s
Test of H0: AB/X,ABY/XY
against H:  AB/AX,ABY/XY
F:    0.0003    DF:   1,  21    P: 0.9856
```

The initial model, $AB/AX, ABY/XY$, is homogeneous and decomposable, and the removal of the edge $[AX]$ satisfies the above criteria.

For large samples, the p-values from the F-test and from the χ^2 test are almost identical. For small samples, the reference χ^2 distribution will be unreliable: generally it will be too liberal, i.e., will tend to reject the null hypothesis too often (Porteous, 1989). For this reason, F-tests should be preferred when available in small samples.

5.4   Exact Conditional Tests

In the previous sections we described tests based on the hierarchical interaction model family for which the distribution of the variables is specified except for a number of unknown parameters. We now describe some tests of more general validity in the sense that they do not require specific distributional assumptions. In this section, we describe a general rationale for tests and some computational aspects, and in the following sections we review some particular tests.

To introduce the topic, consider a simple two-sample problem involving a continuous variable. Suppose we have N independent and identically distributed observations on two variables— a binary factor A, and a continuous response Y. The null hypothesis, $\mathcal{H}_0$, is that of independence (homogeneity), i.e., that

$$f_{Y|A}(y|1) = f_{Y|A}(y|2) = f_Y(y).$$

Under this model, the order statistics $Y_{(1)}, Y_{(2)}, \ldots Y_{(N)}$ are sufficient for $f(y)$, and so similar tests can be constructed by conditioning on the observed order statistics, i.e., that $Y_{(.)} = y_{(.)}$. Under $\mathcal{H}_0$, all $N!$ permutations of y are equally likely. We can represent the observed data in a simple table, as follows:

A	$y_{(1)}$	$y_{(2)}$	$\cdots$	$y_{(N)}$	Total
1	1	0	$\cdots$	0	n_1
2	0	1	$\cdots$	1	n_2
Total	1	1	1	1	N

The columns represent the values of y in ascending order, and for each column, there is a 1 in the row indicating membership of the first or second sample. Each possible permutation of y corresponds to a similar $2 \times N$ table of zeros and ones, with the same row and column sums.

For any test statistic, for example, the difference in sample means

$$t_{\text{obs}} = \bar{y}_1 - \bar{y}_2,$$

we can compute the permutation significance level as

$$p_{\text{obs}} = \Pr\{T \geq t_{\text{obs}}|Y_{(.)} = y_{(.)}, \mathcal{H}_0\} = \kappa/N!,$$

where κ is the number of permutations giving values of the test statistic greater than or equal to t_{obs}.

If the observed responses contain ties, then these will appear in the above table as columns with identical y's. We can represent the observed table more compactly by combining such columns so as to arrive at a table of the form

A	$\tilde{y}_{(1)}$	$\tilde{y}_{(2)}$	$\cdots$	$\tilde{y}_{(C)}$	Total
1	n_{11}	n_{12}	$\cdots$	n_{1C}	n_{1+}
2	n_{21}	n_{22}	$\cdots$	n_{2C}	n_{2+}
Total	n_{+1}	n_{+2}	$\cdots$	n_{+C}	N

where $\tilde{y}_{(1)}, \tilde{y}_{(2)} \cdots \tilde{y}_{(C)}$ are the distinct order statistics, and n_{ij} is the number of observations in sample i with $y = \tilde{y}(j)$.

Similarly, instead of generating all permutations, we can generate all the $2 \times C$ tables of nonnegative integers with fixed column margins $n_{+1}, n_{+2}, \ldots, n_{+C}$ and row margins n_{1+} and n_{2+}. It can be shown that the probability of such a table, $M = \{m_{ij}\}$, is given by the hypergeometric distribution

$$\Pr(M|H_0) = \frac{\prod_{i=1}^{2} n_{i+}! \prod_{j=1}^{C} n_{+j}!}{N! \prod_{i=1}^{2} \prod_{j=1}^{C} m_{ij}!}. \tag{5.2}$$

For any test statistic T, we can compute the permutation significance level

$$p_{\text{obs}} = \Pr(T \geq t_{\text{obs}}|H_0)$$
$$= \sum_{M \in \Upsilon : T(M) \geq t_{\text{obs}}} \Pr(M|H_0).$$

This method of constructing tests is easily generalized in several respects. First, the response variable does not have to be continuous, but can be discrete with, say, C levels. The above description applies precisely, with the distinct order statistics $\tilde{y}_{(1)}, \tilde{y}_{(2)} \dots \tilde{y}_{(C)}$ being simply the factor levels $1, 2 \dots, C$. Secondly, the row variable A does not have to be binary but can be discrete with, say, R levels. Then the process involves generating all $R \times C$ tables of nonnegative integers with fixed column margins $n_{+1}, n_{+2} \dots n_{+C}$ and row margins $n_{1+}, n_{2+} \dots n_{R+}$. The conditional probability of such a table M is then

$$ \Pr(M|H_0) = \frac{\prod_{i=1}^{R} n_{i+}! \prod_{j=1}^{C} n_{+j}!}{N! \prod_{i=1}^{R} \prod_{j=1}^{C} m_{ij}!}. \tag{5.3} $$

Similarly, the row variable could be continuous, but we do not use the possibility in the sequel.

Finally, we can extend the approach to test conditional independence, given a third, discrete variable. Suppose the hypothesis is $A \perp\!\!\!\perp Y \mid S$, where we call S the stratum variable and suppose that it has L levels. The order statistics in stratum k, $Y_{(1,k)}, Y_{(2,k)}, \dots Y_{(n_{++k},k)}$, say, are sufficient for $f_{Y|S=k}(y)$, and so similar tests can be constructed by conditioning on the observed order statistics, i.e., that $Y_{(.,k)} = y_{(.,k)}$ for each $k = 1, \dots, L$. Within stratum k, there are $n_{++k}!$ possible permutations. Under $\mathcal{H}_0$, all $\prod_{k=1}^{L} n_{++k}!$ permutations of y are equally likely.

We summarize the data in an $R \times C \times L$ table, $N = \{n_{ijk}\}$, where R is the number of levels of A, and C is the number of distinct values of Y. The exact tests are constructed by conditioning on the marginal totals $\{n_{i+k}\}$ and $\{n_{+jk}\}$. We can think of the three-way table as a stack of $A \times B$ tables, one for each stratum k of S. The kth slice has fixed row and column totals $\{n_{i+k}\}$ and $\{n_{+jk}\}$.

The sample space, which consists of all $R \times C \times L$ contingency tables with these fixed margins, is denoted Υ. Under the null hypothesis H_0, the probability of a given table $M = \{m_{ijk}\}$ is

$$ \Pr(M|H_0) = \prod_{k=1}^{L} \frac{\prod_{i=1}^{R} n_{i+k}! \prod_{j=1}^{C} n_{+jk}!}{n_{++k}! \prod_{i=1}^{R} \prod_{j=1}^{C} m_{ijk}!}. \tag{5.4} $$

In this fashion, we can in principle compute the exact conditional tests for $A \perp\!\!\!\perp B \mid S$, where A and S are discrete and where B can either be discrete or continuous. Before we describe ways of computing the tests in practice, let us first consider how they fit into the graphical modelling framework. Recall that a decomposable edge deletion test is equivalent to a test for edge removal from a saturated marginal model (see Section 5.2). That is, if the edge to be removed is $[UV]$, then the test is equivalent to a test for

$U \perp\!\!\!\perp V \mid w$ for some $w \subseteq \Delta \cup \Gamma$. When w consists of discrete variables only, and either U or V (or both) are discrete, then the test can be performed as an exact conditional test of the type we have just described. We do this by constructing (conceptually) a new factor S by "crossing" all the factors in w, i.e., with a factor level for each combination of levels for the factors in w.

We now consider three alternative ways of computing the tests.

The direct method is *exhaustive enumeration*. For each table $M \in \Upsilon$, the probability $\Pr(M|H_0)$ and the statistic $T(M)$ are calculated, enabling exact calculation of p_{obs}. Unfortunately, since the number of tables in Υ may be astronomical, exhaustive enumeration is often infeasible.

We mention in passing that Mehta and Patel (1983) have developed a network algorithm for these calculations that can be used for the computation of stratified linear rank tests. It is implemented in the program StatXact (Mehta and Patel, 1991). This algorithm is considerably faster than the one used by MIM.

As an alternative when exhaustive enumeration is not feasible, *Monte Carlo sampling* can be used. Using an algorithm of Patefield (1981), a large fixed number, say K, of random tables is sampled from Υ in such a way that for any $M \in \Upsilon$, the probability of sampling M is $\Pr(M|H_0)$. For the table M_k, define

$$z_k = \begin{cases} 1 & \text{if } T(M_k) \geq t_{\text{obs}} \\ 0 & \text{otherwise.} \end{cases}$$

We estimate p_{obs} as $\hat{p}_{\text{obs}} = \sum_{k=1}^{K} z_k/K$. It is easy to show that $\hat{p}_{\text{obs}}$ is an unbiased estimate of p_{obs}. An approximate 99% confidence interval can be calculated as

$$\hat{p}_{\text{obs}} \pm 2.576 \sqrt{\hat{p}_{\text{obs}}(1 - \hat{p}_{\text{obs}})/K}.$$

By taking K large enough, p_{obs} can be estimated to any desired precision, although greater precision than four decimals is rarely necessary.

The use of Monte Carlo sampling to compute exact tests in multidimensional contingency tables, particularly in connection with ordinal variables, has been strongly advocated by Kreiner (1987).

The third way to estimate p_{obs} is by using *sequential Monte Carlo sampling*. The purpose of this is to stop sampling if it is clear early on that there is little or no evidence against the null hypothesis. Sampling is continued until a prescribed number, say h, of tables M_k with $T(M_k) \geq t_{\text{obs}}$ are sampled, or a maximum number of samples have been taken, whichever comes first. The estimate of p_{obs} is, as usual, $\hat{p}_{\text{obs}} = \sum_{k=1}^{K} z_k/K$, where K

is the number of tables sampled. This stopping rule was proposed by Besag and Clifford (1991). It has the attractive property that p-values close to zero are estimated accurately, whereas in the less interesting region where p-values are large, sampling stops quickly.

We now turn to a detailed description of some specific tests for conditional independence, which can be computed as exact conditional tests. The first type uses test statistics we are already familiar with, since these are based on the standard deviance statistic.

5.5 Deviance-Based Tests

To reiterate, we are considering exact tests for $A \perp\!\!\!\perp B \mid S$, where A and S are discrete, and B can be either discrete or continuous. In the mixed model framework, there are three cases to consider. The first applies when the column variable (B) is discrete. Then the test corresponds to the test of

$$\mathcal{M}_0 : AS, BS$$
$$\text{versus } \mathcal{M}_1 : ABS,$$

for which the deviance has the form

$$G^2 = 2 \sum_{k=1}^{L} \sum_{j=1}^{C} \sum_{i=1}^{R} n_{ijk} \ln\left(\frac{n_{ijk} n_{++k}}{n_{i+k} n_{+jk}}\right).$$

As an example of the use of this test, consider the data shown in Table 5.1, taken from Mehta and Patel (1991). Data were obtained on the site of oral lesions in house-to-house surveys in three geographic regions in India. There are two variables: site of oral lesion (A) and region (B). We are interested

Site	Kerala	Gujarat	Andhra
Labial Mucosa	0	1	0
Buccal Mucosa	8	1	8
Commisure	0	1	0
Gingiva	0	1	0
Hard Palate	0	1	0
Soft Palate	0	1	0
Tongue	0	1	0
Floor of Mouth	1	0	1
Alveolar Ridge	1	0	1

TABLE 5.1. Oral lesion data from three regions in India. Source: Mehta and Patel, 1991.

in determining whether the distribution of lesion sites differs between the three regions. The table is very sparse, so the asymptotic test of $A \perp\!\!\!\perp B$ is unlikely to be reliable. In the following fragment, we first define the data and then estimate the size of the reference set Υ using the approximation given in Gail and Mantel (1977) by using **TestDelete** with the C option.

```
MIM->fact A9B3
MIM->sread AB
DATA->0 1 0 8 1 8 0 1 0 0 1 0 0 1 0 0 1 0 0 1 0 1 0 1 1 0 1 !
Reading completed.
MIM->model AB
MIM->testdelete AB c
Test of H0: B,A
against H: AB
Estimated number of tables:          35200
Likelihood Ratio Test.
LR:  23.2967    DF:   16 Asymptotic P: 0.1060
```

The size of Υ is estimated as $35,200$ and the asymptotic test gives a non-significant result. It is feasible to generate $35,200$ tables, so we proceed to calculate the exact test by using the E option:

```
MIM->testdelete AB e
Test of H0: B,A
against H: AB
Exact test - exhaustive calculations.
No. of tables:       26109
Likelihood Ratio Test.
LR:  23.2967    DF:   16 Asymptotic P: 0.1060
                            Exact P: 0.0356
```

The exact p-value of 0.0356 implies that the distribution of lesion sites differs significantly between regions. For this table, the reference set Υ contained $26,109$ tables. We could also have chosen to use sequential Monte Carlo sampling, which is rather more efficient:

```
MIM->testdel AB q
Test of H0: B,A
against H: AB
Exact test - monte carlo with sequential stopping rule.
Maximum no of more extreme tables:  20
Maximum no of random tables: 1000
No. of tables:       719
Likelihood Ratio Test.
LR:  23.2967    DF:   16 Asymptotic P: 0.1060
                        Estimated P: 0.0278
```

The prescribed maximum number of sampled tables with $T(M_k) \geq t_{obs}$ is 20 by default, and the maximum number sampled is 1000. These settings can be changed using the commands SetMaxExcess and SetMaxSim, respectively.

Another example, using the lizard perching habits data (Section 2.1.1), is as follows. First we define the data, and then we perform an asymptotic χ^2 test for $B \perp\!\!\!\perp C | A$.

```
MIM->fact A2B2C2; sread ABC
32 86 11 35 61 73 41 70 !
MIM->model ABC
MIM->testdelete BC
Test of H0: AC,AB
against H: ABC
Likelihood Ratio Test.
LR:    2.0256    DF:    2 Asymptotic P: 0.3632
```

To evaluate the feasibility of calculating the exact test by exhaustive enumeration, we again use the C option:

```
MIM->testdel BC c
Test of H0: AC,AB
against H: ABC
Estimated number of tables:            4067
Likelihood Ratio Test.
LR:    2.0256    DF:    2 Asymptotic P: 0.3632
```

It is feasible to generate 4067 tables, so we calculate the exact test by using the E option:

```
MIM->testdel BC e
Test of H0: AC,AB
against H: ABC
Exact test - exhaustive calculations.
No. of tables:        4532
Likelihood Ratio Test.
LR:    2.0256    DF:    2 Asymptotic P: 0.3632
                            Exact P: 0.3648
```

We see that there were precisely 4532 tables in Υ. In the example shown, the asymptotic test and the exact test were almost identical.

Now suppose that the response is a continuous variable, say X, and that the initial model is homogeneous. A test of $A \perp\!\!\!\perp X \mid S$ corresponds to

$$\mathcal{M}_0 : AS/SX/X$$
$$\text{versus } \mathcal{M}_1 : AS/ASX/X,$$

for which the deviance difference takes the form

$$d = N \ln \left(\frac{\text{RSS}_0}{\text{RSS}_1} \right), \tag{5.5}$$

where RSS_0 is the sum of squares about the stratum (S) means, and RSS_1 is the sum of squares about the cell means (see Section 4.1.7).

When the initial model is heterogeneous, the test corresponds to

$$\mathcal{M}_0 : AS/SX/SX$$
$$\text{versus } \mathcal{M}_1 : AS/ASX/ASX,$$

and the deviance difference takes the form

$$d = \sum_{k=1}^{L} n_{+k} \ln \left(\frac{\text{RSS}_k}{n_{+k}} \right) - \sum_{k=1}^{L} \sum_{i=1}^{R} n_{ik} \ln \left(\frac{\text{RSS}_{ik}}{n_{ik}} \right), \tag{5.6}$$

where RSS_k is the sum of squares about the mean in stratum k, and RSS_{ik} is the sum of squares about the mean in cell (i, k).

The data in Table 5.2, taken from Mehta and Patel (1991), serve as an illustration. For 28 patients undergoing different chemotherapy regimens, a measure of hematological toxicity, namely the number of days for which the white blood count was less than 500, was registered. There are thus two variables: chemotherapy regimen (A), a discrete variable with 5 levels, and the response variable, say X. There are no strata, i.e., S is null.

We first define the data and then estimate how many tables there are in the reference set Υ:

```
MIM->fact A5; cont X; read AX
DATA->1 0 1 1 1 8 1 10
DATA->2 0 2 0 2 3 2 3 2 8
DATA->3 5 3 6 3 7 3 14 3 14
DATA->4 1 4 1 4 6 4 7 4 7 4 7 4 8 4 8 4 10
DATA->5 7 5 10 5 11 5 12 5 13 !
Reading completed.
MIM->model A/AX/X
MIM->testdel AX c
Test of H0: A/X/X
```

Drug regimen	Hematologic toxicity
1	0 1 8 10
2	0 0 3 3 8
3	5 6 7 14 14
4	1 1 6 7 7 7 8 8 10
5	7 10 11 12 13

TABLE 5.2. Hematologic toxicity data. Source: Mehta and Patel, 1991.

```
against H: A/AX/X
Estimated number of tables:   2.928522E+0011
Likelihood Ratio Test.
LR:  14.6082    DF:    4 Asymptotic P: 0.0056
```

We see that the number of tables in Υ is around 3×10^{11}, so exhaustive enumeration is quite infeasible. Instead, we choose to estimate the permutation significance level using crude Monte Carlo sampling, using the M option:

```
MIM->testdelete AX m
Test of HO: A/X/X
against H: A/AX/X
Exact test - monte carlo estimates.
No. of tables:        1000
Likelihood Ratio Test.
LR:  14.6082    DF:    4 Asymptotic P: 0.0056
                       Estimated P: 0.0228 +/- 0.012147
MIM->model A/AX/AX
MIM->testdelete AX m
Test of HO: A/X/X
against H: A/AX/AX
Exact test - monte carlo estimates.
No. of tables:        1000
Likelihood Ratio Test.
LR:  17.7251    DF:    8 Asymptotic P: 0.0234
                       Estimated P: 0.0469 +/- 0.017227
```

We show both the homogeneous and the heterogeneous variants. The algorithm samples 1000 tables randomly from Υ. In both cases, we see that there is reasonable agreement between the asymptotic results and the exact results.

5.6 Permutation F-Test

We now consider the exact conditional test constructed on the basis of the F-statistic— the so-called permutation F-test. The advantage of this test is that it is valid for arbitrary distributions of the continuous variable.

We illustrate the test using the data shown in Table 5.2.

```
MIM->model A/AX/X
MIM->testdel AX lsm
Test of HO: A/X/X
against H: A/AX/X
Exact test - monte carlo estimates.
```

```
No. of tables:          1000
Likelihood Ratio Test.
LR:  14.6082    DF:    4 Asymptotic P: 0.0056
                          Estimated P: 0.0191 +/- 0.011158
Permutation F Test.
F:    3.9383    DF:    4,  23 Asymptotic P: 0.0141
                          Estimated P: 0.0191 +/- 0.011158
```

We show computation of the exact deviance test and the permutation F-test. Since the F-statistic is a monotonic transformation of the deviance difference with respect to the homogeneous model, as shown in (5.3), the two tests are equivalent in the conditional distribution. We see that the F-test appears to be rather more reliable than the χ^2 test, since the p-value for the former is closer to the exact conditional p-values.

The next two tests that we examine are only appropriate when the response variable is discrete. They are competitors to the contingency table test G^2.

5.7   Pearson χ^2-Test

The Pearson χ^2-test has the form

$$X^2 = \sum_{k=1}^{L}\sum_{j=1}^{C}\sum_{i=1}^{R} \frac{(n_{ijk} - \hat{m}_{ijk})^2}{\hat{m}_{jk}},$$

where $\hat{m}_{ijk} = n_{i+k}n_{+jk}/n_{++k}$ are the expected cell counts under the null hypothesis. The asymptotic distribution under the null hypothesis is the same as the likelihood ratio test G^2, namely χ^2 with $L(R-1)(C-1)$ degrees of freedom.

The option P calculates the Pearson goodness-of-fit test. Continuing the lizard data example:

```
MIM->testdel BC p
Test of HO: AC,AB
against H: ABC
Pearson Goodness-of-fit Test.
X2:    2.0174    DF:    2 Asymptotic P: 0.3647
```

The exact test gives an almost identical result:

```
MIM->testdel BC pe
Test of HO: AC,AB
against H: ABC
Exact test - exhaustive calculations.
No. of tables:          4532
```

```
Pearson Goodness-of-fit Test.
X2:   2.0174   DF:   2 Asymptotic P: 0.3647
                             Exact P: 0.3643
```

The likelihood ratio test gives more of the same:

```
MIM->testdel BC pl
Test of HO: AC,AB
against H: ABC
Likelihood Ratio Test.
LR:   2.0256   DF:   2 Asymptotic P: 0.3632
Pearson Goodness-of-fit Test.
X2:   2.0174   DF:   2 Asymptotic P: 0.3647
```

There is hardly any difference between the tests in this example.

5.8 Fisher's Exact Test

Fisher's exact test is usually associated with 2×2 tables but can readily be extended to $R \times C \times L$ tables. The test statistic is $T = \Pr(N|H_0)$, the hypergeometric probability of the observed table N under H_0. The exact p-value is the probability of observing a table with a $\Pr(N|H_0)$ at least as small as the observed.

Less well-known than the exact test is a closely related asymptotic test (Freeman and Halton, 1951). The statistic

$$FH = -2\ln(\gamma \Pr(N|H_0)),$$

where γ is a constant given as

$$\gamma = (2\pi)^{L(R-1)(C-1)/2} \prod_{k=1}^{L} n_{++k}^{-(RC-1)/2} \prod_{j=1}^{C} n_{+j+}^{(C-1)/2} \prod_{i=1}^{R} n_{i++}^{(R-1)/2}.$$

FH is a monotone transformation of $\Pr(N|H_0)$. The asymptotic distribution of FH under H_0 is the same as the X^2 and G^2 statistics, i.e., χ^2 with $L(R-1)(C-1)$ degrees of freedom.

This test is available through use of the F option. Continuing the lizard data example, we obtain

```
MIM->fact A2B2C2; sread ABC
32 86 11 35 61 73 41 70 !
MIM->model ABC; testdel BC f
Test of HO: AC,AB
against H: ABC
Fishers Test.
FI:   1.9882   DF:   2 Asymptotic P: 0.3700
```

Again, in this example, the results are very close to those given by the X^2 and the likelihood ratio test.

Calculating the exact test for the lesion data (Table 5.1), we obtain similar results as before:

```
MIM->model AB; testdel AB fm
Test of HO: B,A
against H: AB
Exact test - monte carlo estimates.
No. of tables:       1000
Fishers Test.
FH:  19.7208    DF:   16 Asymptotic P: 0.2331
                         Estimated P: 0.0120 +/- 0.008887
```

5.9 Rank Tests

We saw in Section 5.4 that exact conditional tests for $A \perp\!\!\!\perp B \mid S$ can be computed for any test statistic. In the preceding sections, we considered statistics that are based on parametric models— namely, the deviance statistics, the closely related tests due to Pearson and Fisher, and the randomisation F-test. For such statistics, the exact conditional tests can be regarded as a natural part of the modelling process. In the case of discrete data, the exact conditional tests are attractive because their competitors require large samples. When a continuous response is involved and variance homogeneity can be assumed, small-sample tests (F-tests) are available, but the exact tests are still attractive since their validity does not require assumptions of normal errors. With continuous responses with heterogeneous variances, both advantages apply.

We now consider a rather different approach, which is based on test statistics that are not directly derived from models— the so-called rank tests. In this section, we introduce the rationale for these tests and discuss some general issues, deferring a detailed description of some specific tests to the following sections. Although most of the following discussion refers to Wilcoxon's two-sample test, it is also applicable in general terms to other rank tests. A good reference for these tests is Lehmann (1975).

As their name implies, these tests use only ranks, that is to say, the serial number of the observations when the observed data are arranged in serial order. For example, in the simple two-sample case that introduced Section 5.4, the data were arranged as follows:

A	$y_{(1)}$	$y_{(2)}$	$\cdots$	$y_{(N)}$	Total
1	1	0	$\cdots$	0	n_1
2	0	1	$\cdots$	1	n_2
Total	1	1	1	1	N

In rank tests, the columns are numbered 1 to N and test statistics use these numbers, not the original observations. Thus, for example, the Wilcoxon test statistic W is simply the sum of the ranks in one of the rows, say $A = 1$. Assuming for the moment that all the observations are distinct, every set of data of size N will induce ranks of 1 to N so the null distribution of W depends on n_1 and n_2 only. We say that rank tests are unconditionally distribution-free -- unconditionally, in the sense that conditioning on the observed order statistics $y_{()}$ is not necessary. This property was particularly important when computers were not readily available, since it meant that critical values for the exact conditional test could be tabulated in terms of n_1 and n_2 only.

If there are tied observations, ranks are not well-defined. The usual approach is to use the *midranks* for tied observations. If the data are arranged in tied form as

A	$\tilde{y}_{(1)}$	$\tilde{y}_{(2)}$	$\cdots$	$\tilde{y}_{(C)}$	Total
1	n_{11}	n_{12}	$\cdots$	$n_{1C}0$	n_{1+}
2	n_{21}	n_{22}	$\cdots$	n_{2C}	n_{2+}
Total	n_{+1}	n_{+2}	$\cdots$	n_{+C}	N

then data in the jth column would correspond to the ranks $\tau + 1$ to $\tau + n_{+j}$, where $\tau = \sum_{l=1}^{j-1} n_{+l}$. The midrank is the average of these values, i.e., $\tau + (n_{+j} + 1)/2$. Ties complicate the tabulation of critical values since in principle every possible configuration of ties should be tabulated. Since the number of possible configurations is very large, this is infeasible for all but the most minute sample sizes. Corrections to the asymptotic approximations to take ties into account have been developed, but their performance in small samples is generally unclear.

Problems with ties have led to the emphasis of rank tests for continuous data, where ties are hopefully infrequent. It is sometimes argued that for continuous distributions, the probability of ties occurring is zero, and this may serve to allay concerns. However, this is rather like proving that bumblebees cannot fly: ties often do occur, since even continuous variables are measured with finite precision. In any case, it is no longer difficult to compute the exact significance levels, even when ties are present. There is a good case for arguing that in fact rank tests are most appropriate for ordinal categorical data for which there is no metric relation between categories besides that of ordering, so no information is lost when forming ranks. As we see below, such tests often have very superior power to tests not exploiting ordinality.

A general point of criticism against rank tests is that they contribute little to the modelling process; that is to say, they may enable association to be detected, but they do not imply a parametric form for that association. In many applications, it is desirable to achieve a concise description of the system under study. This may be in terms of a small number of model parameters, or may take the form of a data reduction whose justification is model-based. Compare, for example, the Wilcoxon test and the permutation F-test in connection with a continuous response. The tests are equally valid when normality of errors cannot be assumed. The results of performing the F-test can be summarized in terms of the difference in sample means, but for the rank test no such summaries are available. (There is a widespread misapprehension that somehow medians are involved).

Another point of criticism against rank tests is that, despite their apparent simplicity, they can be difficult to interpret. Again, this is due to the fact that they are not model-based. We illustrate this using the Wilcoxon test. Consider a clinical study to compare two treatments, A and B, with respect to an ordinal endpoint measuring treatment effect (categorized as worsening, no change, and slight, moderate, and great improvement). Suppose the results from the study are as follows:

Treatment	Worse	No change	Slightly better	Moderately better	Much better
A	14	12	10	8	6
B	8	9	10	11	12

The standard contingency table test treats the variables as nominal. Here, $G^2 = 4.6018$ on four degrees of freedom, giving $P = 0.3306$. Thus, the test does not detect a treatment difference. In contrast, if we use a Wilcoxon test, we obtain $W = 2225$, with $P = 0.0347$. So Wilcoxon's test finds a significant treatment effect.

So far, so good. Suppose, however, that the study results were not as given above, but instead were as follows:

Treatment	Worse	No change	Slightly better	Moderately better	Much better
A	12	10	6	10	12
B	5	10	20	10	5

We see that treatment B often results in a slight improvement, whereas treatment A appears to lead to improvement in some patients but to no change or worsening in others. Now $G^2 = 13.89$ on four degrees of freedom, corresponding to $P = 0.0076$, which is highly significant. However, the Wilcoxon test gives $W = 2525$, corresponding to a $P = 1.0000$— in other words, no evidence of difference between treatments. So here the nonordinal test is more powerful.

Why is this? The Wilcoxon test is sensitive to departures from independence known as stochastic ordering, i.e., that $F_A(x) \leq F_B(x)$ (or vice versa), where, as usual, F is the distribution function $F(x) = \Pr(X \leq x)$. The first table above was consistent with $F_A(x) \leq F_B(x)$, hence the greater power of the Wilcoxon test for these data. The second table does not exhibit stochastic ordering, and here the Wilcoxon test has low power.

There is another derivation of the Wilcoxon test that does not appeal to stochastic ordering (this is the Mann-Whitney form). Here it is formulated as a test of $\Pr(X > Y) = \frac{1}{2}$ against the alternative $\Pr(X > Y) \neq \frac{1}{2}$ (see Hand, 1992 for a detailed discussion). Notice that here the null hypothesis is not homogeneity but a type of symmetry, $\Pr(X > Y) = \frac{1}{2}$. This means that a negative outcome of the test can be interpreted in different ways. If we know, or are prepared to assume, that stochastic ordering does hold, then we can conclude that there is no evidence against homogeneity. If not, then we conclude that there is no evidence against symmetry. The snag is that it is difficult to examine the stochastic ordering property. Asymptotic tests have been proposed (Robertson and Wright, 1981) but are difficult to compute and are apparently seldom used.

This section has emphasized problematic aspects of rank tests. In their defense, it must be said that they can be valuable, particularly in the preliminary analysis of problems involving ordinal categorical variables. However, a parametric analysis will often be more informative (Agresti, 1984). Exact rank tests are also used when very cautious group comparisons are to be made, for example, in efficacy comparisons in clinical trials. Here, it may be argued that exact model-based tests, for example, the permutation F-test, may often be preferable on the grounds of ease of summarization and interpretation.

5.10   Wilcoxon Test

As we saw in the preceding section, the Wilcoxon test is used to compare discrete or continuous distributions between two populations. In terms of variables, it presupposes a binary (row) factor and a column variable that may be either discrete (ordinal) or continuous.

Suppose the distribution of B given $A = i, S = k$ is $F_{i,k}(x)$, i.e., that $F_{i,k}(x) = \Pr(B < x | A = i, S = k)$. As we just discussed, the statistic tests homogeneity,

$$H_0 : F_{1,k}(x) = F_{2,k}(x) \quad \forall x, \forall k,$$

against the alternative hypothesis that at least one of these distributions is stochastically larger than the other, i.e., that for some k,

$$F_{1,k}(x) < F_{2,k}(x), \quad \forall x,$$

or

$$F_{2,k}(x) < F_{1,k}(x), \quad \forall x.$$

The test statistic is the sum of the ranks of the first population summed over all the strata, i.e., $W = \sum_{k=1}^{L} R_{1k}$, where R_{ik} is the rank sum for the ith treatment group in the kth stratum, given as $R_{ik} = \sum_{j=1}^{C} r_{jk} n_{ijk}$, where r_{jk} is the midrank of an observation in column j, stratum k. Note that W is appropriate when the difference between the two populations is in the same direction in all strata.

In the conditional distribution, under H_0, W has mean

$$\mathrm{E}(W|H_0) = \sum_{k=1}^{L} \mathrm{E}(R_{1k}|H_0)$$

$$= \sum_{k=1}^{L} n_{1+k}/n_{++k} \sum_{j=1}^{C} r_{jk} n_{+jk}$$

and variance

$$\hat{\mathrm{Var}}(W|H_0) = \sum_{k=1}^{L} \frac{n_{1+k} n_{2+k}}{n_{++k}(n_{++k}-1)} \sum_{j=1}^{C} \left[r_{jk} - \frac{\mathrm{E}(R_{1k}|H_0)}{n_{1+k}} \right]^2 n_{+jk}.$$

An asymptotic test of H_0 compares

$$\frac{W - \mathrm{E}(W|H_0)}{\sqrt{\hat{\mathrm{Var}}(W|H_0)}}$$

with the $\mathcal{N}(0,1)$ distribution. The two-sided p-value

$$p = \Pr(|W - \mathrm{E}(W|H_0)| \geq |W_{obs} - \mathrm{E}(W|H_0)| \mid H_0)$$

is calculated.

A simple example with unstratified, continuous data is the following. Diastolic blood pressure (mm Hg) was recorded for four subjects in a treatment group and for 11 subjects in a control group, as shown in Table 5.3. To compare the blood pressure of the two groups, we use the Wilcoxon test:

Group	Blood pressure (mm Hg)
Active	94 108 110 90
Control	80 94 85 90 90 90 108 94 78 105 88

TABLE 5.3. Blood pressure data.

```
MIM->fact A2; cont X; read AX
DATA->1 94 1 108 1 110 1 90
DATA->2 80 2 94 2 85 2 90 2 90 2 90 2 108 2 94 2 78 2 105 2 88 !
Reading completed.
MIM->model A/AX/X; testdel AX we
Test of HO: A/X/X
against H: A/AX/X
Exact test - exhaustive calculations.
No. of tables:            230
Wilcoxon Test.
W:  45.0000  E(W|HO):  32.0000 Asymptotic P: 0.0853
                                    Exact P: 0.0989
```

There is good agreement between the asymptotic and the exact tests.

For stratified tests, ranks (and midranks, if there are ties) can be calculated in two ways. The observations can be ranked within each stratum, giving the stratum-specific scores, or the ranks can be calculated by sorting the observations for the combined strata, giving the stratum-invariant scores. In MIM, stratum-specific scores are used by default, but stratum-invariant scores are also available.

The choice between stratum-specific and stratum-invariant scores is discussed in Lehmann (1975, p. 137-140). Briefly, it depends both on the nature of the data and on power considerations. If the response distributions differ between the different strata, then stratum-specific scores will be indicated. For example, if different laboratories (strata) use different methods to measure a variable, then this may lead to differences in scale or location or both between the different laboratories, and so stratum-specific scores will be indicated. However, if the stratum sizes are small, then the stratum-specific scores may be associated with reduced power compared to the stratum-invariant scores. The extreme example of this would be a matched pair study where each pair is a stratum: here, the stratum-specific scores give a sign test.

Stratified Wilcoxon tests were studied by van Elteren (1960), who proposed two different statistics. The so-called design-free statistic is of the form $W' = \sum_{k=1}^{L} R_{1k}/(n_{1+k}n_{2+k})$, and the locally most powerful test is of the form $W'' = \sum_{k=1}^{L} R_{1k}/(1 + n_{++k})$. The above expressions for the mean and variance of W apply if the r_{jk} are replaced by $r_{jk}/(n_{1+k}n_{2+k})$ and $r_{jk}/(1 + n_{++k})$, respectively.

To compare these statistics, note that with equal allocation to the row factor, the expected rank sum for the kth stratum under the null hypothesis is $E(R_{1k} \mid H_0) = n_{++k}(n_{++k} + 1)/4$. So for the crude Wilcoxon statistic $W = \sum_{k=1}^{L} R_{1k}$, each stratum contributes a quantity roughly proportional to n_{++k}^2. For the locally most powerful version, the contribution is roughly

Centre	Status	Treatment	Response		
			Poor	Moderate	Excellent
1	1	Active	3	20	5
		Placebo	11	14	8
	2	Active	3	14	12
		Placebo	6	13	5
2	1	Active	12	12	0
		Placebo	11	10	0
	2	Active	3	9	4
		Placebo	6	9	3

TABLE 5.4. Data from a multicentre analgesic trial. Source: Koch et al. (1983).

proportional to n_{++k}, and for the design-free version, the contributions are roughly equal. The locally most powerful version seems the most judicious in this respect. If there is much variation between strata in sample size and response distribution $F_{i,k}$, then the different versions may lead to very different conclusions. There does not appear to be any consensus in the literature as to which version is preferable.

An example with stratified categorical data is shown in Table 5.4. These show the results of a multicentre clinical trial comparing an analgesic to a placebo for the relief of chronic joint pain, reported by Koch et al. (1983). There are two centres, and the patients were classified prior to treatment into two diagnostic status groups. Response to treatment is rated as poor, moderate, or excellent. Since the response is associated both with centre and diagnostic status, the key hypothesis is

Response $\perp\!\!\!\perp$ Treatment | (Centre, Diagnostic Status).

We illustrate a test for this as follows:

```
MIM->fact A2B2T2R3
MIM->label A "Centre" B "Diag status" T "Treatment" R "Response"
MIM->sread ABTR
DATA->3 20 5 11 14 8 3 14 12 6 13 5 12 12 0 11 10 0 3 9 4 6 9 3!
Reading completed.
MIM->model ABRT; testdel TR wc
Test of H0: ABT,ABR
against H: ABRT
Estimated number of tables:          98250064
Stratum-specific scores.
Wilcoxon Test.
W: 2668.0000  E(W|H0): 2483.0000 Asymptotic P: 0.0459
MIM->testdel TR lwm
Test of H0: ABT,ABR
against H: ABRT
Exact test - monte carlo estimates.
```

```
Stratum-specific scores.
No. of tables:          1000
Likelihood Ratio Test.
LR:  10.8369    DF:     7 Asymptotic P: 0.1459
                          Estimated P: 0.1810 +/- 0.031364
Wilcoxon Test.
W: 2668.00  E(W|HO): 2483.00 Asymptotic P: 0.0459
                          Estimated P: 0.0550 +/- 0.018579
```

First, we compute the approximate number of tables in the reference set, and the asymptotic test. On this basis, treatment effect is just significant at the 5% level. Since $W > \mathrm{E}(W|H_0)$, the effect of active treatment is larger than that of the placebo, i.e., has a better response. The number of tables in Υ is estimated to be nearly 100 million, so exhaustive enumeration is not possible. We therefore use the Monte Carlo approach. It is seen that the exact p-value is in good agreement with the asymptotic value. For comparison, the ordinary G^2 is also calculated: this test clearly has much less power than the ordinal test.

5.11   Kruskal-Wallis Test

The Kruskal-Wallis test is used to compare discrete or continuous distributions between k populations. It presupposes a nominal (row) factor and a column variable that may be either discrete or continuous.

Suppose the distribution of B given $A = i, C = k$ is $F_{i,k}(x)$, i.e., that $F_{i,k}(x) = \Pr(B < x|A = i, C = k)$. The statistic tests homogeneity

$$H_0 : F_{1,k}(x) = F_{2,k}(x) = \ldots = F_{R,k}(x) \ \forall x, \forall k$$

against the alternative hypothesis that at least one of these distributions is stochastically larger than one of the others, i.e., that $F_{i,k}(x) < F_{j,k}(x), \forall x$ for some k and $i \neq j$. The test statistic is

$$KW = \sum_{k=1}^{L} f_k \sum_{i=1}^{R} \frac{[R_{ik} - n_{i+k}(n_{++k} + 1)/2]^2}{n_{i+k}},$$

where

$$f_k = 12\{n_{++k}(n_{++k} + 1)[1 - \frac{\sum_{j=1}^{C}(n_{+jk}^3 - n_{+jk})}{(n_{++k}^3 - n_{++k})}]\}^{-1}$$

and R_{ik} is the rank sum for row i, stratum k. Under H_0, KW is asymptotically χ^2 distributed with $L(R - 1)$ degrees of freedom.

The test is illustrated using the hematological toxicity data shown in Table 5.2.

```
MIM->fact A5; cont X; read AX
DATA->1 0 1 1 1 8 1 10
DATA->2 0 2 0 2 3 2 3 2 8
DATA->3 5 3 6 3 7 3 14 3 14
DATA->4 1 4 1 4 6 4 7 4 7 4 7 4 8 4 8 4 10
DATA->5 7 5 10 5 11 5 12 5 13 !
Reading completed.
MIM->model A/AX/X; testdel AX km
Test of H0: A/X/X
against H: A/AX/X
Exact test - monte carlo estimates.
No. of tables:        1000
Kruskal-Wallis Test.
KW:   9.4147   DF:    4 Asymptotic P: 0.0515
                        Estimated P: 0.0471 +/- 0.017255
```

In this example, the asymptotic and the exact p-values match closely.

An example with a discrete response variable is the following (Mehta and Patel, 1991). For 18 patients undergoing chemotherapy, tumour regression was recorded and classified on an ordinal scale: no response, partial response, or complete response. The data are shown in Table 5.5, and are analyzed as follows:

```
MIM->fact A5B3
MIM->sread AB
DATA->2 0 0 1 1 0 3 0 0 2 2 0 1 1 4 !
Reading completed.
MIM->model AB; testdel AB ke
Test of H0: B,A
against H: AB
Exact test - exhaustive calculations.
No. of tables:        2088
Kruskal-Wallis Test.
KW:   8.6824   DF:    4 Asymptotic P: 0.0695
                            Exact P: 0.0390
```

The exact p-value of 0.0390 differs somewhat from the asymptotic value.

Drug regimen	Response		
	None	Partial	Complete
1	2	0	0
2	1	1	0
3	3	0	0
4	2	2	0
5	1	1	4

TABLE 5.5. Tumour regression data. Source: Mehta and Patel (1991).

5.12   Jonckheere-Terpstra Test

The Jonckheere-Terpstra test is designed for tables in which both the row and the column variables are ordinal. For $2 \times C$ tables, it is equivalent to the Wilcoxon test.

As before, we consider a test of $A \perp\!\!\!\perp B \mid S$. Suppose the distribution of B given $A = i, S = k$ is $F_{i,k}(x)$, i.e., that $F_{i,k}(x) = \Pr(B < x | A = i, S = k)$. The null hypothesis is homogeneity,

$$H_0 : F_{1,k}(x) = F_{2,k}(x) = \ldots = F_{R,k}(x) \; \forall k,$$

against the alternative hypothesis that these distributions are stochastically ordered, i.e., that either

$$i < j \Rightarrow F_{i,k}(x) \geq F_{j,k}(x), \; \forall x, \forall k$$

or

$$i < j \Rightarrow F_{i,k}(x) \leq F_{j,k}(x), \; \forall x, \forall k.$$

The test statistic is

$$J_t = \sum_{k=1}^{L} \sum_{i=2}^{R} \sum_{j=1}^{C} \{ \sum_{s=1}^{C} w_{ijsk} n_{isk} - n_{i+k}(n_{i+k} + 1)/2 \},$$

where w_{ijsk} are the Wilcoxon scores corresponding to a $2 \times C$ table formed from rows i and j of the table in stratum k, i.e.,

$$w_{ijsk} = \sum_{t=1}^{s-1} (n_{itk} + n_{jtk}) + (n_{isk} + n_{jsk} + 1)/2.$$

The mean of J_t under H_0 is

$$\mathrm{E}(J_t | H_0) = \sum_{k=1}^{L} (n_{++k}^2 - \sum_{i=1}^{R} n_{i+k}^2)/2.$$

The p-value calculated is the two-sided version:

$$p = \Pr(|J_t - \mathrm{E}(J_t|H_0)| \geq |J_{t(obs)} - \mathrm{E}(J_t|H_0)| \mid H_0).$$

As an asymptotic test, $(J_t - \mathrm{E}(J_t|H_0))/\sqrt{\hat{\mathrm{Var}}(J_t|H_0)}$ is compared with the $\mathcal{N}(0,1)$ distribution, where the formidable expression cited in Pirie (1983) is used to estimate the asymptotic variance.

We give two illustrations of the test. The first concerns data reported by Norušis (1988), taken from a social survey in the United States. Table 5.6 shows a cross-classification of income and job satisfaction.

Income (US $)	Job satisfaction			
	Very dissatisfied	Little dissatisfied	Moderately satisfied	Very satisfied
< 6000	20	24	80	82
$6000 - 15000$	22	38	104	125
$15000 - 25000$	13	28	81	113
> 25000	7	18	54	92

TABLE 5.6. Cross-classification of Income and Job Satisfaction. Source: Norušis (1988).

We are interested in testing association between income and job satisfaction. This is shown in the following fragment:

```
MIM->fact A4B4; label A "Income" B "Job satis"; sread AB
DATA->20 24  80 82
DATA->22 38 104 125
DATA->13 28 81 113
DATA->7 18 54 92 !
Reading completed.
MIM->model AB; testdel AB ljm
Test of H0: B,A
against H: AB
Exact test - monte carlo estimates.
No. of tables:       1000
Likelihood Ratio Test.
LR:  12.0369    DF:    9 Asymptotic P: 0.2112
                        Estimated P: 0.2110 +/- 0.033237
Jonckheere-Terpstra Test.
JT: 162647.0 E(JT|H0): 150344.5 Asymptotic P: 0.0047
                        Estimated P: 0.0041 +/- 0.005215
```

We see that the test detects highly significant association. The association is positive since $J_t > E(J_t \mid H_0)$, i.e., increasing satisfaction for increasing income (not surprisingly). For comparison purposes, the likelihood ratio test is also shown. This fails to find association, illustrating the greatly superior power of the ordinal tests.

The second example concerns data reported by Everitt (1977) (see also Christensen, 1990, p. 61ff). A sample of 97 children was classified using three factors: risk of home conditions (A), classroom behaviour (B), and adversity of school conditions (C). The data are shown in Table 5.7.

We test whether classroom behaviour is independent of school conditions, given the home conditions, i.e., $B \perp\!\!\!\perp C \mid A$, using the Jonckheere-Terpstra test. Note that even though B is binary rather than ordinal, the test is still appropriate since the alternative hypothesis here means that the

Home conditions	Adversity of School conditions	Classroom behaviour Nondeviant	Deviant
Not at risk	Low	16	1
	Medium	15	3
	High	5	1
At risk	Low	7	1
	Medium	34	8
	High	3	3

TABLE 5.7. Data on classroom behaviour. Source: Everitt (1977).

conditional probabilities

$$q_{i,k} = \Pr(\text{deviant behaviour}|C = i, A = k)$$

are monotone with respect to i for each k, or in other words, that

$$q_{1,k} \leq q_{2,k} \leq q_{3,k}$$

(or vice versa) for $k = 1, 2$.

```
MIM->fact A2B3C2
MIM->sread BAC
DATA->16 1 7 1 15 3 34 8 5 1 3 3 !
Reading completed.
MIM->model ABC; testdel CB jle
Test of H0: AC,AB
against H: ABC
Exact test - exhaustive calculations.
No. of tables:      1260
Likelihood Ratio Test.
LR:    4.1180    DF:    4 Asymptotic P: 0.3903
                         Exact P: 0.5317
Jonckeere-Terpstra Test.
JT: 435.0000  E(JT|H0): 354.0000 Asymptotic P: 0.1482
                         Exact P: 0.0730
```

For comparison purposes, results from the likelihood ratio test are also shown. We see that this test, which does not use the ordinality of C, detects no evidence of association between classroom behaviour and school conditions. In contrast, the Jonckheere-Terpstra test suggests that there is slight though inconclusive evidence that classroom behaviour depends on school conditions.

Since in the example $J_t > \mathrm{E}(J_t|H_0)$, the direction of association (if any) is positive: that is, the larger the C (the more adverse the school conditions), the larger the B (the more deviant the behaviour). This is consistent with our expectation.

This concludes our description of exact tests. We now turn to some other testing problems. The framework of hierarchical interaction models subsumes some testing situations that have been studied in classical normal-based multivariate analysis. For some of these, sharp distributional results have been derived. In the next three sections, we sketch some of these results and link them to the present framework.

5.13   Tests for Variance Homogeneity

Since variance homogeneity plays an important role in inference, it is natural to focus attention on a test for homogeneity given that the means are unrestricted. An example with $p = q = 2$ is a test of

$$\mathcal{M}_0 : AB/ABX, ABY/XY$$
$$\text{versus } \mathcal{M}_1 : AB/ABX, ABY/ABXY.$$

From (4.17) and (4.18), we find that the deviance difference is

$$d = N \ln |S| - \sum_{j \in I} n_j |S_j|,$$

where, as usual, $S_j = \sum_{k:i^{(k)}=j} (y^{(k)} - \bar{y}_j)(y^{(k)} - \bar{y}_j)'/n_j$ is the MLE of Σ_j under $\mathcal{M}_1$, and $S = \sum_j n_j S_j / N$ is the MLE of $\Sigma_j = \Sigma$ under $\mathcal{M}_0$. Under $\mathcal{M}_0$, d is approximately distributed as χ_v^2, where $v = q(q+1)(\#I - 1)/2$, and $\#I = \Pi_{\delta \in \Delta} \# \delta$ is the number of cells in the underlying contingency table.

This approximation is rather poor for small to moderate sample sizes: generally, the test is too liberal, which is to say that it is apt to reject homogeneity too frequently. Various ways of modifying the test to improve the χ^2 approximation have been studied. Seber (1984, p. 448–451) gives a useful summary. The following test is due to Box (1949). The test statistic is

$$d' = (1 - c)\{(N - \#I) \ln |\tilde{S}| - \sum_{j \in I}(n_j - 1) \ln |\tilde{S}_j|\},$$

where $\tilde{S}_j = (\frac{n_j}{n_j-1})S_j$ and $\tilde{S} = (\frac{N}{N-\#I})S$, and the constant c is given as

$$c = \frac{2q^2 + 3q - 1}{6(q+1)(\#I - 1)}\left\{\sum_j \frac{1}{(n_j - 1)} - \frac{1}{(N - \#I)}\right\}.$$

Under $\mathcal{M}_0$, d' has an asymptotic χ_v^2 distribution: for small samples, the approximation is far superior to the uncorrected deviance d. This test is the multivariate generalization of Bartlett's test for homogeneity.

Both tests, that is, both d and d', are sensitive to non-normality so that rejection of $\mathcal{M}_0$ can be due to kurtosis rather than heterogeneity.

The test is illustrated in Sections 4.1.8 and 4.1.9.

There are hierarchical interaction models that are intermediate between homogeneity and (unrestricted) heterogeneity. One such model (discussed in Section 4.4) is $A/AX, AY/AX, XY$. In a sense, these models supply a more refined parametrization of heterogeneity. Note that by exploiting collapsibility properties, Box's test can sometimes be applied to such models. For example, a test of $A/AX, AY/XY$ versus $A/AX, AY/AX, XY$ can be performed as a test of $A/AX/X$ versus $A/AX/AX$ using Box's test.

5.14   Tests for Equality of Means Given Homogeneity

Another hypothesis that has been studied in depth is that of equality of multivariate means assuming homogeneity. An example with $p = 1$ and $q = 3$ is

$$\mathcal{M}_0 : A/X, Y, Z/XYZ$$
$$\text{versus } \mathcal{M}_1 : A/AX, AY, AZ/XYZ,$$

corresponding to the removal of $[AX]$, $[AY]$, and $[AZ]$ from the graph

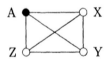

More generally, the test concerns removal of all edges between the discrete and the continuous variables, assuming a homogeneous full model. For this test, a function of the likelihood ratio test has a known small-sample distribution called Wilks' Λ-distribution. We now relate this to the deviance statistic.

Under $\mathcal{M}_1$, the MLEs of the cell means are the sample cell means $\hat{\mu}_i = \bar{y}_i$, and the MLE of the common covariance is the sample "within-cell" SSP, i.e.,

$$\hat{\Sigma}_1 = \sum_j \sum_{k:i^{(k)}=j} (y^{(k)} - \bar{y}_j)(y^{(k)} - \bar{y}_j)'/N.$$

Under $\mathcal{M}_0$, the corresponding estimates are the overall means $\hat{\mu} = \bar{y}$ and the "total" SSP, i.e.,

$$\hat{\Sigma}_0 = \sum_k (y^{(k)} - \bar{y})(y^{(k)} - \bar{y})'/N.$$

The corresponding maximized likelihoods are

$$\hat{\ell}_1 = \sum_i n_i \ln(n_i/N) - Nq \ln(2\pi)/2 - N \ln|\hat{\Sigma}_1|/2 - Nq/2$$

and

$$\hat{\ell}_0 = \sum_i n_i \ln(n_i/N) - Nq \ln(2\pi)/2 - N \ln|\hat{\Sigma}_0|/2 - Nq/2.$$

So the deviance difference is $d = 2(\hat{\ell}_1 - \hat{\ell}_0) = N \ln|\hat{\Sigma}_1^{-1}\hat{\Sigma}_0|$. If we write the corresponding "between-cells" quantity as B,

$$B = \hat{\Sigma}_0 - \hat{\Sigma}_1 = \sum_j (\bar{y}_j - \bar{y})(\bar{y}_j - \bar{y})'/N,$$

and we can re-express d as

$$d = N \ln|\hat{\Sigma}_1^{-1}(\hat{\Sigma}_1 + B)| = N \ln|I + \hat{\Sigma}_1^{-1}B|.$$

The quantity $\hat{\Sigma}_1^{-1}B$ is a generalization of the variance ratio from ANOVA; it tends to unity under $\mathcal{M}_0$.

Wilks' Λ-statistic is defined as

$$\Lambda = |I + \hat{\Sigma}_1^{-1}B|^{-1},$$

and under $\mathcal{M}_0$ this follows a known distribution, the so-called Wilks' Λ-distribution with parameters $(q, N - \#A, \#A - 1)$.

Wilks' Λ is not available in MIM. The rationale is that rather than test for the simultaneous removal of all edges between the discrete and the continuous variables, it would generally seem preferable to decompose this into single edge removal F-tests.

5.15 Hotelling's T^2

This is a special case of the previous section where there is one discrete binary variable. A test statistic due to Hotelling (1931) may be used to test equality of multivariate means assuming equal, unrestricted covariance matrices. This is a generalization of the univariate t-test. Consider a test of

$$\mathcal{M}_0 : A/X, Y, Z/XYZ$$
$$\text{versus } \mathcal{M}_1 : A/AX, AY, AZ/XYZ,$$

where A has two levels. Under $\mathcal{M}_0$, we have that $\mathrm{E}(\bar{y}_1 - \bar{y}_2) = 0$ and $\mathrm{Var}(\bar{y}_1 - \bar{y}_2) = \Sigma(\frac{1}{n_1} + \frac{1}{n_2})$, so we can construct Hotelling's T^2 statistic as

$$T^2 = \left(\frac{1}{n_1} + \frac{1}{n_2}\right)^{-1}(\bar{y}_1 - \bar{y}_2)'\hat{\Sigma}(\bar{y}_1 - \bar{y}_2)$$
$$= (\frac{n_1 n_2}{N})(\bar{y}_1 - \bar{y}_2)'\hat{\Sigma}(\bar{y}_1 - \bar{y}_2).$$

From the previous section, since $\#A = 2$, we can rewrite B as

$$B = \left(\frac{n_1 n_2}{N^2}\right)(\bar{y}_1 - \bar{y}_2)(\bar{y}_1 - \bar{y}_2)'.$$

Hence,

$$|I + \hat{\Sigma}_1^{-1}B| = 1 + T^2/N,$$

and so $d = N\ln(1 + T^2/N)$, or equivalently, $T^2 = N(e^{\frac{d}{N}} - 1)$. The test rejects the null hypothesis at the α level if

$$T^2 \geq \frac{q(N-2)}{(N-q-1)}F^{\alpha}_{q,N-q-1},$$

where $F^{\alpha}_{q,N-q-1}$ is the $(1-\alpha)$-percent point of the F-distribution with q and $N - q - 1$ degrees of freedom.

As before, we comment that in general it seems preferable to test for the removal of the edges separately using F-tests.

6

Model Selection and Criticism

In many applications of statistics, little prior knowledge or relevant theory is available, and so model choice becomes an entirely empirical, exploratory process. Three different approaches to model selection are described in the first three sections of this chapter. The first is a stepwise method, which starts from some initial model and successively adds or removes edges until some criterion is fulfilled. The second is a more global search technique proposed by Edwards and Havránek (1985, 1987), which seeks the simplest models consistent with the data. The third method is to select the model that optimizes one of the so-called information criteria (AIC or BIC). In Section 4 a brief comparison of the three approaches is made.

Section 5 describes a method to widen the scope of the CG-distribution by allowing power transformations of the continuous variables (Box and Cox, 1964). The last two sections describe techniques for checking whether the continuous variables satisfy the assumptions of multivariate normality.

We preface the chapter with some introductory remarks about model selection. Perhaps the first thing to be said is that all model selection methods should be used with caution, if not downright scepticism. Any method (or statistician) that takes a complex multivariate dataset and, from it, claims to identify one true model, is both naive and misleading. The techniques described below claim only to identify simple models consistent with the data, as judged by various criteria: this may be inadequate for various reasons.

For example, if important variables have been omitted or are unobservable, the models selected may be misleading (for some related issues, see Section 1.4). Some problems seem to require multiple models for an adequate description (see the discussion of split models in Section 2.2.6), and for these, the adoption of one grand, all-embracing supramodel may be unhelpful. Finally, the purpose to which the models will be put and the scientific interpretation and relevance of the models ought to play decisive roles in the evaluation and comparison of different models.

The first two model selection approaches described here are based on significance tests; many tests may be performed in the selection process. This may be regarded as a misuse of significance testing, since the overall error properties are not related in any clear way to the error levels of the individual tests (see Section 6.4, however, for a qualification of this statement).

There is also a deeper problem. In statistical modelling we generally choose a model that fits the data well, and then proceed under the assumption that the model is true. The problem with this— the problem of *model uncertainty* — is that the validity of most model-based inference rests on the assumption that the model has *not* been chosen on the basis of the data. Typically, estimators that would be unbiased under a true, fixed model are biased when model choice is data-driven. Estimates of variance generally underestimate the true variance, for example. Similarly, hypothesis tests based on models chosen from the data often have supranominal type I error rates. Chatfield (1995) gives an accessible introduction to model uncertainty. With randomised studies the problem can be circumvented (Edwards, 1999).

To summarize: It is essential to regard model selection techniques as explorative tools rather than as truth-algorithms. In interplay with subject-matter considerations and the careful use of model control and diagnostic techniques, they may make a useful contribution to many analyses.

6.1 Stepwise Selection

This is an incremental search procedure. Starting from some initial model, edges are successively added or removed until some criterion is fulfilled. At each step, the inclusion or exclusion of eligible edges is decided using significance tests. Many variations are possible and are described in this section.

Stepwise selection is performed in MIM using the command `Stepwise`. The standard operation of this command is backward selection; that is to say, edges are successively removed from the initial model. At each step, the eligible edges are tested for removal using χ^2-tests based on the deviance difference between successive models. The edge whose χ^2-test has the largest (nonsignificant) p-value is removed. If all p-values are significant (i.e., all $p < \alpha$, where α is the critical level), then no edges are removed and the procedure stops.

An edge may not be eligible for removal at a step. There can be several reasons for this: first, edges can be *fixed* in the model, using the Fix command. For example,

```
MIM->mod ABCDE
MIM->fix BCD
Fixed variables: BCD
MIM->stepwise
```

initiates stepwise selection starting from ABCDE, in which the edges [BC], [CD], and [BD] are fixed in the model.

Secondly, in the default operation we are describing, the principle of *coherence* is respected. In backward selection, this just means that if the removal of an edge is rejected at one step (the associated p-value is less than the critical level α), then the edge is not subsequently eligible for removal. This rule speeds up the selection process, but this is not the only reason for observing it, as we will explain later.

A third reason for ineligibility comes into play when the procedure runs in decomposable mode. In other words, only decomposable models are considered at each step: thus, at any step, the edges whose exclusion (or inclusion, in forward selection) would result in a nondecomposable model are considered ineligible.

Here is an example of the default mode of operation using the mathematics marks data of Section 3.1.6:

```
MIM->model //VWXYZ
MIM->stepwise
Coherent Backward Selection
Decomposable models, chi-squared tests.
Critical value:   0.0500
Initial model: //VWXYZ
Model: //VWXYZ
Deviance:   0.0000 DF:   0 P:  1.0000
      Edge        Test
  Excluded   Statistic DF         P
      [VW]    10.0999   1     0.0015 +
      [VX]     4.8003   1     0.0285 +
      [VY]     0.0002   1     0.9880
      [VZ]     0.0532   1     0.8176
      [WX]     7.2286   1     0.0072 +
      [WY]     0.5384   1     0.4631
      [WZ]     0.0361   1     0.8494
```

```
            [XY]      18.1640  1      0.0000 +
            [XZ]      11.9848  1      0.0005 +
            [YZ]       5.8118  1      0.0159 +
Removed edge [VY]
Model: //VWXZ,WXYZ
Deviance:   0.0002 DF:   1 P:  0.9880
      Edge        Test
Excluded    Statistic DF         P
      [VZ]       0.0550  1      0.8146
      [WY]       0.5960  1      0.4401
Removed edge [VZ]
Model: //VWX,WXYZ
Deviance:   0.0552 DF:   2 P:  0.9728
      Edge        Test
Excluded    Statistic DF         P
      [WY]       0.5960  1      0.4401
      [WZ]       0.0794  1      0.7782
Removed edge [WZ]
Model: //VWX,WXY,XYZ
Deviance:   0.1346 DF:   3 P:  0.9874
      Edge        Test
Excluded    Statistic DF         P
      [WY]       0.7611  1      0.3830
Removed edge [WY]
Selected model: //VWX,XYZ
```

At the first step, all 10 edges are tested for removal. Of these, six are rejected at the 5% level and are marked with +'s. Since the procedure is in coherent mode, it does not subsequently try to remove these edges. The edge with the largest p-value, $[VY]$, is removed. The formula of the resulting model, $VWXY, WXYZ$, is printed out, together with its deviance, degrees of freedom, and the associated p-value.

At the second step, two edges are tested for removal. Note that $[WZ]$, though not removed or fixed in the model at the first step, is not among the two edges tested. This is because the procedure is in decomposable mode and the removal of $[WZ]$ would result in a nondecomposable model.

However, after $[VZ]$ is removed at the second step, $[WZ]$ is eligible for removal and is indeed removed at the third step.

After the fourth step, no further simplification is possible. The model VWX, XYZ, whose graph is shown in Figure 3.2, is selected.

6.1.1 Forward Selection

Forward selection acts by adding the most significant edges instead of removing the least significant edges. In other words, at each step, the edge with the smallest p-value, as long as this is less than the critical level, is added to the current model.

To compare backward and forward selection, note that backward stepwise methods start with a complex model that is usually consistent with the data and which is then successively simplified. So these methods step through models that are consistent with the data.

In contrast, forward selection methods start with a simple model that is usually inconsistent with the data. This is then successively enlarged until an acceptable model is reached. So forward selection methods step through models that are inconsistent with the data.

This implies that in backward selection, the individual significance tests involve comparison between pairs of models where at least the larger model of the pair (the alternative hypothesis) is valid. In contrast, in forward selection, the significance tests involve pairs of models, both of which are invalid. For this reason, backward selection is generally preferred to forward selection.

Despite this, the two approaches often give quite similar results. Forward selection is often attractive with sparse data, where the simple models give rise to fewer problems concerning the existence of maximum likelihood estimates and the accuracy of the asymptotic reference distributions.

We illustrate forward selection using the mathematics marks data once more. This time we start from the main effects model.

```
MIM->model //V,W,X,Y,Z
MIM->stepwise f
Non-coherent Forward Selection
Decomposable models, chi-squared tests.
Critical value:   0.0500
Initial model: //V,W,X,Y,Z
Model: //V,W,X,Y,Z
Deviance: 202.5151 DF:  10 P:  0.0000
   Edge        Test
  Added    Statistic DF          P
   [VW]     32.1776  1      0.0000 +
   [VX]     31.2538  1      0.0000 +
   [VY]     16.1431  1      0.0001 +
   [VZ]     14.4465  1      0.0001 +
   [WX]     40.8922  1      0.0000 +
   [WY]     23.6083  1      0.0000 +
   [WZ]     18.5964  1      0.0000 +
```

```
            [XY]     61.9249   1      0.0000 +
            [XZ]     51.3189   1      0.0000 +
            [YZ]     40.4722   1      0.0000 +
Added edge [XY]
Model: //V,W,XY,Z
Deviance: 140.5901 DF:   9 P:  0.0000
            Edge       Test
            Added    Statistic DF           P
            [VW]     32.1776   1      0.0000 +
            [VX]     31.2538   1      0.0000 +
            [VY]     16.1431   1      0.0001 +
            [VZ]     14.4465   1      0.0001 +
            [WX]     40.8922   1      0.0000 +
            [WY]     23.6083   1      0.0000 +
            [WZ]     18.5964   1      0.0000 +
            [XZ]     51.3189   1      0.0000 +
            [YZ]     40.4722   1      0.0000 +
Added edge [XZ]
Model: //V,W,XY,XZ
Deviance:  89.2712 DF:   8 P:  0.0000
            Edge       Test
            Added    Statistic DF           P
            [VW]     32.1776   1      0.0000 +
            [VX]     31.2538   1      0.0000 +
            [VY]     16.1431   1      0.0001 +
            [VZ]     14.4465   1      0.0001 +
            [WX]     40.8922   1      0.0000 +
            [WY]     23.6083   1      0.0000 +
            [WZ]     18.5964   1      0.0000 +
            [YZ]      5.9788   1      0.0145 +
Added edge [WX]
Model: //V,WX,XY,XZ
Deviance:  48.3789 DF:   7 P:  0.0000
            Edge       Test
            Added    Statistic DF           P
            [VW]     32.1776   1      0.0000 +
            [VX]     31.2538   1      0.0000 +
            [VY]     16.1431   1      0.0001 +
            [VZ]     14.4465   1      0.0001 +
            [WY]      0.7611   1      0.3830
            [WZ]      0.2445   1      0.6209
            [YZ]      5.9788   1      0.0145 +
Added edge [VW]
Model: //VW,WX,XY,XZ
Deviance:  16.2014 DF:   6 P:  0.0127
            Edge       Test
            Added    Statistic DF           P
```

```
[VX]      9.3269  1      0.0023 +
[WY]      0.7611  1      0.3830
[WZ]      0.2445  1      0.6209
[YZ]      5.9788  1      0.0145 +
Added edge [VX]
Model: //VWX,XY,XZ
Deviance:   6.8745 DF:   5 P:  0.2301
   Edge          Test
  Added     Statistic DF        P
   [VY]      0.1094  1      0.7408
   [VZ]      0.1481  1      0.7003
   [WY]      0.7611  1      0.3830
   [WZ]      0.2445  1      0.6209
   [YZ]      5.9788  1      0.0145 +
Added edge [YZ]
Model: //VWX,XYZ
Deviance:   0.8957 DF:   4 P:  0.9252
   Edge          Test
  Added     Statistic DF        P
   [VY]      0.1094  1      0.7408
   [VZ]      0.1481  1      0.7003
   [WY]      0.7611  1      0.3830
   [WZ]      0.2445  1      0.6209
No change.
Selected model: //VWX,XYZ
```

The first step consists of testing the marginal independence of all variable pairs. It is seen that all pairs are highly correlated. The edge $[XY]$ has the smallest associated p-value (highest correlation) and so is added at the first step. At the sixth step, the same model as before — VWX, XYZ — is selected.

6.1.2   Restricting Selection to Decomposable Models

The two examples above restricted attention to decomposable models only. Stepwise selection restricted to decomposable models was first proposed by Wermuth (1976a). Decomposable models are an attractive subclass for a variety of reasons. Ease of interpretation has been discussed in Section 4.4. Considerable efficiency gains can be achieved by using non-iterative estimates and by exploiting collapsibility properties to avoid unnecessary fitting. A final consideration is that exact tests, F-tests, and sparsity-corrected degrees-of-freedom calculations are only available in decomposable mode.

The decomposable models also enjoy a connectedness property in the following sense. For any nested pair of decomposable models, there exists a

path of decomposable models from the larger to the smaller model, formed
by removing a sequence of edges. In other words, there are no blind alleys
to trap the stepwise process (Edwards, 1984).

Consider a model that has been selected using backward selection restricted
to decomposable models. Hopefully, most of the edges in the graph are
present because their removal was rejected at some stage. However, there
may also be edges that are retained because their removal would have led
to a nondecomposable model. Edges occurring in more than one generator
(see Section 5.2) may be of this type. Sometimes it may be appropriate
to check whether this is the case, either by scrutinizing the output or by
initiating a stepwise selection from the selected model in unrestricted mode
(using the U option).

6.1.3   Using F-Tests

As described in Section 5.3, a decomposable edge deletion test between
homogeneous models can be performed as an F-test, provided at least one
of the nodes of the edge is continuous. This can also be exploited in stepwise selection using the S option. We illustrate this using, once again, the
mathematics marks data.

```
MIM->mod //VWXYZ
MIM->step s
Coherent Backward Selection
Decomposable models, F-tests where appropriate.
DFs adjusted for sparsity.
Critical value:   0.0500
Initial model: //VWXYZ
Model: //VWXYZ
Deviance:   0.0000 DF:   0 P:  1.0000
     Edge         Test
  Excluded    Statistic DF          P
     [VW]       10.0943  1,   83 0.0021 +
     [VX]        4.6533  1,   83 0.0339 +
     [VY]        0.0002  1,   83 0.9883
     [VZ]        0.0502  1,   83 0.8233
     [WX]        7.1057  1,   83 0.0092 +
     [WY]        0.5094  1,   83 0.4774
     [WZ]        0.0340  1,   83 0.8541
     [XY]       19.0283  1,   83 0.0000 +
     [XZ]       12.1098  1,   83 0.0008 +
     [YZ]        5.6667  1,   83 0.0196 +
Removed edge [VY]
Model: //VWXZ,WXYZ
```

```
Deviance:   0.0002 DF:   1 P:  0.9880
     Edge       Test
Excluded    Statistic DF          P
    [VZ]        0.0525  1,  84 0.8193
    [WY]        0.5708  1,  84 0.4521
Removed edge [VZ]
Model: //VWX,WXYZ
Deviance:   0.0552 DF:   2 P:  0.9728
     Edge       Test
Excluded    Statistic DF          P
    [WY]        0.5708  1,  84 0.4521
    [WZ]        0.0758  1,  84 0.7838
Removed edge [WZ]
Model: //VWX,WXY,XYZ
Deviance:   0.1346 DF:   3 P:  0.9874
     Edge       Test
Excluded    Statistic DF          P
    [WY]        0.7384  1,  85 0.3926
Removed edge [WY]
Selected model: //VWX,XYZ
```

At each step, the value of the F-statistic, the associated degrees of freedom, and the associated p-value are written out. The same model as before is selected.

Note incidentally that if we fix all variables except one continuous variable and use F-tests, then we are able to perform a standard stepwise regression selection.

6.1.4 Coherence

The principle of coherence (Gabriel, 1969) states that for any two nested models, say $\mathcal{M}_0 \subseteq \mathcal{M}_1$, if $\mathcal{M}_1$ is rejected (deemed inconsistent with the data), then $\mathcal{M}_0$ must also be rejected. Conversely, if $\mathcal{M}_0$ is accepted (deemed consistent with the data), then $\mathcal{M}_1$ must also be accepted.

In backward selection, the coherence principle can be interpreted as follows. Suppose the current step starts with the model $\mathcal{M}$ and the removal of an edge e is rejected. This implies that the resulting model $\mathcal{M}_1$, say, is inconsistent with the data. The coherence principle says that all submodels of $\mathcal{M}_1$ are also inconsistent with the data. All submodels of $\mathcal{M}$ not containing e are also submodels of $\mathcal{M}_1$. Thus, by fixing e in subsequent models, coherence is respected.

The corresponding argument for forward selection is uncompelling. It would sound something like this: if at some step inclusion of an edge e to the current model $\mathcal{M}$ is rejected (since the improvement in fit is nonsignificant), then e should not be subsequently eligible for inclusion. That would be sensible if $\mathcal{M}$ were consistent with the data, but as mentioned above, this will not generally be the case since in forward selection the models considered are generally too simple, i.e., inconsistent with the data.

This can also be seen from a different viewpoint. Consider a high-dimensional contingency table, and contrast forward selection from the main-effects model with backward selection from the saturated model. Arguments related to Simpson's paradox indicate that marginal independence does not imply conditional independence, whereas in general, conditional dependence does imply marginal dependence.

6.1.5   Other Variants of Stepwise Selection

It is often useful in the process of an analysis to test whether edges can be added to or removed from the current model without actually changing the model. This is the purpose of the O (one step only) option. The tests for edge removal or inclusion are carried out, but the initial model is not changed. After completion (in backward selection), the p-values may be written on the graph as described in Section B.6. Similarly, the commands DeleteLSEdge and DeleteNSEdge remove the least significant edge, and all the nonsignificant edges, respectively.

Another variant is headlong model selection, obtained by using the H option. Edges eligible for removal or inclusion are examined in random order and the first available edge is chosen. That is to say, in backward selection, the first nonsignificant edge is removed, whereas in forward selection, the first significant edge is added. Since the edges are examined in random order, different models may be selected when starting repeatedly from the same initial model.

Headlong selection is obviously faster than the standard method. Furthermore, it may be used to examine whether the selection process depends on the order in which the edges are removed or added. Note that after Stepwise, it is easy to revert to the initial model by using the BackToBase command.

Finally, exact tests may be used in the analysis of sparse contingency tables: they are only available in decomposable mode. The example in Section 2.2.5 illustrates this.

6.2   The EH-Procedure

The previous section described stepwise model selection. This approach is excellent for identifying one simple model consistent with the data. However, it may overlook other models that fit the data just as well.

The present section describes a rather different form of model selection proposed by Edwards and Havránek (1985, 1987) (see also Horáková, 1991 and Edwards, 1993). The approach is based on a search algorithm that seeks the simplest models consistent with the data. The models are chosen from one of two classes: either homogeneous or heterogeneous graphical models. During the search, a sequence of models are fit; these are classified as accepted (i.e., deemed consistent with the data) or rejected (deemed inconsistent). This decision is based on the χ^2 deviance test; that is, if the associated $p \leq \alpha$, the model is rejected; otherwise it is accepted.

The coherence principle described in Section 6.1.4 is applied so that if a model is rejected, then all its submodels are also considered rejected. Similarly, if a model is accepted, then all models that contain it are also considered accepted. These rules determine which models are fitted next.

At any stage in the search process, the current lists of accepted and rejected models divide the model class into three disjoint sets, as sketched schematically in Figure 6.2(a).

The upper set consists of all those models that contain an accepted model as a submodel, and therefore can also be considered to be consistent with the data. These are called *w-accepted*, i.e., weakly accepted. The lower set consists of all those models that are a submodel of one or more of the rejected models, and therefore can also be considered to be inconsistent with the data. These are called *w-rejected*, i.e., weakly rejected. Between these two sets lies a third— the models whose consistency with the data has yet to be determined.

The next step in the search process consists of testing either the minimal or the maximal models in the undetermined set (usually whichever list is the shortest). In Figure 6.2(b), the maximal undetermined models have been fitted: one has been rejected and the other accepted. The lists of accepted and rejected models are updated, and the process is repeated until all models are classified as w-accepted or w-rejected, i.e., the middle set vanishes.

We illustrate the search procedure using the mathematics marks data. First we initialize the process, using the `InitSearch` command, and then we invoke `StartSearch` to perform the search:

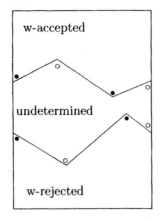   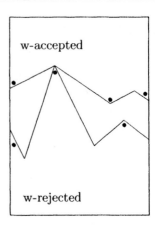

(a)   (b)

FIGURE 6.1. A step in the EH-procedure. In (a), four models (filled circles) have been fit: two have been accepted and two rejected. These generate a partition of the model family into three sets— the w-accepted models, the w-rejected models, and the undetermined models. The minimal and maximal models in the undetermined set are shown as hollow circles. In (b), the two maximal undetermined models have been fit— one has been accepted and the other rejected. These generate a new partition. The process continues in this fashion until the undetermined set is empty.

```
MIM->initsearch
Initializing the EH-procedure ...
Fitting maximum model...
MIM->startsearch
Starting the EH-procedure ...
Maximum no. of models not specified.

------------------------------------------------------------------
EH-procedure complete.

------------------------------------------------------------------
Significance level used: 0.05
12 models tested, out of in all 2^10=1024 possible models.

The minimal accepted model is:
//VWX,XYZ
Deviance:    0.8957 DF:       4  LR:       0.8957 DF:       4 P: 0.9252
------------------------------------------------------------------
```

We see that the search procedure found one single simplest accepted model, $//VWX, XYZ$. This model is thus the simplest model consistent with the data at the 5% level.

The list of maximal rejected models also provides interesting information. This can be obtained using the command Show S:

```
MIM->show s
EH-procedure complete.

-------------------------------------------------------------
The minimal accepted models are:
//VWX,XYZ
Deviance:    0.8957 DF:      4  LR:     0.8957 DF:      4 P: 0.9252

-------------------------------------------------------------
The maximal rejected models are:
//VXYZ,WXYZ
Deviance:   10.0999 DF:      1  LR:    10.0999 DF:      1 P: 0.0015
//VWYZ,WXYZ
Deviance:    4.8003 DF:      1  LR:     4.8003 DF:      1 P: 0.0285
//VWYZ,VXYZ
Deviance:    7.2286 DF:      1  LR:     7.2286 DF:      1 P: 0.0072
//VWXZ,VWYZ
Deviance:   18.1640 DF:      1  LR:    18.1640 DF:      1 P: 0.0000
//VWXY,VWYZ
Deviance:   11.9848 DF:      1  LR:    11.9848 DF:      1 P: 0.0005
//VWXY,VWXZ
Deviance:    5.8118 DF:      1  LR:     5.8118 DF:      1 P: 0.0159

-------------------------------------------------------------
```

We see the largest p-value in the list is 0.0285. This implies that the selection process would have given the same result using a critical level of anywhere between 0.05 and 0.0286.

Another illustration of the use of the EH-procedure follows.

6.2.1 Example: Estrogen and Lipid Metabolism, Continued

The analysis of the lipid data presented in Section 4.1.10 found a simple and well-fitting model,

$$A/AV, AW, X, AY, Z, WXZ, VWXY, AVWY,$$

with a deviance of 32.1893 on 24 degrees of freedom (corresponding to $p = 0.1224$). To find out whether there are other models that are just as simple and well-fitting, we can employ the EH-procedure to find the simplest models that are accepted using a χ^2-test at the level $P = 0.1220$.

```
MIM->critlev .1220
Critical level: 0.122
MIM->initsearch
Initializing the EH-procedure ...
Fitting maximum model...
MIM->startsearch
Starting the EH-procedure ...
Maximum no. of models not specified.
```

```
-----------------------------------------------------------------
EH-procedure complete.
-----------------------------------------------------------------
Significance level used: 0.122
28 models tested, out of in all 2^15=32768 possible models.

The minimal accepted models are:
A/Z,X,AY,AW,AV/XZ,WZ,VXY,AVWY
Deviance:  33.3438 DF:      25  LR:     33.3438 DF:     25 P: 0.1227
A/AZ,AY,AX,AW,AV/AVWYZ,AVWXY
Deviance:   5.5468 DF:       3  LR:      5.5468 DF:      3 P: 0.1359
-----------------------------------------------------------------
```

The procedure has found two minimal models consistent with the data at the $P = 0.1220$ level. The first,

$$A/Z, X, AY, AW, AV/XZ, WZ, VXY, AVWY$$

is a slight simplification of the model previously selected, formed from that by the removal of the edge [WX]. It is nondecomposable, which is why it was not found by the backwards selection procedure. Its graph is shown in Figure 6.2.

The other model selected is relatively complex, with only three degrees of freedom, i.e., only three parameter constraints compared with the full model. It expresses the hypothesis $X \perp\!\!\!\perp Z \mid (A, V, W, Y)$. When tested against the full model, it has a deviance of 5.5468, which is not significant. Its graph is shown in Figure 6.3.

The results confirm that the model found using backward selection is a good choice, since they show there are no other models that are equally simple and equally consistent with the data.

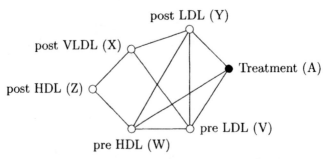

FIGURE 6.2. The first model selected by the EH-procedure.

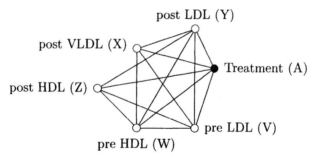

FIGURE 6.3. The second model selected by the EH-procedure.

The hypothesis $X \perp\!\!\!\perp Z \mid (A, V, W, Y)$ can also be tested under simpler models, for example, under the model used to start the selection process:

```
MIM->mod A/Z,X,AY,AW,AV/WXZ,VWXY,AVWY
MIM->step uo
Coherent Backward Selection
Unrestricted models, chi-squared tests.
Single step.
Critical value:   0.0500
Initial model: A/Z,X,AY,AW,AV/WXZ,VWXY,AVWY
Model: A/Z,X,AY,AW,AV/WXZ,VWXY,AVWY
Deviance:  32.1893 DF:  24 P:  0.1224
     Edge        Test
Excluded    Statistic DF         P
     [XV]       9.2271  1      0.0024 +
     [XW]       1.1545  1      0.2826
     [YA]      39.6777  8      0.0000 +
     [YV]     107.0670  3      0.0000 +
     [YW]      13.7958  3      0.0032 +
     [YX]       8.4166  1      0.0037 +
     [ZW]      23.7128  1      0.0000 +
     [ZX]       6.4550  1      0.0111 +
No change.
Selected model: A/Z,X,AY,AW,AV/WXZ,VWXY,AVWY
```

Now the hypothesis is rejected ($P = 0.0111$). That it was accepted before, when tested under the saturated model, can be ascribed to lack of power due to overparametrization.

6.3   Selection Using Information Criteria

Two criteria, AIC and BIC, have been proposed as a basis for model selection. Akaike's Information Criterion (Akaike, 1974) which is derived from information-theoretic arguments, chooses the model with the smallest value of $-2\ln Q - 2p$ where Q is the maximized likelihood under the model and p is the dimension (number of free parameters) of the model. The Bayesian Information Criterion (Schwarz, 1978), chooses the model with the smallest value of $-2\ln Q - \sqrt{N}p$, where N is the number of observations. The justification for this is that it corresponds, asymptotically, to choosing the model with the largest posterior probability. Note that for $N > 8$ and increasingly for larger numbers of observations, the model with the least BIC will tend to be simpler than the model with the least AIC.

The `Select` command uses a brute force approach, viz., each model in the class specified is fitted and the criterion calculated. This is only feasible for problems of small to moderate dimension. The approach is illustrated using the mathematics marks data:

```
MIM->satmod
MIM->select
MIM->pr
The current model is: //VWX,XYZ.
```

Here all the graphical submodels of the initial model are examined and the one with the least BIC is selected. Since there are five variables, the full graph has $5 \times 4/2 = 10$ edges, hence there are $2^{10} = 1024$ possible models. It is not difficult to see that this approach is infeasible for high-dimensional problems.

Stepwise selection can also be based on the information criteria: see Appendix 6.1 for details.

6.4   Comparison of the Methods

In this section we make a brief comparison of the model selection approaches described above. All are heuristic methods that identify simple models consistent with the data, but they differ both in respect to the search method and in respect to the criterion to determine consistency with the data.

The stepwise search method is straightforward and efficient, but suffers from the weakness that it may overlook good models. Its sampling properties seem to be intractable.

The EH-procedure, which uses a more complex search algorithm, selects multiple models and is more global in its approach. That it selects multiple models may be regarded as an advantage in that this reflects inherent model ambiguity. The analyst is forced to consider and compare the models selected, which may often be salutary. It uses tests for overall goodness-of-fit rather than tests between successive models; as a result, the models selected by the EH-procedure will typically be simpler than those selected by stepwise methods. Arguably, this use of overall goodness-of-fit tests is the greatest weakness of the approach. Particularly with high-dimensional contingency tables, such tests are known to be less reliable than tests based on the deviance differences.

Selection by optimizing one of the information criteria is conceptually straightforward, but may be computationally infeasible for larger model families.

The sampling properties of the EH-procedure have been studied by Smith (1992). If one assumes that there is one true model underlying the data, say m_t, then the model family can be partitioned into the models that contain m_t and hence should be accepted, and those that do not and hence should be rejected. Type I error can be defined as false rejection, i.e., rejection of a "true" model, and Type II error can be defined as false acceptance, i.e., acceptance of a "false" model.

Suppose that $(1 - \alpha)$-level tests are employed and consider the upwards-only version of the procedure, which starts by fitting the minimum model in the family, and at each stage fits the minimal undetermined models. In other words, it first fits the main effects model, then models with one edge, then models with two edges that both were rejected at the first step, and so on. As noted by Havránek (1987) and Smith (1992), this procedure is closed in the sense of Marcus, Peritz, and Gabriel (1976), and so controls the familywise type I error rate, i.e.,

$$\Pr(\#\text{false rejections} > 0) < \alpha.$$

Smith (1992) compared the upwards-only and downwards-only versions of the procedure by simulation using a variety of graphical Gaussian models and sample sizes. He found that the upwards-only version had smaller type I error than the downwards-only version, and as expected, the proportion of simulations in which true models were falsely rejected using the upwards-only version was approximately α. Conversely, the downwards-only version had smaller type II error than the upwards-only version. Smith also considered the standard (bidirectional) version and noted that this has properties intermediate between the upwards- and downwards-only versions. Since it is also the fastest, it seems a good compromise.

That the bidirectional version is the fastest can be seen by considering the six-dimensional contingency table analyzed in Section 2.2.4. Here, there are $p(p - 1)/2 = 15$ edges and hence $2^{15} = 32,768$ possible graphical models. The bidirectional version of the search procedure, using $\alpha = 0.05$, after fitting 28 models, selected two: ACE, ADE, BC, F

and AC, ADE, BC, BE, F. It is simple to calculate that there are 768 w-accepted models and 32,000 w-rejected models, so in this example the downwards-only version would involve fitting approximately 768 models and the upwards-only would involve fitting approximately 32,000 models. The figures are not precise since different solutions may be obtained.

6.5 Box-Cox Transformations

Now we leave the topic of model selection and turn to model criticism. An ever-present concern in the application of hierarchical interaction models is the appropriateness of the conditional Gaussian assumption; this is the focus of interest in the remainder of this chapter. In the present section, we extend the usefulness of the models by embedding them in a larger family of distributions. This follows the general approach of Box and Cox (1964).

As usual, we assume we have p discrete variables, Δ, and q continuous variables, Γ, and we write the vector of continuous random variables as $Y = (Y_1, \ldots, Y_q)'$. We assume that for one continuous variable (without loss of generality we can take this to be Y_1), it is not Y that is CG-distributed, but rather $Z = (g_\lambda(Y_1), Y_2, \ldots, Y_q)'$, where $g_\lambda(.)$ is the transformation

$$g_\lambda(y) = \begin{cases} \dfrac{(y^\lambda - 1)}{\lambda} & \text{if } \lambda \neq 0 \\ \ln(y) & \text{if } \lambda = 0 \end{cases},$$

where λ is an unknown transformation parameter. For $g_\lambda(y)$ to be well-defined, y must be positive. When $\lambda \to 0$, $g_\lambda(y) \to \ln(y)$, so that $g_\lambda(y)$ is a continuous function of λ.

Thus, we assume that for some unknown λ, the assumptions of the model are satisfied when Y_1 is replaced with $g_\lambda(Y_1)$. We now describe a technique to find the λ for which this holds. Since Z is CG-distributed, we know that the density of (I, Z) is given by

$$f(i, z) = \exp\{\alpha_i + \beta_i' z - \tfrac{1}{2} z' \Omega_i z\}, \tag{6.1}$$

where α_i, β_i, and Ω_i are the canonical parameters, possibly subject to model constraints. It follows that the density of (I, Y) is given as

$$f(i, y) = (\frac{dz_1}{dy_1}) f(i, z),$$
$$= y_1^{\lambda-1} f(i, z). \tag{6.2}$$

For each fixed value of λ, we transform the observed values of y_1, i.e., $y_1^{(k)}$ to $g_\lambda(y_1^{(k)})$, for $k = 1, \ldots, N$, and fit the model by maximum likelihood to the transformed data. Writing the maximized log likelihood thus obtained as $\ell_z(\lambda)$, we compare different values of λ by examining the profile log likelihood

$$\ell_y(\lambda) = (\lambda - 1) \sum_{k=1}^{N} \ln(y_1^{(k)}) + \ell_z(\lambda), \tag{6.3}$$

obtained from (6.2). This is essentially a one-parameter log likelihood function that could be handled in the usual way: for example, we could in principle find the estimate of λ that maximizes it. But, since we are only interested in simple transformations such as $\ln(y)$, y^{-1}, y^2, or $\sqrt{y}$, it is sufficient to calculate the profile log likelihood over a grid of suitable λ values.

We illustrate this using the digoxin clearance data (Section 3.1.4):

```
MIM->model //XYZ
MIM->fit
Calculating marginal statistics...
Deviance:        0.0000 DF: 0
MIM->boxcox X -2 2 4
Box-Cox Transformation of X:
               -2*loglikelihood -2*loglikelihood
    Lambda     (full model)     (current model)      Deviance

   -2.0000       770.5850          770.5850           0.0000
   -1.0000       726.8188          726.8188           0.0000
    0.0000       702.9458          702.9458           0.0000
    1.0000       711.6995          711.6995          -0.0000
    2.0000       750.0611          750.0611           0.0000
```

Values of $-2\ell_y(\lambda)$ are displayed for $\lambda = -2, -1, 0, 1, 2$. They are calculated for both the full and the current model. In this example, the current model is the full model, so the columns are identical. The minimum value is at $\lambda = 0$, indicating that a log transformation should be made.

It is sometimes useful to note that an approximate $100(1 - \alpha)\%$ confidence interval for λ consists of those values of λ for which the profile log likelihood is within $\frac{1}{2}\chi_{1-\alpha,1}^2$ of its maximum. For example, to construct a 95% confidence interval, we need to know which values of λ give rise to values of $-2\ell_y(\lambda)$ within 3.8 of its minimum, using $\chi_{0.95,1}^2 = 3.8$.

It should also be noted that the choice of λ is sensitive to outliers. Extreme values of $y_1^{(k)}$ will dominate the factor $(\lambda - 1) \sum_{k=1}^{N} \ln(y_1^{(k)})$ in Equation (6.3), often leading to unlikely estimates of λ. So if the technique suggests an implausible value of λ, the first thing to look for is the presence of outliers.

Each column can be used as the basis for choosing a value of λ. To illustrate this, we consider an example described by Box and Cox (1964). This concerns a 3×4 factorial experiment studying survival times (X) after treatment (A) and poison (B). The program fragment

```
mod ab/abx/x; boxcox x -2 2 8
```

gives the following output:

Lambda	-2*loglikelihood (full model)	-2*loglikelihood (current model)	Deviance
-2.0000	118.2175	149.7583	31.5409
-1.5000	115.5179	135.3324	19.8146
-1.0000	113.3732	128.1617	14.7885
-0.5000	111.8172	129.2898	17.4726
0.0000	110.8743	138.7643	27.8900
0.5000	110.5556	155.5891	45.0335
1.0000	110.8574	178.2841	67.4267
1.5000	111.7614	205.4623	93.7008
2.0000	113.2386	236.0609	122.8224

The full model assumes normality of errors and variance heterogeneity, while the current model constrains the variances to be homogeneous. Thus, choice of λ based on the first column will attempt to achieve normality within cells, and a choice based on the second column will attempt to achieve both normal errors and constant variances. Finally, choosing λ to minimize the deviance difference will attempt to stabilize the variances only, ignoring normality of errors. In the present example, a value of $\lambda = -1$, i.e., the reciprocal transformation, seems indicated (see Box and Cox, 1964 for a thorough discussion).

6.6 Residual Analysis

Another way to check the appropriateness of the conditional Gaussian assumption, along with other aspects of the model and the data, is to examine the residuals, i.e., the deviations between the observations and the predicted values under the model. Although residuals can be defined for discrete variables (see, for example, Christensen, 1990, p. 154 ff), we restrict attention to residuals for continuous variables only.

In this section, we examine some different types of residuals and show some simple techniques for examining residuals from different perspec-

tives, including plots against covariates and quantile-quantile plots of the
Mahalanobis distances.

In Section 6.7, a rather different form of residual analysis is presented.
There the focus of interest is on whether there is evidence of second-order
interactions between the continuous variables.

Suppose, as usual, that there are p discrete variables and q continuous
variables, and that we partition the continuous variables into q_1 responses
and $q_2 = q - q_1$ covariates, corresponding to $\Gamma = \Gamma_1 \cup \Gamma_2$. The covariates
can be fixed by design, or we can choose to condition on them in order to
focus attention on the remaining q_1 responses.

If we denote the corresponding observed random variables as (I, X, Y),
then the conditional distribution of Y given $I = i$ and $X = x$ is multi-
variate normal with conditional mean $E(Y|I = i, X = x)$ and covariance
$\text{Var}(Y|I = i, X = x)$, where these are

$$E(Y|I = i, X = x) = (\Omega_i^{11})^{-1}(\beta_i^1 - \Omega_i^{12}x)$$

and

$$\text{Var}(Y|I = i, X = x) = (\Omega_i^{11})^{-1}.$$

If we define a random variable R, the "true" residual, by

$$R = Y - (\Omega_I^{11})^{-1}(\beta_I^1 - \Omega_I^{12}X), \tag{6.4}$$

then since the conditional distribution of R given $I = i$ and $X = x$
is $\mathcal{N}(0, (\Omega_i^{11})^{-1})$, which does not involve x, it follows that $R \perp\!\!\!\perp X \mid I$.
Moreover, if Ω_i^{11} does not depend on i, then $R \perp\!\!\!\perp (I, X)$.

It is useful to derive the independence graph of (I, X, R) obtained when Y
is transformed to R. We call this the *residuals graph*. Suppose, for example,
that the current model is homogeneous with the graph shown in Figure 6.4.
There are three continuous variables, $\Gamma = \{Y_1, Y_2, Y_3\}$, and three discrete
variables, $\Delta = \{A, B, C\}$. Let $\Gamma_1 = \Gamma$, i.e., we set $Y = (Y_1, Y_2, Y_3)'$ and
transform to $R = Y - E(Y|I, X)$. Since the model is homogeneous, $R \perp\!\!\!\perp I$

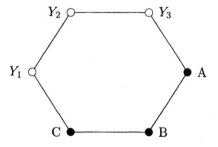

FIGURE 6.4. An independence graph.

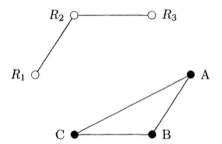

FIGURE 6.5. The residuals graph corresponding to Figure 6.3.

and so in the derived independence graph of (I, R) shown in Figure 6.5, the discrete variables are unconnected with $R = (R_1, R_2, R_3)'$. The covariance of R is the same of that of Y so the subgraph $\mathcal{G}_\Gamma$ remains unchanged. Finally, the subgraph $\mathcal{G}_\Delta$ shows the independence structure of the discrete variables: using collapsibility arguments, we see that we must add the edge $[AC]$ so that the boundary of $\mathcal{G}_\Gamma$ in the original graph is made complete.

For a general homogeneous model, the residuals graph is formed by completing the boundary of every connected component of $\mathcal{G}_{\Gamma_1}$, and removing all edges between $\Delta \cup \Gamma_2$ and Γ_1.

In practice, we cannot observe the true residuals R, but we calculate the observed residuals and we can informally check whether the relations shown in the residuals graph appear to hold. For example, we may plot a residual against a covariate x: if the model is homogeneous, the two should be approximately independent. If, however, the model is misspecified, and in fact the dependence on x is nonlinear, then this may be evident as a tendency to curvature in the plot.

Two types of residual are important. If we write the kth observation as $(i^{(k)}, x^{(k)}, y^{(k)})$, then the *(ordinary) residual* for that observation is

$$\hat{r}^{(k)} = y^{(k)} - \hat{\mu}_{y|i^{(k)}, x^{(k)}},$$

where $\hat{\mu}_{y|i^{(k)}, x^{(k)}}$ is the estimate of $E(Y|I = i^{(k)}, X = x^{(k)})$, obtained from fitting the current model to the data. Note that all observations are used in this fit.

If, alternatively, we exclude the kth observation when estimating the conditional mean so as to obtain a modified estimate $\tilde{\mu}_{y|i^{(k)}, x^{(k)}}$, we can use this to calculate the *deletion residuals*, i.e.,

$$\tilde{r}^{(k)} = y^{(k)} - \tilde{\mu}_{y|i^{(k)}, x^{(k)}}.$$

These will tend to reflect large deviations from the model more dramatically. If the deletion residual for an observation is substantially greater than the (ordinary) residual, then this suggests that the observation has a large influence on the estimates.

Residuals of Y

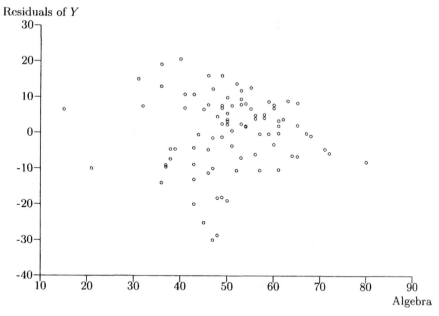

FIGURE 6.6. A residual-covariate plot based on the mathematics marks data. The deletion residual of Analysis (Y) given V, W, X and Z is plotted against Algebra (X). The conditional variance of Y appears to decrease with increasing X.

We can illustrate a covariate plot using the mathematics marks data. The program fragment

```
MIM->model //VWX,XYZ; fit
Deviance:      0.8957 DF: 4
MIM->fix VWXZ; residual -R
Fixed variables: VWXZ
```

calculates the deletion residuals of Analysis (Y) and stores them in R. Figure 6.6 shows a plot of these versus Algebra (X).

In the fragment, the `Residuals` command is used to calculate the residuals and store them in R. The minus sign before R specifies that the deletion residuals are to be calculated; otherwise, the (ordinary) residuals are calculated. Variables that are fixed (using the command `Fix`) are treated as covariates in the calculations. In the fragment, V, W, X, and Z are treated as covariates.

Similarly, if under the model $//VWX, XYZ$ (see Figure 3.2), we want to examine the conditional independence of Mechanics (V) and Analysis (Y) more closely, we can calculate the residuals of V and Y given W, X, and

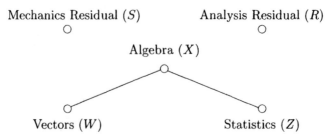

FIGURE 6.7. The residuals of the butterfly graph.

Z. From the residuals graph shown in Figure 6.7, we see that the residuals should be approximately independent.

We examine this by first calculating and storing the residuals as variates R and S, as shown in the following fragment:

```
MIM->model //VWX,XYZ; fit
Deviance:        0.8957 DF: 4
MIM->fix WXZ; resid -RS
Fixed variables: WXZ
```

Figure 6.8 shows a plot of R versus S: the residuals do appear to be independent.

Another useful way to display the residuals is in index plots. Here, residuals are plotted against observation number: this indicates whether there were any trends in the data collection process or similar anomalies.

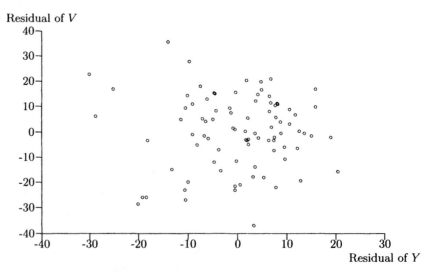

FIGURE 6.8. A plot of the deletion residuals of Mechanics (V) and Analysis (Y).

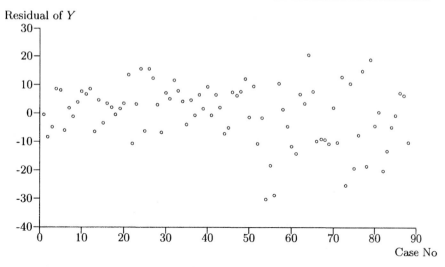

Residual of Y

FIGURE 6.9. A index plot of residuals based on the mathematics marks data. There appears to be a change after observation 52 or thereabouts. A similar finding was described in Section 4.6.4.

The following fragment generates data for the index plot shown in Figure 6.9.

```
MIM->model //VWX,XYZ; fit
Deviance:      0.8957 DF: 4
MIM->fix VWXZ; calc O=obs; label O "Case No"
Fixed variables: VWXZ
MIM->resid -R
```

Another way to examine residuals is to calculate the Mahalanobis distance for each case, i.e.,

$$\hat{d}^{(k)} = \hat{r}^{(k)\prime} \hat{\Omega}_i^{11} \hat{r}^{(k)}$$

for the ordinary residuals, and

$$\tilde{d}^{(k)} = \tilde{r}^{(k)\prime} \tilde{\Omega}_i^{11} \tilde{r}^{(k)}$$

for the deletion residuals. If the model is true, then these are approximately χ^2 distributed with q_1 degrees of freedom. The distances can, for example, be plotted against the corresponding quantiles of the χ^2 distribution (a Q-Q plot); this should be approximately linear.

Mahalanobis distances are obtained using the command Mahalanobis, which calculates the distances and the corresponding quantiles of the χ^2 distribution. For example, the fragment

```
MIM->model //VWX,XYZ; fit; mahal mc
```

Mahal dist

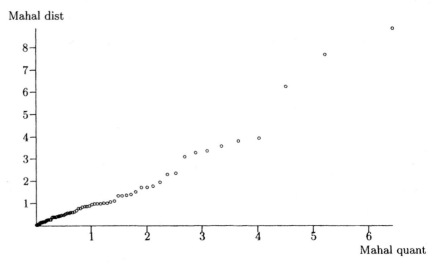

FIGURE 6.10. A Q-Q plot of Mahalanobis' distances based on the mathematics marks data.

calculates the distances and the corresponding quantile and stores them in two variates, m and c: Figure 6.10 shows a plot of m versus c.

We now turn to heterogeneous mixed models. If Ω_i^{11} in Equation (6.4) *does* depend on i, then the variances of the residuals will depend on the discrete variables; this may distort plots of residuals against covariates. In this case, it is useful to employ standardized residuals. These are defined for univariate residuals ($q_1 = 1$), in other words, when we focus on one response variable by conditioning on all the remaining ones. If we define

$$R_s = R\sqrt{\omega_i^{11}},$$

then we see that $R_s \sim \mathcal{N}(0,1)$ so that $R_s \perp\!\!\!\perp (I,X)$. The observed quantities are similarly defined in the obvious way as

$$\hat{r}_s^{(k)} = \hat{r}^{(k)}\sqrt{\hat{\omega}_i^{11}}$$

and

$$\tilde{r}_s^{(k)} = \tilde{r}^{(k)}\sqrt{\tilde{\omega}_i^{11}}.$$

(We could similarly define standardized residuals for $q_1 > 1$, but since this would involve a matrix transformation of $\tilde{e}^{(k)}$, they would be less interpretable in terms of the original variables and thus less suitable for the present purpose.)

Since $\hat{d}^{(k)} = (\hat{r}_s^{(k)})^2$ and $\tilde{d}^{(k)} = (\tilde{r}_s^{(k)})^2$, we can calculate the standardized residuals as

$$\hat{r}_s^{(k)} = \sqrt{\hat{d}^{(k)}}\mathrm{sign}(\hat{r}^{(k)})$$

and similarly, the deletion standardized residuals as

$$\tilde{r}_s^{(k)} = \sqrt{\tilde{d}^{(k)}}\mathrm{sign}(\tilde{r}^{(k)}).$$

To illustrate this, we consider the lipid data (Section 4.1.10) and the model whose graph is shown in Figure 4.3. The conditional distribution of Y given the remaining variables is heterogeneous. To calculate the standardized partial residuals we proceed as follows:

```
MIM->mod A/AV,AW,X,AY,Z/WXZ,VWXY,AVWY; fit; fix VWXZ
Deviance:      32.1893 DF: 24
MIM->resid R; mahal MC
MIM->calc S=sqrt(M)*((R>0)-(R<0))
```

The plot of S versus X is shown in Figure 6.11. No departure from independence is evident.

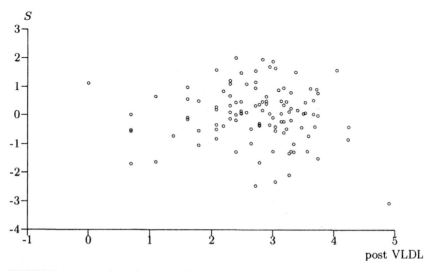

FIGURE 6.11. A plot of standardized residuals.

6.7 Dichotomization

In this section, we introduce an approach to model checking through dichotomization of the continuous variables.

To motivate this, we first observe that the multivariate normal distribution is in some ways analogous to a loglinear model with zero three-factor interactions. This can be seen, for example, through the density (3.10) in which only pairwise interactions between the continuous variables are allowed. Similarly, if $Y = (Y_1, Y_2, Y_3)'$ is multivariate normal distributed, then the conditional distribution of, say, $(Y_1, Y_2)'$ given $Y_3 = y_3$, is again normal, with a conditional mean that depends linearly on y_3, but with a covariance that does not depend on y_3. An analogous property of the no three-factor interaction loglinear model AB, AC, BC is that the conditional cross-product ratio between A and B is constant over levels of C, i.e.,

$$\frac{m_{111}m_{221}}{m_{121}m_{211}} = \frac{m_{112}m_{222}}{m_{122}m_{212}}.$$

These analogies suggest that one type of departure from multivariate normality that is of interest to check is precisely the presence of second-order interactions between variables, and that a way to do this could be to dichotomize the variables and check for the presence of three-factor interactions in the resulting contingency table.

We first describe a method along these lines proposed by Cox and Wermuth (1994) as a test for multivariate normality. We then introduce a similar but more powerful method based on symmetry tests. Finally, we sketch how the methods can be applied to homogeneous mixed models.

Consider a trivariate normal distribution involving random variables V, W, and X, and define three binary random variables A, B, and C by dichotomizing the continuous variables at their means, for example,

$$A = \begin{cases} 1 & \text{if } V < \mu^v \\ 2 & \text{otherwise} \end{cases}$$

and similarly for B and C. Then McFadden (1955) showed that their joint distribution is given explicitly through

$$p_{ijk} = \frac{1}{8} + (-1)^{i+j-2}\nu_{AB} + (-1)^{i+k-2}\nu_{AC} + (-1)^{j+k-2}\nu_{BC},$$

where, for instance,

$$\nu_{AB} = (4\pi)^{-1}\sin^{-1}\rho^{vw}$$

and ρ^{vw} is the marginal correlation between V and W. Moreover, this distribution is such that A, B, and C exhibit no three-factor interaction. Cox and Wermuth (1994) propose therefore that this be used to examine possible divergence from multivariate normality. For each triad of continuous variables, the corresponding binary variables are constructed by dichotomizing the variables at their sample means (or medians) and testing for the presence of three-factor interaction between the binary variables. We illustrate this using the mathematics marks data:

```
MIM->pr s
Empirical counts, means and covariances.
     V   302.293
     W   125.777 170.878
     X   100.425  84.190 111.603
     Y   105.065  93.597 110.839 217.876
     Z   116.071  97.887 120.486 153.768 294.372
  Means  38.955  50.591  50.602  46.682  42.307    88.000
              V       W       X       Y       Z    Count
MIM->fact A2B2C2D2E2
MIM->calc A=1+(V>38.955); calc B=1+(W>50.591); calc C=1+(X>50.602)
MIM->calc D=1+(Y>46.682); calc E=1+(Z>42.307)
MIM->mod AB,AC,BC; fit; test
Calculating marginal statistics...
Deviance:       0.0053 DF: 1
Test of H0: AB,AC,BC
against H:  ABC
LR:    0.0053    DF:   1    P: 0.9418
```

First, the binary variables A to E are calculated by dichotomizing the variables V to Z, and then the likelihood ratio test for zero three-factor interaction is performed. Continuing in this fashion, we obtain Table 6.1.

We see that there is a suggestion of second-order interaction between V (Mechanics), X (Algebra), and Y (Analysis). According to the butterfly graph (Section 3.2), $V \perp\!\!\!\perp Y \mid X$, so this result suggests that the conditional

Continuous variables	Dichotomized variables	Deviance	P-value
V, W, X	A, B, C	0.0053	0.9418
V, W, Y	A, B, D	2.5624	0.1094
V, W, Z	A, B, E	0.5666	0.4516
V, X, Y	A, C, D	8.6575	0.0033
V, X, Z	A, C, E	0.1394	0.7089
V, Y, Z	A, D, E	0.0761	0.7827
W, X, Y	B, C, D	2.5481	0.1104
W, X, Z	B, C, E	0.1359	0.7124
W, Y, Z	B, D, E	0.2972	0.5856
X, Y, Z	C, D, E	0.2827	0.5950

TABLE 6.1. Tests for zero three-factor interaction for each triad of variables.

association between V and Y given X is in fact nonzero, perhaps due to some nonlinearity.

We now consider a modification to Cox and Wermuth's approach. Suppose we have a q-variate normal distribution and that the variables are dichotomized at their means. Lemma 2 of Appendix C states that the joint distribution of the dichotomized variables follows a nonhierarchical loglinear model in which all interactions involving an odd number of factors vanish. In other words, the one-factor interactions (i.e., main effects), three-factor interactions, and so on are all zero. This generalizes the result of McFadden mentioned above, namely that when $q = 3$, the three-factor interaction term vanishes.

Moreover, a goodness-of-fit test for this model is easily derived. As shown in Appendix C, this is essentially a test of multivariate symmetry. For example, we may apply the test to the dichotomized mathematics marks data (using the command SymmTest):

```
MIM->model ABCDE; symmtest
Test of multivariate symmetry:  40.8424  DF:   15  P:   0.0003
```

This indicates that the original variables are asymmetrically distributed about the sample means.

We can also apply the symmetry test to the marginal distribution of each triad of variables in order to compare with the three-factor interaction approach. Table 6.2 shows the results. We would expect the symmetry test to be more powerful since the no three-factor interaction models contain main effects and so are overparametrized. Comparing Table 6.2 with Table 6.1 confirms this clearly.

Continuous variables	Dichotomized variables	Symmetry Test	P-value
V, W, X	A, B, C	3.5598	0.4688
V, W, Y	A, B, D	9.4351	0.0511
V, W, Z	A, B, E	5.3941	0.2492
V, X, Y	A, C, D	21.9078	0.0002
V, X, Z	A, C, E	5.1066	0.2765
V, Y, Z	A, D, E	12.3579	0.0149
W, X, Y	B, C, D	12.9615	0.0115
W, X, Z	B, C, E	2.2890	0.6828
W, Y, Z	B, D, E	10.6461	0.0308
X, Y, Z	C, D, E	13.2029	0.0103

TABLE 6.2. Symmetry tests for each triad of variables.

Continuous variables	Dichotomized variables	Symmetry Test	P-value
V	A	2.2368	0.1348
W	B	0.1819	0.6698
X	C	0.1819	0.6698
Y	D	5.5588	0.0184
Z	E	1.6415	0.2001

TABLE 6.3. Symmetry tests for each variable.

Remarkably, each triad containing D exhibits asymmetry. If we calculate the univariate symmetry tests, we obtain Table 6.3.

This suggests that the asymmetries detected above are due to skewness in the marginal distribution of Y. We can check this by applying the symmetry test to the marginal distribution of the remaining variables, i.e., to the joint distribution A, B, C, and E. The test statistic is 6.5551 on eight degrees of freedom, so there is no evidence of asymmetry. So it does seem that the asymmetries are simply due to skewness in the marginal distribution of Y.

We now describe briefly how these methods can be applied to homogeneous mixed models. Consider such a model and suppose we wish to focus on the multivariate normal assumption for a q_1-vector of responses Y, given (I, X) where X is a q_2-vector of covariates. This is the framework adopted in Section 6.6, where we derived the residuals graph from the original independence graph. If we now dichotomize the variables at their conditional means, or in other words, dichotomize the residuals at zero, we can further derive the graph of (I, X, D), where D is the q_1-tuple of dichotomized variables. For example, Figure 6.12 shows the dichotomized graph corresponding to Figure 6.5.

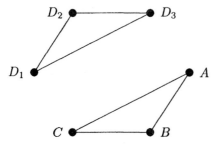

FIGURE 6.12. The dichotomized residuals graph corresponding to Figure 6.5.

To see this, note that whenever we dichotomize a variable in a separating set in a conditional independence relation, the conditional independence is *not* retained, as illustrated below:

In other words, an artificial dependence is introduced. In the notation of Section 6.6, we can in general derive the dichotomized residuals graph from the residuals graph by simply completing every connected component of $\mathcal{G}_{\Gamma_1}$. It is then easy to see how to apply the model checking techniques described above: we simply apply them to every connected component of the dichotomized variables.

7

Directed Graphs and Their Models

Up to now the focus of this book has been on the undirected graphs and models. This chapter describes a variety of other types of independence graph and their associated models. In common for almost all of these is that some or all of the edges are drawn as arrows, indicating direction of influence or, sometimes, causal direction. The first section treats graphs with arrows only, the so-called DAGs: such graphs have a long history, starting in path analysis (Wright, 1921) and extending to categorical data (Goodman, 1973, and Wermuth and Lauritzen, 1983). The next section describes graphs of more recent origin that have both lines (i.e., undirectional edges) and arrows, the so-called *chain graphs*. These are appropriate when the variables can be grouped into blocks, so that the variables within a block are not ordered, but there is a clear ordering between the blocks. In the remaining sections we examine more briefly some other types of graphs. These include *local independence graphs*, which appear to be useful in analyzing stochastic processes; *covariance graphs*, in which marginal rather than conditional independences are represented; and *reciprocal graphs*, which capture the independence properties of simultaneous equation systems.

To introduce these graphs, we can perform a small thought experiment. We suppose that we are market researchers who are studying the prospects of a new instant noodle product. We are interested in finding out who, if anyone, likes noodles, and to do this we interview a representative sample of people, recording their race, gender, and response to the question "Do you like noodles?" Let us suppose that the results are as shown in Table 7.1.

If we apply the joint loglinear models described in Chapter 2, we find that the simplest model consistent with these data is the one shown in Figure 7.1.

Race	Gender	Do you like noodles?	
		Yes	No
Black	Male	32	86
	Female	35	121
White	Male	61	73
	Female	42	70

TABLE 7.1. The fictive noodles data.

But this model is obviously inappropriate. How can we suppose that race and gender are conditionally independent *given the response*? Surely the respondents' race and gender, characteristics determined decades before, cannot be affected by whether or not they are partial to noodles. Race and gender might be marginally independent, but they can hardly be conditionally independent given the response.

The problem arises because we have not taken the ordering of the variables into account. Here race and gender are clearly prior to the response. If we reanalyse the data using directed graphs and the associated models (described below), then we obtain the graph shown in Figure 7.2. This resembles the previous graph, except that the edges are replaced by arrows pointing towards the response. As we shall see, directed graphs have different rules for derivation of conditional independence relations. Now the missing edge between between race and gender means that they are marginally independent, not conditionally independent given the response.

The example illustrates another simple but very important point. This is that the arrows need not represent causal links. Here, for example, it would not be appropriate to regard gender and race as *causes*. They are not manipulable, in the sense that they cannot be regarded as an intervention or treatment. It is not possible to think of a given person with a different race: if "you" were of a different race, "you" would be a completely different person. It may be legitimate to call gender and race *determinants* of the response, if this term is understood in a purely descriptive sense (for example, indicating merely that males and females have, in general, different responses). But the term *cause* seems to imply a type of relation that is not appropriate here. See Chapter 8 for further discussion.

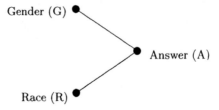

FIGURE 7.1. The undirected graph showing $G \perp\!\!\!\perp R|A$.

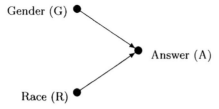

Gender (G)

Answer (A)

Race (R)

FIGURE 7.2. The directed graph showing $G \perp\!\!\!\perp R$.

The graph shown in Figure 7.2 has only directed edges (arrows), and so falls within the class of directed graphs described in the following section. But if we supposed that gender and race were *not* marginally independent, then we would be hard put to say which is prior to which. It would be natural to put them on an equal footing, by connecting them with a line, not an arrow. This would be a simple chain graph, as described in Section 7.2. Removing the line between gender and race in the chain graph setting would also lead to Figure 7.2, so this graph can also be considered a simple chain graph: the two classes of graphs intersect, and the graph shown in Figure 7.2 lies in the intersection.

7.1   Directed Acyclic Graphs

The graphs we first consider are *directed*, that is, contain only directed edges (drawn as *arrows*, not as lines). So we can again write a graph as a pair $\mathcal{G} = (\mathcal{V}, \mathcal{E})$, where $\mathcal{V}$ is a set of vertices and $\mathcal{E}$ is a set of edges, but where now we identify edges with *ordered* pairs of vertices. If there is an arrow from v to w then we write this as $v \rightarrow w$, or equivalently as $[vw] \in \mathcal{E}$. Note that $[vw]$ is not the same as $[wv]$.

If $v \rightarrow w$ or $w \rightarrow v$, we say that v and w are *adjacent* and write $v \sim w$. By a *path* we mean a sequence of vertices $\{v_1, \ldots, v_k\}$ such that $v_i \sim v_{i+1}$ for each $i = 1, \ldots, k-1$. In contrast, for a *directed path* from v_1 to v_k we require that $v_i \rightarrow v_{i+1}$ for each $i = 1, \ldots, k-1$. When the first and last vertices coincide, i.e., $v_1 = v_k$, the directed path is called a *directed cycle*.

We restrict attention to directed graphs with no directed cycles. These are usually known as *directed acyclic graphs*, or *DAGs* for short. (As Andersen et al. (1997) point out, it would be more logical to call them acyclic directed graphs: but then the acronym would be difficult to pronounce. Here we keep to the conventional expression.)

Figure 7.3 shows two directed graphs. The first is acyclic, but the second is not.

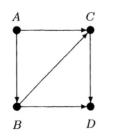 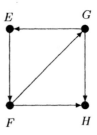

FIGURE 7.3. Two directed graphs. The first is a DAG, the second is not.

If $v \to w$, then v is called a *parent* of w and w is called a *child* of v. The set of parents of w is denoted pa(w) and the set of children as ch(w).

If there is a directed path from v to w, then v is called an *ancestor* of w and w is called a *descendent* of v. The set of ancestors of w is denoted an(w) and the set of descendents as de(w).

These four definitions (of parents, children, ancestors, and descendents) can easily be extended to apply to sets of nodes. For example, for a set $S \subseteq V$ we define pa(S) = $\{\bigcup_{v \in S} \text{pa}(v)\} \backslash S$, that is to say, as the set of nodes not in S that are parent to a node in S. The other definitions are extended similarly. Furthermore, we define $\text{an}^+(S) = S \cup \text{an}(S)$ to be the *ancestral set* of S.

It is not difficult to show that the absence of any directed cycles is equivalent to the existence of an ordering of the nodes $\{v_1, \dots, v_n\}$ such that $v_i \to v_j$ only when $i < j$. In other words, there exists a numbering of the nodes so that arrows point only from lower-numbered nodes to higher-numbered nodes. Of course, the numbering is not necessarily unique. A DAG with n nodes and no edges is compatible with all $n!$ orderings, and a complete DAG is compatible with only one. The first DAG in Figure 7.3 is compatible with one ordering, namely $A \to B \to C \to D$. Figure 7.2 is compatible with two orderings, $G \to R \to A$ and $R \to G \to A$.

Although from a graph-theoretic point of view the natural perspective is to consider which orderings are compatible with a given DAG, from the perspective of an applied modeller the natural starting point is an a priori ordering of the variables. So we assume that subject-matter knowledge tells us that the variables can be labelled $v_1, \dots, v_n$ such that v_i is prior to v_{i+1} for $i = 1, \dots, n-1$. Corresponding to this ordering, we can factorize the joint density of $\{v_1, \dots, v_n\}$ as

$$f(v_1)f(v_2|v_1) \dots f(v_n|v_{n-1}v_{n-2} \dots v_1). \tag{7.1}$$

In constructing a DAG, an arrow is drawn from v_i to v_j, where $i < j$, unless $f(v_j|v_{j-1} \dots v_1)$ does not depend on v_i, in other words, unless

$$v_i \perp\!\!\!\perp v_j \mid \{v_1 \dots v_j\} \backslash \{v_i, v_j\}. \tag{7.2}$$

This is the key difference between DAGs and undirected graphs. In both types of graph a missing edge between v_i and v_j is equivalent to a conditional independence relation between v_i and v_j; in undirected graphs, they are conditionally independent given all the remaining variables, whereas in DAGs, they are conditionally independent given all *prior* variables. Thus in Figure 7.2 the missing arrow between G and R means that $G \perp\!\!\!\perp R$, not that $G \perp\!\!\!\perp R|A$.

Having constructed the DAG from (7.2), we can rewrite the joint density (7.1) more elegantly as

$$\Pi_{v \in \mathcal{V}} f(v|\text{pa}(v)) \tag{7.3}$$

and the pairwise conditional independence relations corresponding to a missing arrow between v_i and v_j as

$$v_i \perp\!\!\!\perp v_j \mid \text{an}(\{v_i, v_j\}). \tag{7.4}$$

These expressions do not make use of any specific vertex ordering.

7.1.1 Markov Properties of DAGs

Markov properties on directed acyclic graphs have been the subject of much recent research, including Kiiveri et al. (1984), Pearl and Paz (1986), Pearl and Verma (1987), Smith (1989), Geiger and Pearl (1988, 1993) and Lauritzen et al. (1990).

Up to now we have only used DAGs to represent pairwise independences, as in (7.2); this is the DAG version of the pairwise Markov property. The natural question arises: Can we deduce any stronger conditional independence relations from a DAG? In other words, is there an equivalent of the global Markov property for DAGs? For example, in Figure 7.4 there is no arrow from B to D. The pairwise Markov property states that $B \perp\!\!\!\perp D|\{A, C\}$; but does it also hold that $B \perp\!\!\!\perp D|C$? Intuitively, this would seem likely.

For undirected graphs, we saw that a simple criterion of separation in the graph-theoretic sense was equivalent to conditional independence in the statistical sense. A similar result is true of DAGs, though the graph-theoretic property, usually called *d-separation*, is alas somewhat more difficult to grasp.

There are actually two different formulations of the criterion. The original formulation is due to Pearl (1986a, 1986b) and Verma and Pearl (1990a,

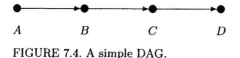

$A \qquad B \qquad C \qquad D$

FIGURE 7.4. A simple DAG.

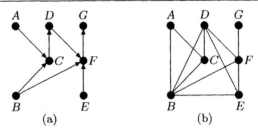

FIGURE 7.5. (a) shows a DAG $\mathcal{G}$ and (b) shows its moral graph $\mathcal{G}^m$, which is formed by marrying parents in $\mathcal{G}$ and then deleting directions. In $\mathcal{G}$ we see that pa(C) = $\{A, B\}$ and pa(F) = $\{B, D, E\}$.

1990b); shortly after, Lauritzen et al. (1990) gave an alternative formulation. We give both here, since each has its own advantages. The later one is conceptually simpler, but the original one builds on concepts that are useful in other contexts (for example, see Section 8.3.3).

We first look at the later version of the criterion. To do this, we need to define something called a *moral graph*. Given a DAG $\mathcal{G} = (\mathcal{V}, \mathcal{E})$, we construct an undirected graph $\mathcal{G}^m$ by marrying parents and deleting directions, that is,

1. For each $v \in \mathcal{V}$, we connect all vertices in pa(v) with lines.

2. We replace all arrows in $\mathcal{E}$ with lines.

We call $\mathcal{G}^m$ the moral graph corresponding to $\mathcal{G}$. Figure 7.5 shows a DAG and its moral graph.

Now suppose that we want to check whether $v_i \perp\!\!\!\perp v_j | S$ for some set $S \subseteq \mathcal{V}$. We do this in two steps. The first step is to consider the ancestral set of $\{v_i, v_j\} \cup S$ (see Section 7.1), that is, an$^+(\{v_i, v_j\} \cup S) = \mathcal{A}$, say. From (7.3), since for $v \in \mathcal{A}$, pa(v) $\in \mathcal{A}$, we know that the joint distribution of $\mathcal{A}$ is given by

$$\Pi_{v \in \mathcal{A}} f(v|\text{pa}(v)), \tag{7.5}$$

which corresponds to the subgraph $\mathcal{G}_{\mathcal{A}}$ of $\mathcal{G}$. This is a product of factors $f(v|\text{pa}(v))$, that is, involving the variables $v \cup \text{pa}(v)$ only. So it factorizes according to $\mathcal{G}_{\mathcal{A}}^m$, and thus the global Markov properties for undirected graphs (see Section 1.3) apply. So, if S separates v_i and v_j in $\mathcal{G}_{\mathcal{A}}^m$, then $v_i \perp\!\!\!\perp v_j | S$.

This is the required criterion. We illustrate its application to Figure 7.5(a). Suppose we want to know whether $C \perp\!\!\!\perp F | D$ under this graph. To do this, we first form in Figure 7.6(a) the subgraph $\mathcal{G}_{\mathcal{A}}$ corresponding to $\mathcal{A} = \{A, B, C, D, E, F\}$, and then in Figure 7.6(b) its moral graph. In (b), D does not separate C from F, so $C \not\perp\!\!\!\perp F | D$.

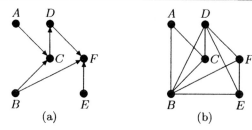

FIGURE 7.6. Applying the d-separability criterion.

The criterion is easily extended to sets of variables, in the following sense. The directed version of the *global Markov property* states that for three disjoint sets S_1, S_2, and S_3, $S_1 \perp\!\!\!\perp S_2 | S_3$ whenever S_3 separates S_1 and S_2 in $\mathcal{G}_{\mathcal{A}}^m$, where $\mathcal{A} = \mathrm{an}^+(S_1 \cup S_2 \cup S_3)$.

The pairwise and global Markov properties are equivalent under very general conditions (see Lauritzen, 1996, p. 51). In other words, when we construct a DAG using the pairwise properties, the criterion can be used to derive stronger conditional independence relations. Furthermore, *all* such conditional independencies can be derived. That is to say, all conditional independencies that hold for all densities in the model can be derived using the criterion. (There may be specific densities that obey additional conditional independencies not implied by the graph.)

We now turn to the original version of the criterion. This focusses on individual paths between vertices. In *undirected* graphs, the existence of a path between V and W, say, indicates that they are marginally dependent. If we are interested in the conditional dependence of V and W given a set S, then if the path does *not* contain a node in S, it (still) indicates conditional dependence. If it does contain such a node, then it is not clear whether conditional independence applies or not. However, if *all* paths between V and W contain a node in S, then S separates V and W, so that these are conditionally independent given S.

A similar argument applies to DAGs, though here it is crucial to distinguish a certain type of configuration on a path. We call a node on a path a *collider* if it has converging arrows. Consider the DAGs shown in Figure 7.7: both have paths from V to W. We examine various (in)dependence relations between V and W that are associated with these paths, keeping in mind that when these graphs are imbedded in larger graphs, the independences we find here may vanish, but the dependences will still hold.

In Figure 7.7(a), the path contains no colliders, and we have that $V \not\perp\!\!\!\perp W$, but that $V \perp\!\!\!\perp W | X$ and $V \perp\!\!\!\perp W | Y$. We can say that the path indicates that V and W are marginally dependent, but that the path can be *blocked* by conditioning on the noncolliders X or Y.

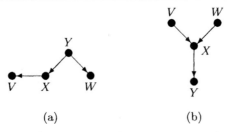

FIGURE 7.7. Two DAGs showing a path between V and W. In (a), there are no colliders, in (b) there is one collider (X).

In Figure 7.7(b) the opposite is true. The path contains a collider, and we have that $V \perp\!\!\!\perp W$, but that $V \not\!\perp\!\!\!\perp W|X$ and $V \not\!\perp\!\!\!\perp W|Y$. So the path does not indicate marginal dependence, since it contains a collider; however, if we condition on the collider or on a descendent of the collider, the path *does* indicate dependence between V and W.

Putting these ideas together, we say that a path between V and W can be *active* or *blocked*. Being active means that it indicates a dependence between V and W. A path is blocked if either (i) it has a noncollider that is conditioned on, or (ii) it has a collider that is not conditioned on (and none of its descendents are conditioned on either).

We are now ready to state the d-separation criterion in its original formulation (Pearl, 1986a, 1986b; Verma and Pearl, 1990a, 1990b). We seek to define the d-separation of sets S_1 and S_2 by S_3. We consider paths between a vertex in S_1 and S_2. We say that S_3 *blocks* such a path if either (i) the path has a noncollider, say x, such that $x \in S_3$, or (ii) the path has a collider, say y, such that $y \notin S_3$ and $de(x) \cap S_3 = \emptyset$. The criterion states that S_3 d-separates S_1 and S_2 if S_3 blocks all paths between S_1 and S_2.

7.1.2 Modelling with DAGs

In a sense, modelling with DAGs is quite straightforward. Since the conditional densities $f(v_j|v_{j-1} \ldots v_1)$ can be specified freely, any appropriate univariate response models can be used. That is to say, for each j, we can model the dependence v_j on the prior variables $v_1 \ldots v_{j-1}$ using any model in which $v_1 \ldots v_{j-1}$ are included as covariates; if only a subset are included, then v_j depends only on this subset. There is, of course, a huge variety of univariate response models that may be applied: for example, generalized linear models (McCullagh and Nelder, 1989) and regression models for ordinal response variables (Agresti, 1984). Different types of model may be used for each step. Furthermore, standard techniques from univariate models for model criticism, residual analysis, and so on can be applied. This makes for great flexibility of modelling, but it makes it impossible to cover in a book such as this — the class of possible models is huge.

One aspect deserves emphasis. This is that the choice of model at each step is quite independent (logically and statistically) of the choice of model at all other steps. In contrast to undirected graphs, in which the decision whether or not to include an edge depends on which other edges are present or absent, with DAGs the ordering of the variables introduces a great simplification. Here the decision to include an arrow $v \to w$ depends only on the presence or absence of other edges pointing at w; the presence or absence of all other edges in the DAG is quite immaterial.

In the following section we illustrate the use of DAGs to model a study involving longitudinal measurements of a discrete, ordinal response.

7.1.3 Example: Side Effects of Neuroleptics

Lingjærde et al. (1987) describe a randomised, double-blind parallel study comparing two antidepressant drugs. Fifty patients were treated with each drug. The patients were assessed at the end of a placebo washout period prior to active treatment, and again after one, two, and three weeks on the active drug. The present analysis concerns a single rating scale item relating to severity of a side effect. The severity levels are: not present, mild, moderate, and severe. The data form a contingency table with $2 \times 4^4 = 512$ cells.

This example illustrates several aspects that occur quite frequently in contingency table analysis. Firstly, there is the clear temporal sequence of the measurements, which we address by adopting the DAG modelling approach just described. Secondly, the underlying contingency table is very sparse, with only 100 observations distributed over 512 cells, so asymptotic tests cannot be used. Thirdly, the response is ordinal rather than nominal; that is to say, there is a natural ordering of the categories: not present, mild, moderate, and severe, so we should use tests sensitive to ordinal alternatives.

Since only discrete variables are involved, joint loglinear models can be used to model the conditional distributions. This is explained in more detail in Section 4.2. In effect, to condition on a set of variables in a loglinear model, all interactions between these variables must be included in any model considered. So for $f(v_j|v_{j-1} \ldots v_1)$ we apply a model to the marginal table for $\{v_1, \ldots, v_j\}$ that includes all interactions between $\{v_1, \ldots, v_{j-1}\}$. We do this first for $j = 2$, then for $j = 3$, and so on.

We return to the example to illustrate this process. The variables are labelled as follows: treatment group (G), baseline score (A), score after one week (B), score after two weeks (C), and score after three weeks (D). Since patients were allocated at random to treatment group after the baseline period, A is prior to G, so the order of the variables is $\{A, G, B, C, D\}$. First we define the variables and read in the data:

```
MIM->fact A4B4C4D4G2
MIM->label G "Drug" A "Baseline" B "Week 1" C "Week 2" D "Week 3"
MIM->setblocks A|G|B|C|D
MIM->read GDCBA
1 1 1 1 1 1 1 1 2 2 1 1 1 1 1 1 1 1 1 1 1 1 1 1 1 1 1 1 2 1 1 1 1
1 1 1 1 1 2 1 1 2 1 2 2 1 1 1 1 1 1 1 1 1 1 2 2 3 1 1 1 1 1 1 1
1 1 1 1 1 2 2 1 1 1 1 1 1 1 1 1 1 1 1 4 2 1 1 1 1 1 1 1 1 1 1 1
1 1 1 1 2 1 1 1 3 1 1 1 1 1 1 1 1 1 1 1 1 3 1 1 1 1 2 3 1 1 1 2
3 3 1 1 3 1 1 1 1 1 2 3 2 1 1 1 1 1 2 2 2 1 1 1 1 1 1 1 1 1 1 2
1 1 1 1 1 1 1 1 1 2 1 2 1 2 1 1 1 1 1 1 1 1 1 1 1 1 1 1 1 1 1 1
2 1 1 1 1 1 1 1 1 1 1 1 1 1 2 2 1 1 1 1 1 2 1 2 1 1 1 1 1 1 1 1
1 2 1 1 1 3 3 2 1 1 1 1 1 1 1 1 1 1 1 2 2 2 3 2 2 1 1 1 1 2 1 2 3
2 2 3 3 3 1 2 2 2 2 1 2 1 2 2 1 2 2 3 3 1 2 1 2 2 2 2 3 3 3 3 2 1
3 2 1 2 2 1 1 1 2 1 1 2 2 2 2 1 2 1 2 1 1 1 1 2 2 1 1 1 2 1 1 1 1
2 2 2 2 1 2 3 3 2 1 2 2 2 2 2 2 1 1 1 1 2 1 1 1 1 2 1 2 1 1 2 1 3
1 2 2 3 3 3 1 2 1 1 1 1 2 1 1 1 1 2 2 3 3 1 2 1 3 1 1 2 2 1 2 1 2
2 2 2 1 2 2 3 1 1 2 2 1 3 3 2 3 3 1 1 2 1 1 1 1 2 3 3 2 3 2 1 1 1
1 2 1 1 2 1 2 2 3 2 2 2 1 2 2 1 2 2 2 4 1 2 1 2 2 1 2 2 2 2 1 2 2
2 2 1 2 2 3 3 2 2 3 1 1 2 2 3 1 1 1 2 3 2 3 2 2 3 2 2 2 2 1 1 1 1
2 2 1 1 1 !
MIM->satmod
```

The SetBlocks command defines the order of the variables, and turns block mode on. The SatModel command, in block mode, sets the current model to the full DAG shown in Figure 7.8.

The first step involves modelling the dependence of the treatment allocation G on the pretreatment score A, but since we know that the allocation was random, we know that $G \perp\!\!\!\perp A$, and so we can delete the edge $[AG]$:

```
MIM->Delete AG
```

We proceed, therefore, to analyze the dependence of the first week score (B) on G and A. Since the table is very sparse and the response is ordinal, we

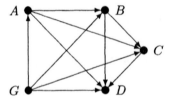

FIGURE 7.8. The full model.

base our analysis on exact tests for conditional independence that are sensitive to ordinal alternatives. Two tests in particular are used. For tests of the type $X \perp\!\!\!\perp Y \mid Z$, where X is binary and Y ordinal, we use the stratified Wilcoxon test, and where both X and Y are ordinal, we use the stratified Jonckheere-Terpstra test (Section 5.12). For comparison purposes, the standard likelihood ratio tests are also computed. Both asymptotic and exact conditional tests are shown.

```
MIM->TestDelete AB ljm
Test of HO: BG,AG
against H: ABG
Exact test - monte carlo estimates.
Stratum-specific scores.
No. of tables:        1000
Likelihood Ratio Test.
LR:  12.6779    DF:    8 Asymptotic P: 0.1234
                          Estimated P: 0.1358 +/- 0.021234
Jonckheere-Terpstra Test.
JT: 555.5000  E(JT|HO): 400.5000 Asymptotic P: 0.0022
                          Estimated P: 0.0016 +/- 0.002503
MIM->TestDelete GB lwm
Test of HO: AG,AB
against H: ABG
Exact test - monte carlo estimates.
Stratum-specific scores.
No. of tables:        1000
Likelihood Ratio Test.
LR:   9.4349    DF:    5 Asymptotic P: 0.0929
                          Estimated P: 0.1442 +/- 0.021770
Wilcoxon Test.
W:  1600.0000  E(W|HO):  1836.0000 Asymptotic P: 0.0070
                          Estimated P: 0.0102 +/- 0.006222
```

First, the deletion of edge $[AB]$, corresponding to a test of $A \perp\!\!\!\perp B|G$, is examined. Since both A and B are ordinal, a stratified Jonckheere-Terpstra test is used. We see that this test rejects the null hypotheses very decisively, with an estimated p-value of 0.0022. In contrast, the standard likelihood ratio test, which does not exploit ordinality, does not detect any association.

Next, the deletion of edge $[GB]$ is examined. Since G is binary and B is ordinal, we use the stratified Wilcoxon test. Similar results are obtained. Neither $[AB]$ nor $[GB]$ can be deleted.

In the next stage of the analysis, we examine the dependence of the score after two weeks' treatment (C) on G, A, and B:

```
MIM->TestDelete BC ljm
Test of H0: ACG,ABG
against H: ABCG
Exact test - monte carlo estimates.
Stratum-specific scores.
No. of tables:        1000
Likelihood Ratio Test.
LR:  48.3514    DF:    17 Asymptotic P: 0.0001
                          Estimated P: 0.0000 +/- 0.000000
Jonckheere-Terpstra Test.
JT: 579.0000  E(JT|H0): 400.0000 Asymptotic P: 0.0000
                          Estimated P: 0.0000 +/- 0.000000
MIM->TestDelete AC ljm
Test of H0: BCG,ABG
against H: ABCG
Exact test - monte carlo estimates.
Stratum-specific scores.
No. of tables:        1000
Likelihood Ratio Test.
LR:  18.8328    DF:    15 Asymptotic P: 0.2214
                          Estimated P: 0.3988 +/- 0.030348
Jonckheere-Terpstra Test.
JT: 131.0000  E(JT|H0): 135.5000 Asymptotic P: 0.7898
                          Estimated P: 0.8310 +/- 0.023227
MIM->Delete AC
MIM->TestDelete GC lwm
Test of H0: BC,ABG
against H: BCG,ABG
Exact test - monte carlo estimates.
Stratum-specific scores.
No. of tables:        1000
Likelihood Ratio Test.
LR:  11.2178    DF:     6 Asymptotic P: 0.0819
                          Estimated P: 0.1527 +/- 0.022293
Wilcoxon Test.
W:  1148.5000  E(W|H0):  1244.0000 Asymptotic P: 0.0327
                          Estimated P: 0.0384 +/- 0.011913
```

The deletion of the edge $[BC]$ is tested first and is rejected strongly by both the likelihood ratio and the ordinal test. The deletion of the next edge, $[AC]$, is then tested and accepted by both tests. The edge $[AC]$ is then deleted, and the deletion of the third edge, $[GC]$, is tested. The ordinal test rejects the deletion with an estimated p-value of 0.0421.

The order in which the edges are examined is not arbitrary; the possible effect of treatment on the response is of primary interest. By choosing to test for the deletion of $[GC]$ last, we seek to maximize the power for this

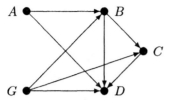

FIGURE 7.9. After the second stage.

test. After $[AC]$ is deleted, the test for the deletion of $[GC]$ becomes a test of $G \perp\!\!\!\perp C|B$ rather than of $G \perp\!\!\!\perp C|(A, B)$, with consequent increase in power.

We have now arrived at the model shown in Figure 7.9.

In the third stage, we examine the dependence of the score after three weeks' treatment (D), on the treatment group G, and the previous scores A, B, and C.

```
MIM->TestDelete CD ljm
Test of HO: ABDG,ABCG
against H: ABCDG
Exact test - monte carlo estimates.
Stratum-specific scores.
No. of tables:        1000
Likelihood Ratio Test.
LR:  38.0929    DF:   26 Asymptotic P: 0.0594
                        Estimated P: 0.1498 +/- 0.022122
Jonckheere-Terpstra Test.
JT: 193.5000  E(JT|HO): 143.0000 Asymptotic P: 0.0037
                        Estimated P: 0.0093 +/- 0.005955
MIM->TestDelete BD ljm
Test of HO: ACDG,ABCG
against H: ABCDG
Exact test - monte carlo estimates.
Stratum-specific scores.
No. of tables:        1000
Likelihood Ratio Test.
LR:  28.6759    DF:   24 Asymptotic P: 0.2326
                        Estimated P: 0.9370 +/- 0.015059
Jonckheere-Terpstra Test.
JT: 156.0000  E(JT|HO): 133.0000 Asymptotic P: 0.1198
                        Estimated P: 0.1224 +/- 0.020317
MIM->Delete BD
MIM->TestDelete AD ljm
Test of HO: CDG,ABCG
against H: ACDG,ABCG
```

```
Exact test - monte carlo estimates.
Stratum-specific scores.
No. of tables:        1000
Likelihood Ratio Test.
LR:  20.3345    DF:    14 Asymptotic P: 0.1199
                          Estimated P: 0.1940 +/- 0.024508
Jonckheere-Terpstra Test.
JT: 194.5000  E(JT|H0): 170.0000 Asymptotic P: 0.2111
                          Estimated P: 0.2132 +/- 0.025385

MIM->Delete AD
MIM->TestDelete GD lwm
Test of H0: CD,ABCG
against H: CDG,ABCG
Exact test - monte carlo estimates.
Stratum-specific scores.
No. of tables:        1000
Likelihood Ratio Test.
LR:  10.1274    DF:     7 Asymptotic P: 0.1815
                          Estimated P: 0.3129 +/- 0.028739
Wilcoxon Test.
W:  1183.5000  E(W|H0):  1292.0000 Asymptotic P: 0.0237
                          Estimated P: 0.0195 +/- 0.008563
```

So $[BD]$ and $[AD]$ are deleted, and $[CD]$ and $[GD]$ are retained. We have arrived at the the DAG shown in Figure 7.10.

This graph has a particularly simple structure. If we rename A, B, C, and D as R_0, R_1, R_2, and R_3, then the DAG can be characterized by (i) the marginal independence of G and R_0, and (ii) the transition probabilities $\Pr\{R_t = r | (R_{t-1} = s, G = g)\}$ for $t = 1, 2, 3$. An obvious question is whether these transition probabilities are constant over time.

The present form of the data is

$$(G^{(i)}, A^{(i)}, B^{(i)}, C^{(i)}, D^{(i)})$$

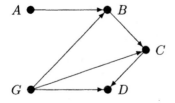

FIGURE 7.10. The final DAG.

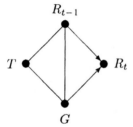

FIGURE 7.11. Time-homogeneity as a graph.

for $i = 1, \ldots, 100$. To test the time-homogeneity hypothesis, we need to transform the data to the form

$$(G^{(i)}, R_{t-1}^{(i)}, R_t^{(i)}, t)$$

for $t = 1, 2, 3$ and for $i = 1, \ldots, 100$. Suppose we have done this and the variables are denoted G (treatment group, as before), Q (i.e., R_{t-1}), R (i.e., R_t) and T (i.e., time $= 1$, 2, 3). In this framework, time-homogeneity can also be formulated as a test of conditional independence: namely that $R_t \perp\!\!\!\perp T \mid (R_{t-1}, G)$, as shown in Figure 7.11. We leave the details of this analysis to the interested reader.

7.2 Chain Graphs

Although problems with complete causal orderings seem to be fairly unusual in applications, partial orderings are often available. For example, an epidemiological study might involve the following characteristics of a sample of individuals:

1. Familial characteristics, such as parental genotype.
2. Genetic characteristics, such as individual genotype.
3. Demographic characteristics, such as sex and ethnic group.
4. Social and economic factors, such as occupation, socioeconomic status, and educational background.
5. Lifestyle characteristics, such as tobacco use, diet, and physical exercise.
6. Biological characteristics, such as elevated cholesterol and body mass index.

Clearly, the familial characteristics are antecedent to the demographic characteristics, which themselves are antecedent to the lifestyle and biological characteristics. It may be reasonable to assume that the socioeconomic factors are antecedent to the lifestyle and biological characteristics, although this is clearly a nontrivial assumption.

A common study design in the health sciences is a *panel study*, in which a number of individuals are followed over a period of time, and information collected at predetermined points of time during this period. Clinical trials often use similar longitudinal designs: after randomised allocation to treatment, patients are followed up over some appropriate interval, and information is collected at predetermined time-points. For such studies, the ordering of the measurement times gives rise to an ordering between blocks of variables, but not to a complete ordering between all variables.

To capture this type of prior information, some work has combined the undirected graphs and DAGs into a single framework, the so-called *block-recursive* or *chain* graphs. Some key references are Lauritzen and Wermuth (1989), Wermuth and Lauritzen (1990), and Frydenberg (1989).

These graphs are based on what is known as a *dependence chain*, that is, we suppose that subject-matter knowledge gives us a partitioning of the variables into an ordered list of *blocks*, say $V = B_1 \cup B_2 \ldots \cup B_k$. A chain graph consistent with this block structure has lines (i.e., undirected edges) within the blocks, and arrows between variables in different blocks, pointing from the lower-numbered to the higher-numbered block. Variables in the same block are assumed to be concurrent, that is, their association structure is taken to be symmetric, without ordering. Figure 7.12 is an example of such a graph.

Corresponding to the block structure we assume that the joint density $f(V_1, \ldots, V_n)$ factorizes as

$$f(B_1)f(B_2|B_1)\ldots f(B_k|B_1 \cup B_2 \ldots \cup B_{k-1}).$$

If a line is missing between two vertices v and w in the same block B_i, or an arrow is missing from $v \in B_j$ to $w \in B_i$, for $j < i$, then this means that

$$v \perp\!\!\!\perp w \mid B_1 \cup B_2 \ldots \cup B_i, \tag{7.6}$$

this being a version of the pairwise Markov property for chain graphs, as explained in the following section.

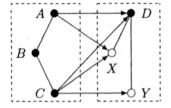

FIGURE 7.12. A chain graph. The two blocks, $\{A, B, C\}$ and $\{D, X, Y\}$, are shown in boxes. Edges between boxes are directed (arrows) and edges within boxes are undirected (lines).

Up to now we have described how chain graphs are constructed from the perspective of an applied modeller. We now turn to a more graph-theoretic perspective.

As before, we write a graph as a pair $\mathcal{G} = (\mathcal{V}, \mathcal{E})$, where $\mathcal{V}$ is a set of vertices and $\mathcal{E}$ is a set of edges, where again we identify edges with *ordered* pairs of vertices. Whenever both $(v, w) \in \mathcal{E}$ and $(w, v) \in \mathcal{E}$, we say that $v \leftrightarrow w$ and draw a line (i.e., an undirected edge) between v and w. Whenever $(v, w) \in \mathcal{E}$ and $(w, v) \notin \mathcal{E}$ we say that $v \rightarrow w$ and draw an arrow from v to w. If $v \rightarrow w$, $w \rightarrow v$, or $v \leftrightarrow w$ we say that v and w are *adjacent* and write this as $v \sim w$.

By a *path* we mean a sequence of vertices $\{v_1, \ldots, v_k\}$ such that either $v_i \rightarrow v_{i+1}$ or $v_i \leftrightarrow v_{i+1}$ for each $i = 1, \ldots, k-1$. In other words, a path can have lines and/or arrows, but it must follow the direction of the arrows. If all its edges are lines, it is called *undirected*; otherwise (that is, if it contains one or more arrows), it is called *directed*. As usual, a path is called a *cycle* when the first and last vertices coincide.

It is simple to characterize chain graphs without referring to the dependence chain structure, namely as graphs that *contain no directed cycles*. So, for example, one cycle in Figure 7.13 is $\{X, Y, Z, X\}$: this is undirected. It is easy to see that all cycles in Figure 7.13 are like this. But if there were a line instead of an arrow between K and Z, $\{K, X, Z, K\}$ would be a directed cycle, and so the graph would not be a chain graph.

It is easy to see that the class of chain graphs includes undirected graphs and DAGs as special cases. If all the edges of a chain graph are undirected, then obviously it is undirected. If all its edges are directed, then it is a DAG.

More insight into the structure of the graphs can be obtained by considering the connected components of a chain graph after deleting all arrows. Write these as $C_1, \ldots, C_r$, say. We call the C_i the *components* of the chain graph $\mathcal{G}$. They contain only undirected edges, and if two components are connected, they are connected by arrows. Moreover all arrows between any two components must have the same direction (otherwise it would be easy

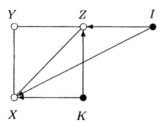

FIGURE 7.13. Another chain graph.

to construct a directed cycle). So we can construct a directed graph whose nodes are the components of $\mathcal{G}$ and where two nodes are connected by an arrow whenever the two components are connected (by arrows in the same direction) in $\mathcal{G}$. Clearly this graph is a DAG; we call it the component DAG.

We can exploit this result in various ways. For example, it follows that the components can be ordered and numbered, so that arrows point only from lower-numbered to higher-numbered components. It also follows that the numbering is not necessarily unique.

If we compare the blocks with the components, we see that the latter give rise to a (possibly) finer partition of the variables than that given by the blocks. That is, each block is a union of one or more components. Two variables in a block will be in the same component whenever they are connected (i.e., there is a path between them in $\mathcal{G}$). For example, the components of Figure 7.13 are $\{I\}$, $\{K\}$ and $\{X, Y, Z\}$: the graph could have been generated from several different block structures:

1. $B_1 = \{I, K\}$ and $B_2 = \{X, Y, Z\}$,
2. $B_1 = \{I\}$, $B_2 = \{K\}$, and $B_3 = \{X, Y, Z\}$, or
3. $B_1 = \{K\}$, $B_2 = \{I\}$, and $B_3 = \{X, Y, Z\}$

Since the components can be identified from the graph, it follows that it is not necessary to draw the boxes on the graph (see Figure 7.14). All the mathematical and statistical properties can be derived from the graph alone, without specification of the block structure. However, the block structure gives us information about alternative models, in particular about missing edges; for example, in Figure 7.2 we do not know whether, if Gender and Race were connected, it would be by a line or an arrow. Knowledge of the block structure is necessary for this. The block structure encodes a priori information about the system, whereas the components are model specific.

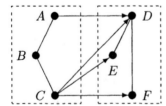 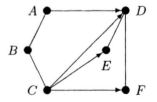

FIGURE 7.14. Unchaining a graph. To the left is shown a chain graph with the blocks drawn, and to the right is the same graph without the boxes. The boxes are superfluous. As Hank Williams might have put it: "Take these chains from my graph and set me free."

7.2.1 Markov Properties of Chain Graphs

Markov properties of chain graphs were studied by Frydenberg (1989); a thorough exposition is given in Lauritzen (1996). There are actually a variety of Markov properties; we give only a simplified version here. See also Section 7.5 for an alternative use of chain graphs.

The first property was mentioned in the previous section: it is called the pairwise Markov property relative to the dependence chain $B_1, \ldots, B_k$. It states: if a line is missing between two vertices v, w in the same block B_i, or an arrow is missing from $v \in B_j$ to $w \in B_i$, for $j \leq i$, then this means that

$$v \perp\!\!\!\perp w \mid B_1 \cup B_2 \ldots \cup B_i. \tag{7.7}$$

In other words, the immediate interpretation of a missing edge between two variables is that they are independent, conditional on all prior and concurrent variables. Prior and concurrent, that is, relative to the later of the two variables, and understood as all the variables in the previous blocks as well as the concurrent block.

We see that this combines the undirected and the DAG frameworks. In the undirected framework, a missing edge denotes independence given the concurrent variables, and in the DAG framework, a missing arrow denotes independence given the prior variables.

As with undirected graphs and DAGs, the question arises: Can we deduce strong conditional independence statements from a chain graph? The global Markov property for chain graphs is due to Frydenberg (1989). It is a generalisation of the global Markov property for DAGs, described in Section 7.1.1, involving separation in a derived undirected graph. To state it precisely we need some new graph-theoretic definitions.

Firstly, we define the boundary of a set $S \subseteq \mathcal{V}$ as

$$\mathrm{bd}(S) = \{v \in \mathcal{V} \setminus S : v \leftrightarrow w \text{ or } v \rightarrow w \text{ for some } w \in S\}.$$

Secondly, given a chain graph $\mathcal{G} = (\mathcal{V}, \mathcal{E})$, we define its moral graph, $\mathcal{G}^m$, to be the undirected graph resulting from the following operations:

1. Complete the boundary of each component of $\mathcal{G}$ with lines.
2. Replace all arrows in $\mathcal{E}$ with lines.

Thirdly, we define the ancestral set $\mathrm{an}^+(S)$ of a set $S \subseteq \mathcal{V}$ as follows. If there is a path from v to w, then v is called an *ancestor* of w. The set of ancestors of nodes in S is denoted $\mathrm{an}(S)$, and $\mathrm{an}^+(S) = S \cup \mathrm{an}(S)$ is called the *ancestral set* of S. Note that $\mathrm{an}^+(S)$ can be found by using the DAG notion of ancestral set on the component DAG. Namely, if we take S^* as the union of the components containing a node in S, regard this is a set of nodes (= components) in the component DAG, and find an$^+(S^*)$ in this DAG, then we obtain an$^+(S)$, as required.

We can now state the global Markov property for chain graphs:

For any three disjoint vertex sets S_1, S_2, and S_3, $S_1 \perp\!\!\!\perp S_2|S_3$ whenever S_3 separates S_1 and S_2 in $\mathcal{G}_\mathcal{A}^m$, where $\mathcal{A} = $ an$^+(S_1 \cup S_2 \cup S_3)$.

For example, consider the graph shown in Figure 7.15(a). Suppose that we want to determine whether $F \perp\!\!\!\perp D|E$. We perform the following steps.

1. We find the set $\mathcal{A} = $ an$^+(\{D, E, F\})$. Here $\mathcal{A} = \{A, B, C, D, E, F, G\}$.

2. We construct $\mathcal{G}_\mathcal{A}$, the subgraph on $\mathcal{A}$.

3. To moralize $\mathcal{G}_\mathcal{A}$ we examine the its components: $\{A, B, C\}$, $\{D, E\}$, and $\{F, G\}$. The boundary of $\{F, G\}$ is $\{B, D, E\}$, the boundary of $\{D, E\}$ is $\{C\}$, and $\{A, B, C\}$ has no boundary. Accordingly we complete $\{B, D, E\}$ with lines and change the remaining arrows to lines. This results in $\mathcal{G}_\mathcal{A}^m$, shown in Figure 7.15(b).

4. Since E does not separate F from D in this graph (there being paths between D and F not going through E, for example $\{D, G, F\}$, it follows that $F \not\perp\!\!\!\perp D|E$.

7.2.2   Modelling with Chain Graphs

In contrast to DAGs, there are not too many model families that have the properties we require. We need multivariate response models for B_i given $B_1 \cup \ldots \cup B_{i-1}$, in which arbitrary sets of conditional independence relations of the form (7.6) can be specified. As emphasized in Section 4.5,

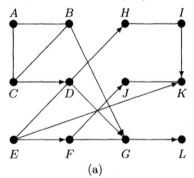 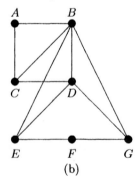

(a)   (b)

FIGURE 7.15. A chain graph $\mathcal{G}$ is shown in (a). Is $F \perp\!\!\!\perp D|E$? To determine this we find the set $\mathcal{A} = $ an$^+(\{D, E, F\})$, construct the subgraph on $\mathcal{A}$, and then moralize this. The result, $\mathcal{G}_\mathcal{A}^m$, is shown in (b). Since E does not separate D from F in $\mathcal{G}_\mathcal{A}^m$, we conclude that $F \not\perp\!\!\!\perp D|E$.

the undirected CG-distribution models can be used except when there are discrete responses and continuous covariates, by including all interactions between the covariates in all models considered. When there are discrete responses and continuous covariates, the CG-regression models of Section 4.5 can be used, but maximum likelihood estimation for these models is computationally more difficult.

The modelling challenge is to find appropriate models for each block, i.e., for B_i given $B_1 \cup \ldots \cup B_{i-1}$, for each i. As with DAGs, one great simplifying principle applies: the choice of model for each block is quite independent of the models chosen for the other blocks. If we restrict ourselves to graphical models, then we can say that the decision to include an arrow from $v \in B_j$ to $w \in B_i$ depends on which other arrows pointing to a variable in B_i are present, and on which lines between variables in B_i are present— but only on these.

Note the close connection between chain graphs and the concept of collapsibility described in Section 4.2. If a joint model $\mathcal{M}$ is collapsible onto a, then it is equivalent to a chain model with two chain components, a (model $\mathcal{M}_a$) and b (model $\mathcal{M}_{b|a}$), and the models with covariates described in Sections 3.2, 3.2.1, and 4.1.10 can thus be regarded as simple, two-block chain models.

7.2.3 Example: Membership of the "Leading Crowd"

As a very simple example of chain graph modelling with MIM, we repeat an analysis given in Goodman (1973a, 1973b) and Fienberg (1980). Each of $3,398$ schoolboys was interviewed at two points of time and asked about his percieved membership in the "leading crowd" (in = 1, out = 2), and his attitude to the "leading crowd" (favourable = 1, unfavourable = 2). This can be regarded as a very simple panel study. The code fragment

```
MIM->fact A2B2C2D2
MIM->label A "Member 1" B "Attitude 1" C "Member 2" D "Attitude 2"
MIM->sread ABCD
MIM->458 140 110 49 171 182 56 87 184 75 531 281 85 97 338 554 !
MIM->setblocks AB|CD; satmod; step
```

shows how the block structure is set up using the SetBlocks command. Stepwise model selection is then initiated, starting from the saturated model. The selected model is shown in Figure 7.16. The interpretation of the model is that

1. Membership and attitude at the first interview are associated.

2. Membership at the second interview is affected both by membership at the first interview and by attitude at the first interview.

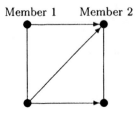

FIGURE 7.16. The Leading Crowd

3. Attitude at the second interview is affected by concurrent membership and previous attitude.

Further examples of modelling with chain graphs are to be found in Cox and Wermuth (1996), Mohamed et al. (1998), Neil-Dwyer et al. (1998), Caputo et al. (1999), Ruggeri et al. (1998), and Stanghellini et al. (1999).

7.3 Local Independence Graphs

Broadly speaking, the undirected graphs, DAGs, and chain graphs described above are suitable for different kinds of data. Undirected graphs are suitable for cross-sectional data, and represent patterns of symmetric associations. In contrast, DAGs represent directed associations. So when variables represent short, nonoverlapping events, it is natural to assume that the direction of influence flows from the earlier events to the later. Chain graphs combine these types of association, and similar remarks apply. Variables are regarded as either concurrent or ordered; the former are connected with lines, and the latter with arrows.

So DAGs and chain graphs can incorporate the time dimension, but only when this is discretized. It is unclear whether graphs are able to represent continuous time systems, in which several processes may be in interplay through time; in such a system, one process may both influence and be influenced by another process.

In this section we give a brief sketch of some recent work by Didelez (1999) that sheds some light on this question. This work, based on previous results of Schweder (1970) and Aalen (1987), applies the ideas of graphical modelling to multivariate stochastic processes. To avoid introducing the formidable mathematical machinery necessary for a proper treatment, we restrict ourselves to a simple case.

We are interested in modelling a Markov process $Y(t) = (Y_1(t), \ldots, Y_K(t))$, where t is time and where the $Y_i(t)$ take on a finite number of values: here

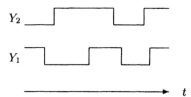

Y_2

Y_1

t

FIGURE 7.17. A two-dimensional binary Markov process.

we suppose that their state spaces are $\{0,1\}$, i.e. $Y_i(t) \in \{0,1\}$ for all i and t. Figure 7.17 represents a realisation of a two-dimensional binary Markov process. This could represent, for example, the time-course of a recurrent illness and the presence of a virus implicated in the illness.

We assume the process is governed by transition intensities

$$h(y; y^*; t) = \lim_{\delta t \to 0} Pr(Y(t + \delta t) = y^* | Y(t) = y)/\delta t$$

for $y \neq y^*$; these represent, as it were, the instantaneous probabilities of changes of state. We need to assume that the probability of simultaneous transitions is zero; that is, when y and y^* differ in more than one component, $h(y; y^*; t) = 0$ for all t. Such a process is called *composable* (Schweder, 1970).

Composable Markov processes are governed by the component transition intensities

$$h_j(y; y_j^*; t) = h(y; y^*; t)$$

for $j = 1, \ldots, K$, where $y_i^* = y_i$ for $i \neq j$. For example, a two-dimensional composable process has possible states $\{(0,0), (0,1), (1,0), (1,1)\}$, and is governed by the eight transition intensities

$$
\begin{array}{ll}
(0,0) \to (1,0) : & h_1(0,0;1;t) \\
(0,0) \to (0,1) : & h_2(0,0;1;t) \\
(1,0) \to (0,0) : & h_1(1,0;0;t) \\
(1,0) \to (1,1) : & h_2(1,0;1;t) \\
(0,1) \to (1,1) : & h_1(0,1;1;t) \\
(0,1) \to (0,0) : & h_2(0,1;0;t) \\
(1,1) \to (0,1) : & h_1(1,1;0;t) \\
(1,1) \to (1,0) : & h_2(1,1;0;t)
\end{array}
$$

since all the other transition intensities are zero.

It is easy to state the condition for independence of $Y_1(t)$ and $Y_2(t)$: this is that

$$h_1(0,0;1;t) = h_1(0,1;1;t),$$

$$h_1(1,0;0;t) = h_1(1,1;0;t),$$
$$h_2(0,0;1;t) = h_2(1,0;1;t), \text{ and}$$
$$h_2(0,1;0;t) = h_2(1,1;0;t).$$

In words, this says that the transition intensity in the Y_1 subprocess does not depend on $Y_2(t)$, and that the transition intensity in the Y_2 subprocess does not depend on $Y_1(t)$. But notice that these conditions can be separated. It is quite feasible that

$$h_1(0,0;1;t) = h_1(0,1;1;t),$$
$$h_1(1,0;0;t) = h_1(1,1;0;t)$$

holds, but not

$$h_2(0,0;1;t) = h_2(1,0;1;t),$$
$$h_2(0,1;0;t) = h_2(1,1;0;t);$$

in this case, we say that Y_1 is locally independent of Y_2, but Y_2 is not locally independent of Y_1.

More formally, we say that Y_i is locally independent of Y_j with respect to $(Y_1, \ldots Y_K)$ when $h_j(y; y_j^*; t)$ is a constant function of y_j for every y_s, where $s = \{1 \ldots K\} \backslash \{i, j\}$. That is, for every fixed value y_s of the remaining subprocesses in $(Y_1, \ldots Y_K)$, the transition intensities $h_i(y; y_i^*; t)$ on the ith component do not depend on y_j. We use the notation $Y_i \perp\!\!\!\perp^l Y_j | (Y_1, \ldots Y_K)$ to represent that Y_i is locally independent of Y_j with respect to $(Y_1, \ldots Y_K)$.

This concept of local independence, due to Schweder (1970), plays the same fundamental role in the present framework as conditional independence does in standard graphical modelling. We notice immediately that the relation is asymmetrical: that is, $Y_i \perp\!\!\!\perp^l Y_j | (Y_1, \ldots Y_K)$ does not imply $Y_j \perp\!\!\!\perp^l Y_i | (Y_1, \ldots Y_K)$. Thus graphs representing the local independence structure must have directed edges, i.e., both single-headed (unidirectional) and double-headed (bidirectional)arrows are allowed.

An example of a local independence graph is shown in Figure 7.18. It represents seven local independence relations, namely $Y_1 \perp\!\!\!\perp^l Y_3 | (Y_1, \ldots Y_4)$, $Y_1 \perp\!\!\!\perp^l Y_4 | (Y_1, \ldots Y_4)$, $Y_2 \perp\!\!\!\perp^l Y_1 | (Y_1, \ldots Y_4)$, $Y_2 \perp\!\!\!\perp^l Y_4 | (Y_1, \ldots Y_4)$, $Y_3 \perp\!\!\!\perp^l Y_1 | (Y_1, \ldots Y_4)$, $Y_3 \perp\!\!\!\perp^l Y_4 | (Y_1, \ldots Y_4)$, and $Y_4 \perp\!\!\!\perp^l Y_1 | (Y_1, \ldots Y_4)$.

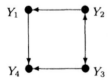

FIGURE 7.18. A local independence graph.

Didelez (1999) studied the Markov properties of local independence graphs: in particular, pairwise, local, and global properties framed in terms of local as well as conditional independence. A modified concept of separation is involved; we refrain from explaining this here. Applied to Figure 7.18, for example, we obtain that $Y_1 \perp\!\!\!\perp^l Y_3|(Y_1, Y_2, Y_3)$ but $Y_2 \not\perp\!\!\!\perp^l Y_4|(Y_1, Y_2, Y_4)$.

Armed with these results, the graphs appear to provide a powerful tool for understanding complex multivariate processes. It is interesting to note that continuous time systems *can* be represented by graphs, but that these are not DAGs or chain graphs.

7.4 Covariance Graphs

Cox and Wermuth (1993, 1996) suggested the use of undirected graphs to display the marginal independence structure of a set of variables, by connecting two vertices by an edge whenever the two variables are marginally dependent. By convention, edges are drawn as dashed lines, and the graphs are called *covariance graphs*. Their treatment was informal; the Markov interpretation of the graphs was unclear in the sense that local and global Markov properties were not given.

A formal treatment was provided by Kauermann (1996), who studied pairwise, local, and global Markov properties for these graphs and obtained general conditions for their equivalence. A family of models corresponding to the graphs is available for the Gaussian case only. These models are dual to graphical Gaussian models, in that they constrain a set of elements of the covariance matrix Σ to be zero, rather than a set of elements of the precision matrix Ω. Kauermann shows that the maximum likelihood estimates have the dual property to those of graphical Gaussian models, namely that $\hat{\omega}^{xy}$ only differs from its counterpart in the sample precision matrix for those x, y for which σ^{xy} is constrained to be zero.

Fitting these models in MIM is straightforward, using the following trick: use StatRead to read the sample precision matrix instead of the covariance matrix, and interpret the fitted covariances as fitted precisions and vice versa. The quantity shown as the deviance has the same asymptotic properties as the true deviance, and so can be used for this purpose.

In illustration we supplement the analysis of the anxiety and anger data shown in Section 3.1.5. We wish to fit the dual model shown Figure 7.19; this represents the hypothesis that $X \perp\!\!\!\perp Y$ and $W \perp\!\!\!\perp Z$. First we obtain the sample precision matrix, using the original data:

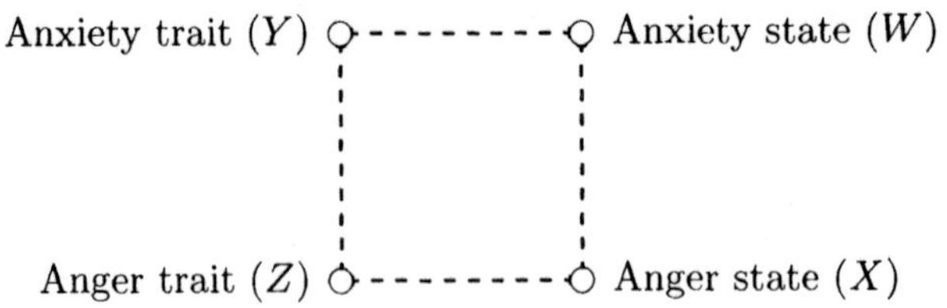

Anxiety trait (Y) ⚪- - - - - - - -⚪ Anxiety state (W)

Anger trait (Z) ⚪- - - - - - -⚪ Anger state (X)

FIGURE 7.19. A covariance graph, expressing that $X \perp\!\!\!\perp Y$ and $W \perp\!\!\!\perp Z$

```
MIM->pr t
Empirical discrete, linear and precision parameters.
    W    0.05600
    X   -0.02142  0.04065
    Y   -0.02676 -0.00107  0.05764
    Z    0.00187 -0.01236 -0.01426  0.03484
Linear   0.20728 -0.09730  0.36695  0.36090-19.64269
              W         X        Y        Z Discrete
```

Then we enter it using StatRead:

```
MIM->StatRead WXYZ
684
18.8744  15.2265  21.2019  23.4217
 0.05600
-0.02142  0.04065
-0.02676 -0.00107  0.05764
 0.00187 -0.01236 -0.01426  0.03484 !
Reading completed.
MIM->satmod; delete XY,WZ
MIM->fit; print fg
Deviance:     214.7895 DF: 2
Fitted counts, means and covariances.
      W    0.056
      X   -0.021    0.041
      Y   -0.027    0.013   0.058
      Z    0.010   -0.012  -0.014   0.035
  Means 18.874   15.226  21.202  23.422  684.000
             W        X       Y       Z    Count
Fitted discrete, linear and precision parameters.
      W   26.897
      X   11.052   32.976
      Y   10.072    0.000  23.678
      Z   -0.000    8.394   6.680  34.414
Linear 889.491 907.301 848.5541075.472-36889.176
             W        X       Y       Z Discrete
```

The model fits very poorly. The fitted covariances are shown as precisions and vice versa.

7.5   Chain Graphs with Alternative Markov Properties

Cox and Wermuth (1993, 1996) suggest the use of chain graphs with a different Markov interpretation to that described in Section 7.2. To motivate this, consider the recursive linear system

$$X_1 = \epsilon_1$$
$$X_2 = \epsilon_2$$
$$X_3 = bX_1 + \epsilon_3$$
$$X_4 = cX_2 + \epsilon_4,$$

where b and c are fixed constants, and ϵ_1, ϵ_2, and $(\epsilon_3, \epsilon_4)'$ are independent stochastic terms with $\epsilon_1 \sim \mathcal{N}(0, \sigma_1)$, $\epsilon_2 \sim \mathcal{N}(0, \sigma_2)$, and $(\epsilon_3, \epsilon_4)' \sim \mathcal{N}(0, \Sigma)$, say. The system satisfies three conditional independence relations: $X_1 \perp\!\!\!\perp X_2$, $X_4 \perp\!\!\!\perp X_1 | X_3$, and $X_3 \perp\!\!\!\perp X_2 | X_1$. If we try to represent the system in a chain graph, the nearest we can come is the graph shown in Figure 7.20, but according to Section 7.2, this graph represents *different* relations, namely that $X_1 \perp\!\!\!\perp X_4 | (X_2, X_3)$, $X_2 \perp\!\!\!\perp X_3 | (X_1, X_4)$, and $X_1 \perp\!\!\!\perp X_2$. This is because the Markov interpretation described in Section 7.2 requires conditioning on the concurrent as well as the prior variables.

Since the system just described is relatively simple and straightforward, it would be useful if the conditional independences it engenders could be represented graphically. Cox and Wermuth (1993, 1996) introduced some modified chain graphs for this purpose. Visually, the modification consists of drawing the arrows (and sometimes also the lines) with dashes instead of solid lines. The chain graphs with dashed arrows and solid lines correspond to block-recursive linear systems; we now describe a component model in such a system.

The conditional distribution of the variates in block B_i given those in the previous blocks, is defined as follows. Each variate is given as a linear function of its parents in the graph, plus an error term. The errors are independent across blocks, while their within-block correlation structure is given by the graphical Gaussian model corresponding to $\mathcal{G}_{B_i}$. (Note that it

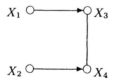

FIGURE 7.20. A chain graph. If it is defined in terms of the standard Markov properties then it represents the relations $X_1 \perp\!\!\!\perp X_4 | (X_2, X_3)$, $X_2 \perp\!\!\!\perp X_3 | (X_1, X_4)$, and $X_1 \perp\!\!\!\perp X_2$. In terms of the alternative Markov properties, it represents instead $X_1 \perp\!\!\!\perp X_4 | X_2$, $X_2 \perp\!\!\!\perp X_3 | X_1$, and $X_1 \perp\!\!\!\perp X_2$.

is assumed that for $i > 1$, all variables in B_i are continuous). Thus the graph represents the model in that the arrows define the linear models for the variates and the lines define the error structure. Furthermore, a property of the system is that the concurrent variables need not be included in the conditioning sets; that is, if an arrow is missing from $v \in B_j$ to $w \in B_i$, for $j < i$, then this means that

$$v \perp\!\!\!\perp w \mid B_1 \cup B_2 \ldots \cup B_{i-1}. \tag{7.8}$$

Comparing with (7.6), we see that the variables in the concurrent block are omitted from the conditioning set.

Andersson, Madigan and Perlman (2000) study chain graphs with these alternative Markov properties, formulating pairwise, local, and global properties and stating conditions for their equivalence. The global property, like that of the standard chain graphs of Section 7.2, involves separation in an undirected graph, but the derivation of this graph appears to be much more complex.

We note in passing that it is straightforward to fit the component models in MIM, in a two-stage process:

1. Fit the linear models for the variates in B_i as follows. Construct an undirected model with all interactions between the covariates, and all edges from the variates in B_i to their parents, but omitting all edges between variates in B_i. (This omission enforces conditional mean linearity; see Section 4.3.) Fit this model and calculate the residuals.

2. Fit the graphical Gaussian model to the residuals.

Cox and Wermuth (1993, 1996) also describe corresponding models in which the errors follow a dual graphical Gaussian model (Section 7.4), and draw these with dashed instead of solid lines. These models correspond more closely to standard simultaneous equation models (which specify zeroes in the error covariance matrix rather than the precision matrix), but the Markov interpretation of the graphs seems unclear (at least to the present author). The component models are again easily fit in MIM, only this time fitting the dual model to the residuals in the way described in Section 7.4.

7.6 Reciprocal Graphs

Koster (1996) studied the Markov properties of simultaneous equation systems models and found that, for a (slightly) restricted subclass of these models, their Markov properties can be represented as reciprocal graphs. These graphs generalize chain graphs by permitting double-headed arrows between undirected path components. The double-headed arrows

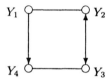

FIGURE 7.21. A reciprocal graph.

represent feedback loops, such as those arising in nonrecursive equation systems. Recall that with chain graphs, arrows only connect vertices in different undirected components; the same is true for reciprocal graphs for double-headed arrows. Figure 7.21 shows such a graph.

Koster derives pairwise, local, and global Markov properties, and states conditions for their equivalence. He also studies the applicability of the results to LISREL models (Jöreskog and Sörbom, 1989). These results clarify when it is legitimate to interpret LISREL path diagrams as conditional independence graphs; see also Koster (1999).

8

Causal Inference

The study of cause and effect relationships is central to much empirical research: consider, for example, studies to identify the causal risk-factors for a disease, or studies that attempt to predict the effect of policy changes. Graphical models may be very useful in such endeavours, by helping to reveal the associational structure of the data. In so doing they may often, as Cox and Wermuth (1996) put it, "point towards explanations that are potentially causal." Since the graphs, particularly the directed graphs, resemble causal networks, it is natural to interpret them in causal terms; perhaps this even happens unconsciously. The key question is therefore: When are such causal interpretations justified? Or in other words, is it ever legitimate to claim that an analysis provides *evidence* of a causal connection? In this chapter we attempt to clarify this and related issues.

Traditionally, statisticians have been reluctant to get involved in discussions about causality, the argument being that the question whether or not an analysis can be given a causal interpretation is a problem for the subject-matter specialist, not for the statistician. According to this view, statistics deals with correlations and associations, and the interpretion of these as causal relations is not a statistical issue. Statisticians should be aware of the dictum "Association does not imply causality" and advocate caution in drawing causal conclusions. In short, they should be applied mathematicians and not amateur philosophers.

It is certainly essential to distinguish between association and causality. Failure to do this is at the root of much misinterpretation of data. For example, a recent Danish study found that, on average, students who had held a year free between school and university got lower grades in their first-year examinations. It is very easy to summarize this as "If you go straight to university you will get better grades" — which implies that the relation is causal. Needless to say, it may well be nothing of the sort.

In recent years the statistical community has shown a more active interest in causal issues. Several types of argument can be made to support this change in attitude.

Firstly, statisticians and subject-matter specialists ought to — and fortunately often do — share responsibility for formulating the problem and for interpreting the analyses. So even though the assessment of causality involves many subject-matter aspects, statisticians share responsibility for any causal inferences. Consequently they need to understand issues arising in the interpretation of analyses, not just those arising in the analyses themselves.

Secondly, although association does not in general imply causality, it may do so sometimes. For example, results from well-conducted randomised studies are widely regarded as providing good evidence of causal relations. One goal of this chapter is to explain why this is so, and another is to examine whether it is possible to make valid causal inferences from other types of study.

Thirdly, there is a growing awareness that some statistical techniques, particularly those used to predict the effect of interventions, build on implicit causal assumptions. Typically, a statistical model specifies the joint distribution of variables in some system. If there is an external intervention to the system, such a model cannot tell us how the distribution is affected. It is *not* enough to calculate a conditional distribution. To see this, suppose that $Y = (Y_1, \ldots, Y_4)'$ follows a multivariate normal distribution, that is, $Y \sim \mathcal{N}(\mu, \Sigma)$. Suppose that we set Y_4 to k: how do we expect the system to react? We can calculate the conditional distribution of (Y_1, Y_2, Y_3) given $Y_4 = k$, using standard expressions given in (3.19) and (3.19). But suppose that the variables were generated by the recursive equation system

$$Y_1 := a_1 + \epsilon_1,$$
$$Y_2 := a_2 + b_2 Y_1 + \epsilon_2,$$
$$Y_3 := a_3 + b_3 Y_1 + c_3 Y_2 + \epsilon_3,$$
$$Y_4 := a_4 + b_4 Y_1 + c_4 Y_2 + d_4 Y_3 + \epsilon_4,$$

with independent errors $\epsilon_j \sim \mathcal{N}(0, \eta_j)$, for suitable parameters $a_1, \ldots, \eta_4$. This is distributionally identical to before. But a property of the recursive system is that setting Y_4 to k does not affect (Y_1, Y_2, Y_3) at all. The point is that if we want to predict how a system will respond to an external intervention, we need a *causal* model, not just a distributional one.

So, whenever we use a model to predict the effect of an intervention, we implicitly give the model a causal interpretation. As an illustration of how routinely this is done, consider the standard definition of a regression coefficient as the amount the response changes when the covariate is changed by one unit. If we suppose this to be true when we actively change the covariate, as opposed to just observing a change, then we are implicitly giving the model a causal interpretation. This may or may not be warranted. Most statistics textbooks fail to address such issues explicitly; presumably this leads to many wrongly interpreted analyses and poorly based predictions. See Box (1966) for a very clear description of the problem.

The structure of this chapter is as follows. The first section describes briefly some philosophical aspects. This is intended primarily as background to the following sections, which outline two statistical approaches to causal inference. The first of these is the framework associated with Neyman, Rubin, Holland, and others, often called Rubin's causal model. The second is an approach based on directed graphs, due to Pearl and others. The last section compares the two frameworks; they are found to differ somewhat in focus but not in substance. Some operational consequences are sketched.

8.1 Philosophical Aspects

What do we mean when we say that one event causes another? A very influential account was given by an eighteenth-century Scottish philosopher, David Hume. He wrote, "We may define a cause to be an object followed by another, and where all the objects, similar to the first, are followed by objects similar to the second." Or, in other words, "where, if the first object had not been, the second never had existed." So, according to the first definition, an event C causes an event E if (i) C and E occur, and (ii) there is a law or regularity that implies that whenever C occurs, E also occurs. The second definition states that C causes E if (i) C and E occur, and (ii) if C had not occurred, E would not have occurred. Hume emphasizes that the causal relation is not directly observed; all that is perceived is the conjunction of the two events. An observer infers that the relation is causal when the two events are an instance of some causal law or regularity. Hume also emphasizes that the two events must be contiguous in space and time, and that the cause C occurs before the effect E.

Hume's definitions taken together imply that C is necessary and sufficient for E. Often this is an oversimplification, in the sense that E may have alternative causes, and C may only cause E under certain conditions. For example, consider a fire (E) caused by an electric shortcircuit (C). The fire could have had other causes; for example, an oil stove could have overturned (A), and the shortcircuit only caused the fire since inflammable materials were nearby (B). Mackie (1965) extended Hume's criteria so as to accommodate this sort of complexity, by introducing the so-called *INUS* condition. This requires C to be an insufficient but necessary part of an unnecessary but sufficient set of conditions for E. So, in this example, C is a cause of E since B and C when occurring alone are insufficient for E, but when occurring together become sufficient. So C is a necessary part of a sufficient set of conditions, $C \cup B$. Furthermore, this set is unnecessary, in the sense that A could also cause E. Some difficulties with this account are described by Kim (1971).

Recall that Hume's second definition of causality says that C causes E if (i) C and E occur, and (ii) if C had not occurred, E would not have occurred. This last statement is an example of a *counterfactual conditional*, that is to say, a statement that involves conditioning on something that is counter to fact. Since C did occur, the event "C did not occur" is counter to fact. Some examples of counterfactual conditionals are: "If I had taken poison this morning I would now be dead," "If kangaroos had no tails, they would topple over" and "If pigs had wings, police cars would be obselete." Philosophers often regard such statements as problematic. It is, for example, difficult to see how the truth of such statements may be ascertained, since they refer to a contingency that did not occur. They may also be virtually meaningless. For example, consider the statement "If the Rocky mountains were 500 feet higher, Denver would be warmer in the winter." It is unclear whether Denver would even exist if the Rockies were higher. However, the statement could have meaning in the context of a theory relating the settlement of the American West to climate and physical geography. If such a theory predicted the existence of Denver under differing geographical conditions, then the statement would begin to make some sense.

Another way of expressing this is to say that counterfactual statements have meaning only when an underlying causal model is implicitly assumed. The counterfactual world in which C does not occur differs from the actual world in which it does, only in the nonoccurrence of the event C and its effects. Suppose, for example, that a stone causes a horse to stumble and its rider to fall. In the counterfactual world, the stone was not there, the horse did not stumble, and the rider did not fall. Here the plausibility of the counterfactual statement rests on a continuity assumption, namely that in the absence of the stone the horse would continue to run smoothly and its rider would not fall off.

So counterfactuals are closely related to causal statements. Lewis (1973) and others have developed theories of causality based on counterfactuals, invoking the concept of parallel worlds.

Hume's second definition implies that causes must be, at least potentially, manipulable, in the sense that it must be conceivable that the cause did *not* occur. So, for example, in most contexts gender cannot be considered a cause, because it gives no meaning to suppose that a given person could have a different gender; if they did, they would be a different person. Similar remarks apply to quantities like country, age, and the passage of time. Accounts of causality based on manipulability have sometimes been criticised. For example, we can say that a tornado causes damage without being able to manipulate it. But such criticism seems to presume that manipulation can be performed only by humans. This is an anthropomorphism; no one would deny that causal laws apply on the sun or any other places where humans cannot tread.

One aspect of the Humean account may seem surprising to modern statisticians. The requirement of constant conjunction means that causality is deterministic; that is, whenever the cause occurs (and possibly some extra conditions are fulfilled) the effect necessarily also occurs. This does not cover the meaning of a statement like "Smoking causes lung cancer," which clearly does not imply that everyone who smokes gets lung cancer.

Determinism may be defined as the view that it is universally true that, given the initial conditions and the laws of nature, only one outcome is possible. In other words, given the past, the future is fixed. Historically, the belief in determinism appears to have arisen in response to the success of Newtonian physics in predicting planetary motion, which suggested the idea of a "clockwork universe." Largely due to the prestige enjoyed by quantum theory, which is inherently indeterministic, determinism as an explicit doctrine has few proponents today. However, as Anscombe (1971) remarked, it remains part of the weltanschauung of western culture, surviving, for example, in deterministic conceptions of causality.

Recently, various authors have proposed theories of probabilistic causality; notably Good (1961), Suppes (1970), and Eells (1991). It is beyond the scope of this book to compare these accounts in any detail: we suffice with a very brief description of Eells's theory. This deals primarily with causal statements relating to *types of events*, rather than specific events. For example, in the statement "Smoking causes lung cancer" the terms "smoking" and "lung cancer" refer not to specific events but rather to types of events. In Eells's terminology, causal statements referring to event types are called *type-level* statements, while statements referring to specific events are called *token-level* statements.

Roughly speaking, Eells's account says that C has a causal influence on E if two criteria are satisfied: firstly, that C occurs before E, and secondly, that for some carefully chosen set of conditions $K_1, \ldots, K_n$,

$$\Pr(E|C, K_1, \ldots, K_n) \neq \Pr(E|\bar{C}, K_1, \ldots, K_n), \tag{8.1}$$

where $\bar{C}$ denotes the nonoccurrence of C. In other words, the occurrence of C influences the probability of the occurrence of E. This is equivalent to the conditional dependence statement $C \not\perp E|K_1, \ldots, K_n$. Eells supposes that C and E are discrete (binary) event types, but clearly the formulation in terms of conditional dependence can be applied to other types, for example events corresponding to real-valued measurements.

Crucial to Eells's approach is the view that these probability statements are relative to a population of singular events, as we now explain. Consider tossing a fair coin. We can suppose that the probability of heads for a general toss is $\frac{1}{2}$. But if we consider a *particular* toss, that is, with a specific coin tossed in a specific way at a specific height over a specific surface, at the point in time when the coin breaks contact with the thumb tossing it, the outcome may be inevitable, so the probability of heads would be zero or one. So the token-level probabilities differ in general from the corresponding type-level probabilities. According to Eells, expressions such as (8.1) implicitly involve sampling from a population of token-level events, in the sense that the probabilities manifest themselves as limiting frequencies in repeated sampling from the population.

Furthermore, the statements are relative to what Eells called a *causal background context*. This is related to the vague phrase "carefully chosen set of conditions" used above. The reason why an arbitrary set of conditions is not appropriate in (8.1) is that it is possible that the inequality vanishes if further conditions are added. For example, $\Pr(E|C) \neq \Pr(E|\bar{C})$ might hold, but this might reflect a spurious association due to a common cause K, so that $\Pr(E|C, K) = \Pr(E|\bar{C}, K)$. So we must require that the inequality does not vanish if any further conditions are added, and for this to make sense there must be a pool of possible conditions to draw from, and we must also assume that there are no further conditions (or hidden variables) outside this pool that could make the association spurious. This could be called a kind of *complete world* assumption.

The above discussion has been in terms of *events* and *conditions*. But what exactly is meant by these terms? In everyday speech we distinguish between events and processes, where the former are thought of as occurrences of short duration, and the latter as occurrences of long duration. When an event A causes an event B, it is reasonable to assume that B occurs after A and therefore does not also cause A. This is not the case with processes, since clearly two processes overlapping in time can each influence the other. But whether we regard a given occurrence as an event or a process seems arbitrary, or rather, context dependent. Consider, for example, a plate falling onto the floor and breaking. We can regard this as a simple event, which did or did not happen, or as a complex process, as when we view it in slow motion. This complex process can be broken down into a large number of simpler events. So before causal explanations can be given, the phenomenon being explained is broken down into conceptual entities at some appropriate level of detail, that is to say, events.

Of course, not all events are complex. For example, they can be simple changes of state, or states or other characteristics at a given time. Similarly, the role of conditions in the above account is primarily to define relevant subpopulations. So conditions are typically states or other characteristics at the time the cause occurs; note that they can sometimes be concomitant causes in their own right.

It seems likely that causality for processes needs a different philosophical framework. One such has been proposed by Salmon (1993). For a statistical account, see Eerola (1994); see also Section 7.3.

Other accounts of the philosophy of causality, as seen from various statistical perspectives, can be found in Holland (1986), Eerola (1994), Sobel (1996), and Pearl (2000). The philosophical literature is vast; Sosa and Tooley (1993) give a good introduction to a variety of modern views.

As we have seen in this section, the central philosophical issue related to causality is What do we mean by a cause? We now turn to some statistical accounts of causal inference. The primary issue here is, How do we assess a causal effect? The questions are of course related, since if there is a causal effect then there must be a cause.

8.2 Rubin's Causal Model

This section describes a framework for causal inference that was developed in a series of papers by Rubin, Holland, Rosenbaum, Stone, and others. The basic idea was introduced by Neyman (1923) in the context of randomised experiments. Current interest in the topic is due to work by Rubin (1974), who extended Neyman's model to nonrandomised experiments; for this reason, it is often referred to as Rubin's Causal Model. The discussion paper by Holland (1986) is an accessible introduction to the topic that gives further references and also draws connections to philosophical accounts.

The basic idea is that of a number of cases or *units*. These are the objects of study on which causes may act. Causes are called *treatments*, reflecting the origin of the ideas in designed experiments. For simplicity we suppose that there are only two treatments, written a and c (active and control). We also suppose that the response Y is real-valued.

The key to the framework is the following notation. The response of the unit i to a is written $Y(i, a)$, and the corresponding response to c is written $Y(i, c)$. This implicitly assumes that each unit is potentially exposable to both treatments, and that the response depends only on (i) which unit it is, and (ii) which treatment is applied to the unit. We examine more closely assumptions (i) and (ii).

Assumption (i) implies that the notion of a "unit" must include all relevant circumstances. Suppose, for example, that we are investigating the effect of an injection of a drug a into a patient. The response may depend not only on the drug and the patient, but also on, say, the blood pressure of the patient when the injection is made, the time of day, and a host of other factors. Then the relevant notion of "unit" is the collection of factors {patient, blood pressure, time of day, ... }. If, after the injection of a, the patient is also injected with a control drug c, then some of these factors will have changed. It follows that only one of $Y(i, a)$ and $Y(i, c)$ can be observed, the other being hypothetical. In other words, the framework is counterfactual. The two worlds (the actual and the counterfactual) are assumed to differ only in which treatment is applied, and any effects this may have.

Assumption (ii) implies that the response to the unit does not depend on which treatments were applied to the other units — in other words, that there is no interference between units. This is known as the stable unit treatment value assumption, or SUTVA for short.

It is also assumed that each treatment event is the same, at least approximately, for all units. Subtle issues may be involved here. Consider a study in which patients are allocated to one month's treatment with either tablet A or tablet B, to be taken, say, three times daily. Some patients may fail to follow this regimen, for various reasons. In that case, we may choose either to regard the treatment event as the allocation to a treatment group or as following the prescribed regimen. The former leads to the so-called intent-to-treat analysis, which compares the randomised treatment groups, and the latter leads to the so-called per protocol analysis, in which non-compliant patients are omitted from the comparison. Note that the latter construct is more complex; often compliance will depend on the treatment received, making the per protocol analysis subject to bias and difficult to interpret.

8.2.1   Estimating Causal Effects

Since the response is taken to be real-valued, we can define the causal effect on the unit i to be the difference

$$C(i) = Y(i, a) - Y(i, c).$$

This is often called the unit-level causal effect. Of course, it cannot in general be observed directly, since either $Y(i, a)$ or $Y(i, c)$ is unobserved. Nevertheless, if certain assumptions are made, we can make inferences about it.

One such assumption is *temporal stability*. If we can assume that the response does not depend on when the treatment is applied to the unit and whether the other treatment has previously been applied, then we can simply apply each treatment to i. This enables us to observe $Y(i, a)$ and $Y(i, c)$ directly and so calculate the unit-level causal effect. This is a great simplifying assumption, which seldom seems to be appropriate in statistical applications.

Alternatively we might assume *unit homogeneity*. This states that the units are entirely homogeneous; in other words, for all units i, $Y(i, a) = y_a$ and $Y(i, c) = y_c$, say. Then, if we apply a to one unit i_1 and c to another unit i_2, we obtain the causal effect as $Y(i_1, a) - Y(i_2, c)$. Of course, this is quite unrealistic.

So unless very strong assumptions can be made, the unit-level causal effects are inaccessible. The statistical way around this is to consider group effects. Two approaches to this can be adopted, one using the concept of average causal effect, the other using assumptions of additivity. The average causal effect C may be defined as the average value of $C(i)$ over the population. We have that $C = \mathrm{E}(Y_a) - \mathrm{E}(Y_c)$, where $\mathrm{E}(Y_a)$ and $\mathrm{E}(Y_c)$ are the average responses over the population if every unit were exposed to treatment a or c, respectively. Of course, $\mathrm{E}(Y_a)$ and $\mathrm{E}(Y_c)$ are counterfactual and cannot be observed. What we can observe are the average responses over the units exposed to treatment a and c, respectively. We write these as $\mathrm{E}(Y_a | Z = a)$ and $\mathrm{E}(Y_c | Z = c)$, where Z is the treatment allocation. Following Holland (1986), we define the *prima facie causal effect* to be the estimate of the average causal effect based on these quantities, that is,

$$C_{PF} = \mathrm{E}(Y_a | Z = a) - \mathrm{E}(Y_c | Z = c).$$

The term *prima facie* means that it is what on first view appears to be the causal effect; we could also call it the *apparent* causal effect.

In general the prima facie causal effect is biased. This will occur, for example, if $C = 0$, and Z is set to a whenever $Y_a > Y_c$. Then it is easy to see that $\mathrm{E}(Y_a | Z = a)$ will overestimate $\mathrm{E}(Y_a)$ and $\mathrm{E}(Y_c | Z = c)$ will understimate $\mathrm{E}(Y_c)$, and so C_{PF} will overestimate C.

The key to avoiding this bias lies in the treatment allocation mechanism. If units are allocated treatments using a known random mechanism, then the prima facie causal effect is unbiased. To see this in a simple case, consider an experiment with fixed numbers n_a of treated units and $n_c = N - n_a$ of controls, allocated at random. Then

$$E(Y_a|Z = a) = \frac{1}{n_a} \sum_i Y(i,a) \Pr[Z(i) = a]$$

$$= \frac{1}{n_a} \sum_i Y(i,a)(\frac{n_a}{N})$$

$$= E(Y_a),$$

so $E(Y_a|Z = a)$ and similarly $E(Y_c|Z = c)$ are unbiased, and therefore $E(C_{PF}) = C$. It is not difficult to generalize this result to all randomised designs in which $Z \perp\!\!\!\perp I$, that is, the treatment allocation is independent of the unit.

Another common study design is stratified randomisation, in which randomisation is performed within each stratum. This design is often used with multicentre clinical studies, where the different clinical centres serve as strata. For such designs $Z \perp\!\!\!\perp I|S$, where S is the stratum variable. Within each stratum the prima facie causal effect is unbiased for the average causal effect, which we can write as $E(C^s_{PF}) = C^s$ for each stratum s. We might be primarily interested in these stratum causal effects, and leave it at that; or we might be interested in the overall average causal effect $C = \sum_s (\frac{n_s}{N})C^s$, in which case the adjusted prima facie effect $C^*_{PF} = \sum_s (\frac{n_s}{N})C^s_{PF}$ would be appropriate. It is easy to see that this is unbiased for C.

The unit-treatment additivity, or constant effect, assumption states that the unit-level causal effect is constant for all units, i.e., $C(i) = C$ for all units i. The first thing to note is that this assumption is *not* sufficient to remove bias in C_{PF}. Generally $E(Y_a|Z = a)$ and $E(Y_c|Z = c)$ will be biased estimates of $E(Y_a)$ and $E(Y_c)$, and so C_{PF} will also be biased. So from the present perspective additivity is not useful, and in most contexts is rather implausible; nevertheless, it underlies much statistical practice, for example in the analysis of variance. It is of course more convenient to parametrize a model with one parameter (here called C) for treatment effect, rather than refer to an average of counterfactual quantities. It can also be noted that although the additivity assumption is not directly verifiable (since $Y(i,a)$ and $Y(i,c)$ cannot both be observed), it can certainly be falsified. For when Z is random, additivity implies that the observed response distributions, i.e., that of $\{Y(i,a) : Z(i) = a\}$ and that of $\{Y(i,c) : Z(i) = c\}$, should differ only by a location shift.

8.2.2   Ignorability

So far, we have been able to derive unbiased estimates of the causal effect only in randomised studies, that is, in studies where the allocation of treatment to unit is under the complete control of the experimenter. In many fields, such studies cannot be performed. For example, in brain science it is not currently possible to alter the concentration of a neurotransmitter at one site without altering many other things at the same time. Similarly, in studies of the influence of lifestyle factors on health, one cannot dictate to the subjects in the study which lifestyle they should follow. Is it not possible to estimate causal effects in such studies also? The answer is yes, as long as the treatment allocation is ignorable, in the sense we now describe. The criterion is due to Rubin (1978); see also Rosenbaum and Rubin (1983, 1984) and Rosenbaum (1984). We give an informal account here.

The basic idea is straightforward. For any set of covariates X, we know that

$$E(C_{PF}) = \sum_x \Pr(X = x)[E(Y_a|X = x, Z = a) - E(Y_c|X = x, Z = c)].$$

If we can find a set of covariates X such that $(Y_a, Y_c) \perp\!\!\!\perp Z|X$, then it follows that $E(Y_a|X = x, Z = a) = E(Y_a|X = x)$ and similarly that $E(Y_c|X = x, Z = c) = E(Y_c|X = x)$, and so we obtain that

$$E(C_{PF}) = \sum_x \Pr(X = x)[E(Y_a|X = x) - E(Y_c|X = x)]$$
$$= E(Y_a) - E(Y_c)$$
$$= C,$$

that is, C_{PF} is unbiased. We say that the treatment allocation Z is *strongly ignorable given covariates* X if $(Y_a, Y_c) \perp\!\!\!\perp Z|X$. For technical reasons that we skip here, we also require that $0 < \Pr(Z = a|X = x) < 1$ for all x. This is the so-called SITA (strongly ignorable treatment allocation) condition. It is of course counterfactual, since Y_a and Y_c cannot both be observed, and so it is not possible to check directly whether or not the condition holds.

The SITA condition can be represented graphically, as in Figure 8.1, and sometimes this may be helpful when judging whether it is a reasonable assumption to make. We suppose that treatment allocation arises in a fashion that depends on X and possibly some other, unmeasured covariates U_z. Similarly, we suppose that the response-pair (Y_a, Y_c) depends on X and

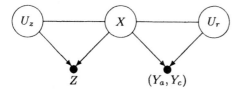

FIGURE 8.1. The SITA condition. U_z and U_r are sets of unobserved covariates, X is the set of observed covariates, Z is the treatment allocation and (Y_a, Y_c) is the counterfactual response-pair.

possibly some other, unmeasured covariates U_r. The sets U_z and U_r are distinct and separated by X. So no unmeasured covariate is a determinant of both the treatment allocation and the response-pair. If these conditions are fulfilled, then the assignment is ignorable. If they are not (so that there is an edge from some node in U_z to (Y_a, Y_c), from some node in U_r to Z, or from some node in U_z to some node in U_r) then the assignment is not ignorable. More precise versions of this sort of reasoning are given in Sections 8.2.4 and 8.3.3.

8.2.3   Propensity Score

Closely related to ignorable treatment assignment is the construct called the *propensity score*. This is the probability of receiving a given treatment given the observed covariates, $\Pr(Z = a | X = x) = s(x)$, say. Rosenbaum and Rubin (1983) show that if the treatment assignment Z is ignorable given covariates X, then $Y_a \perp\!\!\!\perp Z | s(X)$ and $Y_c \perp\!\!\!\perp Z | s(X)$, from which it follows as before that

$$\mathrm{E}(Y_a | Z = a) = \sum_s \Pr(s(X) = s)\mathrm{E}(Y_a | s(X) = s, Z = a)$$

$$= \sum_s \Pr(s(X) = s)\mathrm{E}(Y_a | s(X) = s)$$

$$= \mathrm{E}(Y_a)$$

and similarly $\mathrm{E}(Y_c | Z = c) = \mathrm{E}(Y_c)$, so that $\mathrm{E}(C_{PF}) = C$, as required.

In effect, we divide the observations into strata on the basis of their probability of receiving the active treatment. We can regard observations in the same stratum as being assigned to treatment at random; in other words, we can consider the study to be a "quasi-randomised" experiment.

Rosenbaum (1984) develops a rationale for randomisation inference conditional on the propensity score. See also Rosenbaum (1995) and D'Agostino (1998).

So if we know the propensity score and that treatment allocation is ignorable, then we know that our estimate of causal effect is unbiased. Conditioning on the propensity score may often be technically easier than conditioning on the X (for example, if X consists of 10 binary variables, then there would be 2^{10} strata).

A limiting factor to the usefulness of these results is that it is rare in observational studies that the propensity score is known. We might assume that we know its functional form, and estimate it from the data— but this assumption, however convenient, may often be unwarranted (like the assumption of ignorability).

8.2.4   Causal Hypothesis Testing

So far in this section we have described the framework for causal inference developed by Rubin, Holland, and Rosenbaum in the papers cited. The focus in this work is on unbiased estimation of causal effects. We now turn to causal hypothesis testing, following Stone (1993). He modified the basic framework slightly, by including unobserved variables that in a sense replace the individual units. Since Pearl's account is also based on this framework, it is convenient to describe it now.

Let U be a set of unobserved covariates, and as before let I represent the unit, X the observed covariates, and Z the treatment allocation. We suppose that (X, U, Z) constitutes a *complete set of causal determinants* of Y, by which we mean that $Y \perp\!\!\!\perp I | (X, U, Z)$; in other words, given knowledge of X, U, and Z for a unit, no further information about that unit is relevant for the distribution of Y. Implicitly, we envisage a graph in which the variables adjacent to Y are X, U, and (possibly, this being at issue) Z. This is represented in Figure 8.2.

In philosophical terms, when we modify the causal framework in this way, passing from unit-level quantities such as $Y(i, a)$ to group-level ones such as $f_{Y|X,U,Z}(y|x, u, z)$, we are in effect making a transition from singular to general causal statements. In Eells's terminology, probability statements at the latter level are relative both to a causal background context, which consists of a set of conditions (here corresponding to U and X), and to a population of singular events.

Comparing Figures 8.2 and 8.1, we can identify the set of variables U in Figure 8.2 with U_r in Figure 8.1. The variables U_z in Figure 8.1 are omitted from the present account since they are not part of the complete set of determinants for Y.

Using the observed data, we can test whether $Y \perp\!\!\!\perp Z | X$; we call this the hypothesis of *no association*. When does this correspond to a test of causality? To answer this we can consider various causal hypotheses.

The most immediate, which we call *no unit-level causality*, states that Z is not a member of a complete set of determinants for Y, which corresponds to

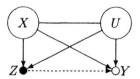

FIGURE 8.2. A framework for causal inference: Z is the treatment allocation, X and U are sets of observed and unobserved covariates, respectively, and Y is the response.

the relation $Y \perp\!\!\!\perp Z \mid (X, U)$. In Figure 8.2 this corresponds to the absence of the edge between Z and Y. We use the expression *unit-level* since if for some units (corresponding to some specific values of X and U) the distribution of Y differs over values of Z, then the hypothesis does not hold.

When is no association equivalent to no unit-level causality? A trivial condition is that of *covariate sufficiency*, i.e., that U is empty. If we can assume that our set of observed variables is so rich that further information about the unit is irrelevant to the distribution of Y for that unit, then clearly the hypotheses of no association and of no unit level causality coincide. This assumption is very strong and almost always implausible. Similarly, $Y \perp\!\!\!\perp U \mid (X, Z)$ would also ensure the equivalence of no association and no causality; this, however, would imply that U can be omitted from the complete set of causal determinants for Y.

So, in statistical applications at least, knowledge of unit-level causality is inaccessible, and some weaker hypothesis is necessary. Consider the hypothetical distribution of Y that would arise if the whole population received treatment $Z = z$. If, for each $X = x$, this distribution is the same for all z, then we say there is *no distribution effect*. Taking, for ease of notation, X and U to be discrete, and using f as a generic density symbol, we require that for each x,

$$\sum_u f(y|x, u, z) \Pr(x, u) \tag{8.2}$$

does not depend on z.

Note that the factor $\Pr(z|x, u)$ is omitted from (8.2) because we are not *conditioning on* $Z = z$ but rather *setting* Z to z. The distinction should become clearer in the next section.

Now we may ask: When is no association equivalent to no distribution effect? To answer this, we need the concept of *confounding*. Roughly speaking, we say that there is confounding if there are unobserved variables that affect the response and that are not independent of the treatment given the observed covariates. In the present context, all the variables in U are assumed to be in the complete set and thus affect the response, so no confounding corresponds simply to $U \perp\!\!\!\perp Z \mid X$. Observe that

$$f(y|z, x) = \sum_u f(y|x, u, z) \Pr(u|z, x).$$

If $U \perp\!\!\!\perp Z \mid X$, then $\Pr(u|z, x) = \Pr(u|x)$, giving

$$f(y|z, x) = \sum_u f(y|x, u, z) \Pr(u|x),$$

from which the equivalence of no association and no distributional effect follows directly.

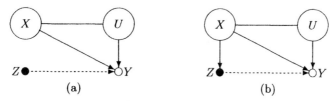

FIGURE 8.3. The causal framework for randomised studies. Z is the treatment allocation, X and U are sets of observed and unobserved covariates, respectively, and Y is the response. The first graph, (a), represents simple randomisation, in which $Z \perp\!\!\!\perp (X, U)$, and the second graph, (b), represents stratified randomisation, in which $Z \perp\!\!\!\perp U | X$. Both designs satisfy the criterion for no confounding, namely that $Z \perp\!\!\!\perp U \mid X$. So for both designs a test of no association, i.e., of $Y \perp\!\!\!\perp Z | X$, is a valid test of no causal effect of Z on Y (where this is understood as a test of no distribution effect).

For randomised studies, treatments will be assigned in such a way that either $Z \perp\!\!\!\perp (X, U)$ (simple randomisation) or $Z \perp\!\!\!\perp U | X$ (stratified randomisation, or randomised blocks), in this way ensuring that there is no confounding. So for such studies, a test of no association is equivalent to a test for no distributional effect. This is illustrated in Figure 8.3.

The requirement of no confounding can be slightly weakened, since the ignorability conditions also lead to the equivalence of no association and no distributional effect. To see this, observe that $Y_a \perp\!\!\!\perp Z | X$ implies that $f(y_a | x)$ is identical to $f(y_a | x, z)$ for each z; in particular, it is identical to $f(y_a | x, Z = a)$ and thus to $f(y | x, Z = a)$. In the same way, $Y_c \perp\!\!\!\perp Z | X$ implies that $f(y_c | x) = f(y | x, Z = c)$. The assumption of strongly ignorable treatment assignment states that $(Y_a, Y_c) \perp\!\!\!\perp Z | X$. It follows that $f(y_a | x) = f(y_c | x)$ is equivalent to $f(y | x, Z = a) = f(y | x, Z = c)$. In other words, no distribution effect is equivalent to no association.

To confirm that the ignorability conditions are indeed weaker than no confounding, we show that the latter implies the former. Now

$$U \perp\!\!\!\perp Z | X \Rightarrow (X, U) \perp\!\!\!\perp Z | X \Rightarrow g(X, U, z) \perp\!\!\!\perp Z | X$$

for any function g and fixed z. Since (X, U, Z) is a complete set of causal determinants for Y, we can regard Y as given by $g(X, U, Z)$ for some unknown g, and hence that $Y_a = g(X, U, a)$ and $Y_c = g(X, U, c)$. The result follows. Note that this proof supposes that Y is deterministic given (X, U, Z) but is easily modified when Y is stochastic. Note also that it does not involve *conditioning* on $Z = z$ but rather *setting* Z to z. Both points should become clearer in the next section.

A •————————→• B

FIGURE 8.4. A Simple Causal Model.

8.3 Pearl's Causal Graphs

We now move to an account of causal inference rooted in directed acyclic graphs (DAGs). This has been developed in a series of papers by Pearl and various of his coworkers; some of these results build on work by other authors, notably Strotz and Wald (1960), Robins (1986), Pearl and Verma (1991), and Spirtes et al. (1993). The results described in this section are discussed more fully in Pearl (1995a, 2000). Lauritzen (1999) gives a good overview, with special focus on the intervention calculus sketched below in Sections 8.3.3 and 8.3.4.

8.3.1 A Simple Causal model

Consider the graph shown in Figure 8.4. We want the graph to represent a simple causal system, so that the arrow from A to B can be taken to mean that A causes B. It can be argued that the key feature distinguishing causal models from associational models is the ability to describe the behaviour of the system under *intervention*. Accordingly, we introduce a new symbol to denote intervention. We define $\Pr(B = j|\mathrm{do}(A = i))$ to mean the probability that B is j when A has been set to i by an action external to the system. Similarly, $f(Y|\mathrm{do}(X = x))$ means the density of Y when X has been set to x. For notational convenience, we focus on discrete variables in this section, but the results are essentially independent of variable type.

In expressions involving both intervention and conditioning, the convention is that conditioning takes place after intervention, so that, for example, $\Pr(A = a|B = b, \mathrm{do}(C = c))$ is equivalent to

$$\Pr(A = a, B = b|\mathrm{do}(C = c))/\Pr(B = b|\mathrm{do}(C = c)).$$

Armed with this new symbol, consider Figure 8.4 again. Since A is prior to B, setting B to j does not affect A, so

$$\Pr(A = i|\mathrm{do}(B = j)) = \Pr(A = i) \neq \Pr(A = i|B = j).$$

And since we want the model to describe the behaviour under intervention, we want it to hold that

$$\Pr(B = j|\mathrm{do}(A = i)) = \Pr(B = j|A = i) \neq \Pr(B = j).$$

So the arrow in Figure 8.4 expresses an asymmetry with respect to intervention, an asymmetry that illustrates the importance of distinguishing between intervention and conditioning.

Suppose we sample A and B from some population and use the data to estimate a joint distribution as $\hat{p}_{ij} = n_{ij}/N$. We want to estimate the postintervention distribution of B as

$$\hat{\Pr}(B = j|\mathrm{do}(A = i)) = \hat{\Pr}(B = j|A = i) = \hat{p}_{ij}/\hat{p}_{i+}. \qquad (8.3)$$

When is this justified?

The problem is that there could be an unobserved second cause of B, say C, that also influences A, so that the observed association between A and B is partly, or perhaps completely, due to this common influence of C. This is shown in Figure 8.5. We say that C *confounds* the effect of A on B. If there is such a confounder, then setting A to i would affect the distribution of B differently, and not in accordance with the above expression.

Note that C need not be a *cause* of A and B in the sense of being potentially manipulable, but could be a *determinant* of A and B — that is to say, a condition that affects the occurrence of A and B.

To justify using the expression (8.3), we need to be able to assume that the occurrence of B, though dependent on the prior state of A, does not also depend on the process determining this state. When this assumption is correct, the probability of the event $B = j$ occurring after the event $A = i$ will not depend on how $A = i$ arose, that is, whether due to an intervention or a passive observation. So the estimate (8.3) will be appropriate. The presence of a confounder would violate the assumption, since it would give rise to an association between B and the process giving rise to A.

So, to interpret the graph shown in Figure 8.4 as a causal graph, we need to assume that it gives, as it were, the whole story — that there are no confounding variables present in reality but absent from the graph. This is a stronger assumption than is usually required in statistical modelling. Normally, we only require that the models we use be (approximately) *true*, but to interpret them as causal models we also need them to be (approximately) *complete* in the sense just described. Needless to say, this may often be challenging.

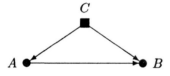

FIGURE 8.5. The effect of A on B is confounded by C.

8.3.2   Causal Graphs

Formalizing the above considerations, Pearl (1995a) specifies assumptions under which a system represented by a given DAG, say $\mathcal{G}$, can be interpreted causally. These are as follows.

Causal Graph Assumptions: *For each i, the process generating V_i takes the form of a functional mechanism $V_i := f_i(pa_i, \varepsilon_i)$, where pa_i denotes the parents of V_i in $\mathcal{G}$, and ε_i is a stochastic term, representing an error or disturbance. Furthermore, the $\{\varepsilon_1, \ldots, \varepsilon_n\}$ are mutually independent.*

It follows from these assumptions that the joint distribution factorises according to $\mathcal{G}$ (see Section 7.1), that is,

$$\Pr(V_1 \ldots V_n) = \Pi_{i=1\ldots n} \Pr(V_i|pa_i). \tag{8.4}$$

Note that in practice we do not actually need to *identify* the functions $f_i()$; we just need to know that such functions exist. Instead of the functional mechanism assumption, we could require that each pa_i constitutes a complete set of causal determinants for V_i. However, the functional mechanism formulation makes it clear that the data generation process is stable, in the sense that the f_i remain unaffected when other elements in the system are subject to external intervention.

Similarly, we do not need to know the distribution of the error terms ε_i; we just need to know that they are mutually independent. In applications, this requirement can often be problematic. To model this, if, say, ε_i and ε_j cannot be assumed independent, we can add a new node with arrows to V_i and V_j to $\mathcal{G}$. This new node represents an unmeasured variable. For example, in Figure 8.4 the assumptions imply the existence of functions f_a and f_b, and random variables ε_a and ε_b such that

$$A := f_a(\varepsilon_a), \ B := f_b(A, \varepsilon_b) \text{ and } \varepsilon_a \perp\!\!\!\perp \varepsilon_b.$$

When there is confounding, the occurrence of B depends on the process determining A, so that $\varepsilon_a \not\perp\!\!\!\perp \varepsilon_b$. To model this, we can add a third node with arrows to A and B, resulting in Figure 8.5. The wider implication is that realistic causal graphs for data from observational studies will generally include unmeasured variables.

The key property of the system we have described is this. If we intervene in the data generating process by setting V_k to v, say, then we replace the functional mechanism $V_k := f_k(pa_k, \varepsilon_k)$ by a new one $V_k := v$, but the stability properties mean that the rest of the system remains intact. In other words, the remaining conditional distributions $\Pr(V_i|pa_i)$ for $i \neq k$ are unaffected. In (8.4) this removes the factor $\Pr(V_k|pa_k)$, so that the joint

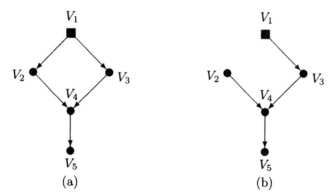

FIGURE 8.6. Intervention surgery in causal graphs. The original graph is shown in (a). Intervening on V_2 corresponds to the removal of all arrows pointing into V_2, as shown in (b).

distribution becomes

$$\Pr(V_1 \ldots V_n | \mathrm{do}(V_k = v)) = \begin{cases} \Pi_{i \neq k} \Pr(V_i | pa_i) & \text{if } V_k = v \\ 0 & \text{otherwise.} \end{cases} \quad (8.5)$$

In graphical terms, the arrows pointing into V_k are removed from the graph.

It should be noted that some authors (for example, Spirtes et al., 1993, and Lauritzen, 1999) prefer to regard this property of stability of conditional distributions encoded in (8.5) as primitive, instead of deriving it from the functional mechanisms f_i and error terms ε_i. An advantage of this approach is that it makes it clear that not all variables need be manipulable.

By applying (8.5) we can calculate the effects of interventions: this is known as the *intervention calculus*. In illustration of this, consider the graph shown in Figure 8.6(a). The square node symbol indicates that V_1 is unmeasured. The graph implies that the preintervention distribution factorizes as follows:

$$\Pr(V_1 \ldots V_5) = \Pr(V_1) \Pr(V_2 | V_1) \Pr(V_3 | V_1) \Pr(V_4 | V_2, V_3) \Pr(V_5 | V_4).$$

Suppose that there is interest in estimating the causal effect of V_2 on V_4. Can this be done using the observed data? If the assumptions described above hold, then the conditional distribution $\Pr(V_4 | V_3, V_2)$ is unaffected by interventions on V_2 (and V_3). So we can use the observed data to estimate the conditional distribution $\Pr(V_4 | V_3, V_2)$, and interpret this as an estimate of the postintervention distribution $\Pr(V_4 | V_3, \mathrm{do}(V_2))$.

In other words, if the assumptions hold, then the estimates from the observed conditional distribution can be interpreted causally. Depending on the purpose of the analysis, we might be primarily interested in the V_3-specific causal effects of V_2 on V_4, $\Pr(V_4|V_3, \mathrm{do}(V_2))$, or alternatively in the overall average causal effect $\Pr(V_4|\mathrm{do}(V_2)) = \sum_{V_3} \Pr(V_3) \Pr(V_4|V_3, \mathrm{do}(V_2))$.

The same remarks apply when we use constrained parametric models to estimate $\Pr(V_4|V_2, V_3)$. For example, we might base this estimate on a log-linear model for $\{V_2, V_3, V_4\}$ with no three-factor interaction, or if V_4 were real-valued, on an additive model.

We could also have derived these results directly from (8.5). Intervention on V_2 results in the distribution

$$\Pr(V_1, V_3, V_4, V_5|\mathrm{do}(V_2)) = \Pr(V_1) \Pr(V_3|V_1) \Pr(V_4|V_2, V_3) \Pr(V_5|V_4)$$

corresponding to Figure 8.6(b). After a little manipulation we obtain that $\Pr(V_4|V_3|\mathrm{do}(V_2)) = \Pr(V_4|V_2, V_3)$, as required.

In Pearl's terminology, we have just shown that the causal effect $\Pr(V_4|\mathrm{do}(V_2))$ is *identifiable*. That is to say, when dealing with a causal graph with unmeasured variables, we call a causal effect $\Pr(Y|\mathrm{do}(Z = z)$ identifiable if it can be estimated from the observed data. One of the main goals of Pearl (1995a) is to find conditions for identifiability. A simple example of a unidentifiable causal effect is shown in Figure 8.5, where

$$\Pr(B = b|\mathrm{do}(A = a)) = \sum_c \Pr(C = c) \Pr(B = b|A = a, C = c),$$

but since C is unobserved, this quantity is unavailable.

8.3.3   The Back-Door Criterion

One general condition for identifiability is the *back-door criterion*, so called because it involves paths from the effect Y to the cause Z going through the parents of Z — as it were, via the back door. A set of variables S is said to satisfy the back-door criterion relative to (Z, Y) if (i) no node in S is a descendant of Z, and (ii) S blocks every path between Z and Y that contains an arrow into Z. (See Section 7.1.1 for the definition of blocking).

The following general result is proved in Pearl (1995a). If there exists a set S, consisting of measured variables, that satisfies the back-door criterion relative to (Z, Y), then $\Pr(Y|\mathrm{do}(Z = z))$ is identifiable and is given by

$$\Pr(Y|\mathrm{do}(Z = z)) = \sum_s \Pr(Y|Z, S = s) \Pr(s).$$

This is illustrated in Figure 8.7(a). The sets $\{V_3, V_4\}$ and $\{V_4, V_5\}$ fulfill the back-door criterion relative to (Z, Y). So, for example, even when V_1, V_2,

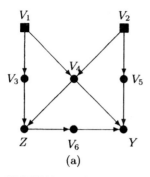 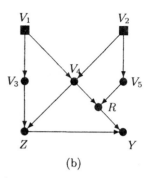

(a)   (b)

FIGURE 8.7. The back-door and SITA conditions. In Figure (a), the sets $\{V_3, V_4\}$ and $\{V_4, V_5\}$ block all back-door paths between X and Y, but $\{V_4\}$ does not block the path $(Z, V_3, V_1, V_4, V_2, V_5, Y)$. Figure (b) represents the same system in Rubin's framework. The variable R contains the counterfactual response-pair (Y_a, Y_c).

V_5, and V_6 are unmeasured, $\Pr(Y|\mathrm{do}(Z = z))$ is identifiable and is given by

$$\Pr(Y|\mathrm{do}(Z = z)) = \sum_{v_3, v_4} \Pr(Y|Z, V_3 = v_3, V_4 = v_4) \Pr(v_3, v_4).$$

An immediate corollary of the criterion is that if all the parents of Z are measured, then $\Pr(Y|\mathrm{do}(Z = z))$ is identifiable.

In the causal graph context, the back-door criterion is equivalent to the SITA condtion described in Section 8.2 (Galles and Pearl, 1998). We illustrate this using the example shown in Figure 8.7(a). Figure 8.7(b) represents the same system in Rubin's framework. We explain how this is derived from Figure 8.7(a). First, since Rubin's framework incorporates only one posttreatment variable (the response), the variable V_6 is marginalised over. So the node V_6 is removed and an arrow from Z to Y is added. Then a variable $R = (Y_a, Y_b)$, representing the potential responses when $Z = a$ and $Z = b$, is added to the graph. Arrows are drawn to R from the parents of Y, that is, from V_4 and V_5. Note that no arrow is drawn from Z to R. This is because R contains responses to both $Z = a$ and $Z = b$, so the actual state of Z is not relevant. Finally, an arrow is drawn from R to Y. Note incidentally that here we know the form of the functional mechanism for Y, namely, $Y := I[Z = a]Y_a + I[Z = c]Y_c$, where $I[\,]$ is the indicator function.

To see the close connection between the SITA condition and the back-door criterion, recall that the former was that $R \perp\!\!\!\perp Z|S$ for some set of measured variables S. From Figure 8.7(b), using the d-separation criterion, we see that $R \perp\!\!\!\perp Z|S$ holds for two sets S; namely, $\{V_3, V_4\}$ and $\{V_4, V_5\}$. So the two criteria are identical.

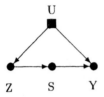

FIGURE 8.8. An example of the front-door criterion for identifiability. The causal effect $\Pr(Y|do(Z = z))$ is identifiable.

This example illustrates how the potential response-pair R in Rubin's framework can be characterized graphically. The parents of R are the parents of Y after marginalising over variables on paths from Z to Y. For example, if there was an arrow from V_1 to V_6 in Figure 8.7(a), then Figure 8.7(b) should include an arrow from V_1 to R.

The back-door criterion is best regarded as supplying a precise condition for lack of confounding in the causal graph framework. Greenland, Pearl, and Robins (1999) compare it to traditional epidemiological criteria for confounding, stressing limitations of the latter when there are multiple potential confounders. They also make the important but uncomfortable point that adjusting for a set of variables that do *not* fulfill the back-door criterion may result in an *increase* in bias. So partial or incorrect adjustment for confounders may be worse than no adjustment at all. The same point is made by Spirtes et al. (1998).

8.3.4 The Front-Door Criterion

Another condition for identifiability is the *front-door criterion*. This has no counterpart in the Rubin framework, and at first sight seems somewhat counterintuitive. We refer to Pearl (1995a) for the general formulation of the criterion, and suffice with a single example. Figure 8.8 shows a causal graph with four variables, resembling Figure 8.5 in the sense that an unmeasured variable U influences both the cause Z and the effect Y. In addition, a measured variable S mediates the effect of Z on Y; it is crucial that S is *not* directly affected by U. Without S, we have seen that $\Pr(Y|do(Z = z))$ is not identifiable, but when S is present the front-door criterion is fulfilled, so that $\Pr(Y|do(Z = z))$ becomes identifiable. It is given as

$$\sum_{s} \Pr(s|z) \sum_{z'} \Pr(y|z', s) \Pr(z').$$

A hypothetical application of this result is the following. Z is smoking, S is the presence of tar deposits in the lungs, Y is lung cancer, and U is an unmeasured genetic disposition to smoking and lung cancer. Provided that the effect of smoking on lung cancer is mediated by the tar deposits, and the tar deposits are not directly affected by the genetic disposition, the causal effect of smoking on cancer is identifiable. For further insight into this example and into the estimate of $\Pr(Y|\mathrm{do}(Z = z))$, see Pearl (1995b).

The front-door and back-door criteria do not exhaust all the possibilities for identifiability — causal effects can sometimes be identifiable even though neither criterion is fulfilled. See Pearl (2000) and Lauritzen (1999).

8.4 Discussion

In this section we briefly compare the approaches described in the previous sections, and then sketch some operational consequences.

8.4.1 Comparison of the Two Approaches

The major difference between the two frameworks is that Rubin's account is based on a population of units, and derives the group-level results carefully from these. In contrast, Pearl's account describes the group level only. Rubin's account posits the existence of a priori fixed potential or counterfactual outcomes at the unit level, whereas Pearl's account is explicitly stochastic at the group level.

At the group level, the two frameworks are very closely related. To see this, note that we can equate $f(Y_a)$ and $f(Y_c)$ in Rubin's account with the post-intervention distributions, $f(Y|\mathrm{do}(Z = a))$ and $f(Y|\mathrm{do}(Z = c))$, of Pearl's account. Similarly, if (X, U, Z) is a complete set of causal determinants for Y, then $f(Y_a|U = u, X = x)$ and $f(Y_c|U = u, X = x)$ in Rubin's account correspond respectively to $f(Y|U = u, X = x, \mathrm{do}(Z = a))$ and $f(Y|U = u, X = x, \mathrm{do}(Z = c))$ in Pearl's account.

In Pearl's account, the notion of causal effect is identified with the post-intervention distributions $f(Y|\mathrm{do}(Z))$, whereas in Rubin's account, it is defined as $\mathrm{E}(Y_a) - \mathrm{E}(Y_c)$. The former is more general. Rubin's definition is appropriate when Y is real-valued and Z binary, whereas Pearl's is applicable for general variable types. Note also that it is natural in Pearl's framework to define absence of causal effect in terms of the equality of postintervention distributions — $f(Y|\mathrm{do}(Z = a)) = f(Y|\mathrm{do}(Z = c))$, for example. This is precisely the property of no distributional effect described in Section 8.2.4.

Pearl's causal graphs permit one to work with multiple causes and effects, whereas Rubin's framework relates a single cause to a single effect. But Rubin's account is more general in another respect, since it does not assume any particular type of dependence structure, for example corresponding to a DAG. The SITA conditions are thus of wide applicability.

Rubin's framework focusses on identifying when unbiased estimates of causal effect can be obtained. But there will usually also be interest in increasing the precision of the estimates, by modelling the response in terms of the covariates; this is integral to Pearl's framework.

In summary, the difference between the two accounts seems to be more a matter of focus and notation than of substance.

8.4.2   Operational Implications

Suppose we want to study whether some event A causes another event B. Moreover, if it does, we want to estimate the magnitude of the effect. How should we go about this?

The obvious answer is: if possible, intervene. Twiggle A, as it were, and observe whether there is an effect on B. If A is a possibly defective light-switch, and B is a light, then we can turn A on and off a few times. Note that it is hardly necessary to randomise here; if we are convinced that we decide spontaneously when to turn it on and off, then this decision cannot be related to B except through the effect of the switch. In other words, volition corresponds to randomisation. In more complex circumstances, for example in allocating drugs to patients, such spontaneous decisions are hardly possible — it would be difficult to avoid unconscious bias, caused for example by tending to allocate the healthier patients to the new, and perhaps risky, drug — so randomisation is necessary.

If it is impractical, unethical, or for any other reason infeasible to perform a randomised intervention, then we must make do with passively observing the allocation of A. The validity of any causal inferences then rests on the validity of assumptions concerning the allocation process. For example, if it is known that there is a dependence between the allocation process and the response, but it can be assumed that this dependence is mediated by one or more measured variables, then valid causal inference is possible. Alternatively, if the putative effect is known to be mediated by other measurements (for example, by current in the wire from the switch to the light, or by the concentration profile of the drug in the blood), which are not associated with the allocation process, then this may also allow valid inference.

Thus, causal inferences from an observational study will be reliable only when there is good understanding of which factors determine the treatment allocation and which factors (apart from the putative cause under study) affect the response. Furthermore, these factors should, as far as possible, be recorded in the study.

In some contexts this may be difficult or impossible. Sometimes a useful approach may hinge on careful choice of the study context, as described by Rosenbaum (1999). Suppose, for example, that the objective is to study the economics effects of unwed motherhood. Since unmarried mothers cannot be assumed to be similar (in regard to their economic behaviour or situation) to unmarried women who do not give birth, differences between these groups cannot be interpreted as effects of unwed motherhood. Bronars and Grogger (1994) compared, instead, unwed mothers who gave birth to twins, with unwed mothers who gave birth to single children, since the event of giving birth to twins rather than a single child can reasonably be assumed to be independent of the mother's economic circumstances. In this way the allocation to comparison group, though not randomised, is nonetheless close; we may regard it as a natural experiment.

To illustrate his results, Pearl (1995a) describes an observational study where it is assumed that the data are generated according to a causal graph with both observed and latent variables. Under this assumption, the intervention calculus may suggest whether and how valid causal inferences may be made. But in most settings, the underlying structure is not known with any certainty. Sometimes it may be reasonable to consider and compare various alternative, plausible causal graphs, and use the framework to find out when and how causal inference is valid. There is a world of difference between (i) presenting a causal graph as the gospel truth and any conclusions from the analysis as unassailable, and (ii) carefully comparing alternative graphs expressing well-grounded scientific hypotheses, and cautiously weighing the evidence for any conclusions.

When asked how to justify causal inferences from observational studies, R.A. Fisher is reported to have replied, "Make your theories elaborate." (Cochran, 1965) By this he meant that detailed, specific theories about the system under study should be developed. Being detailed and specific, they can be compared against reality and alternative theories; if they survive critical scrutiny they win credence and so provide support for causal inference. The graphical models described in this book seem well-suited to the task of developing and testing complex, specific models. Under the assumptions described in this chapter, the graphs can be interpreted causally. But reasonable care must be taken to ensure that these assumptions are warranted. If a careless, cavalier approach is adopted, the interpretations may be very misleading.

Appendix A

The MIM Command Language

The examples in this book have used MIM 3.1, a 32-bit Windows application. A freeware version can be downloaded from the MIM Web Site at `www.hypergraph.dk`, where further information on the program can be found.

A.1 Introduction

All commands may be abbreviated to a few letters—the first letter followed by some other letters (not necessarily the letters immediately following). The parsing routine searches through a command list until it finds one containing the given letters in the right order. Thus, `CGFit` can be abbreviated to `CG`, `PrintFormat` to `PF`, and so on. Trial and error is a good way to find a convenient abbreviation.

There may be several commands on one line, separated by semicolons. Thus the sequence

```
MIM->model ab/abx,aby/xy
MIM->show p
MIM->fit
```

(which specifies a model, describes its properties, and then fits it to the data) may be shortened to

```
MIM->mod ab/abx,aby/xy;sh p;fit
```

if so desired. Some commands have operands, which normally follow on the same line. If necessary, ampersand (&) may be used as a continuation symbol; this enables the operands to be given on the following line(s). All text following the ampersand on the same line is ignored. For example,

```
MIM->model ab/ &
MIM->bx,ay & Hoho
MIM->/xy
```

is equivalent to

```
MIM->model ab/abx,ay/xy
```

Two characters, hash (#) and percent (%), are useful for commenting command lines. Any text following the hash sign (#) on the same line is completely ignored. Any text following the percent sign (%) on the same line is stored internally, but otherwise ignored. The comment lines stored may be displayed using the **Show D** command. For example,

```
MIM->% We begin by declaring three variables: a,b and c.
MIM->fact a2b2c2
MIM->show d
 We begin by declaring three variables: a,b and c.
```

These features are particularly useful on input files (Section A.12.1).

A.2   Declaring Variables

Variable names consist of single letters from A to Z and a to z. Note that these are distinct; for example, X and x refer to different variables. This very short naming convention is convenient in connection with model formulae, but it can be inconvenient in practical work. To remedy this, labels can be attached to variables, as we describe shortly.

Discrete variables are declared using the command **Factor**. The number of levels they can assume must be specified as an integer following the variable name; for example,

```
MIM->fact a 2 b 2 c 3
```

The blanks between the variable names and the integers can be omitted.

Continuous variables are declared using the command **Continuous**. For example,

```
MIM->cont w x y z
```

Again, the blanks between the variable names can be omitted. The command **Variate** is synonymous.

The **Label** command is used to define variable labels with up to 12 characters. The labels should be enclosed in quotes ("). For example,

```
MIM->label w "Sepal Length" x "Sepal width"
```

The labels can be shown on independence graphs. Information on declared variables can be obtained using the command **Show**, as illustrated below.

```
MIM->show v
```

Var	Label	Type	Levels	In Data	In Model	Fixed	Block
g	Gender	disc	2	X	X	.	1
y	Year	disc	2	X	X	.	1
s	Smoking	disc	2	X	X	.	2
a	Alcohol	disc	2	X	X	.	2
w	Work	disc	2	X	X	.	2
b	BMI	cont	.	X	X	.	3
f	FEV (lung)	cont	.	X	X	.	4
k	Cholesterol	cont	.	X	X	.	4
h	Hypertension	disc	2	X	X	.	4

The command **ValLabel** is used to label the levels of factors. The syntax is

```
ValLabels var level "label" level "label"
```

where **var** is a factor name. For example,

```
Factor A2
ValLabel A 1 "Level One Label" 2 "Level Two Label"
```

defines a binary factor A, with "Level One Label" and "Level Two Label" as labels. Value labels may have at most 255 characters. They may be displayed using the **Show L** command.

Factors with more than two levels can be defined as being ordinal (i.e., having ordered categories) using the command **Ordinal**, or nominal (i.e., with unordered categories) with the command **Nominal**. The syntax of the commands is

```
Ordinal <fset>
Nominal <fset>
```

where `fset` is a list of factors. If `fset` is blank, a list of the ordinal or nominal factors is written out. Per default, factors are nominal. These commands are useful in connection with the analysis of contingency tables with ordinal classifying factors.

A.3   Undirected Models

Formulae for undirected models are entered using the `Model` command. Some examples:

```
MIM->fact a2b2c2d2
MIM->model abc,bcd

MIM->cont wxyz
MIM->model //wxy,xyz

MIM->fact a2b2c2; cont x
MIM->model abc/abx/x

MIM->fact a2b2; cont wxyz
MIM->model a,b/aw,bw,x,y,z/aw,bw,wxyz
```

The first two examples illustrate pure models (i.e., those involving only discrete or only continuous variables).

The syntax of the formula is checked and the formula is stored in a concise form. For example,

```
MIM->cont wxyz
MIM->model //wx,xy,wy,xy,xz,yz; print
```

results in the following output:

```
The current model is //wxy,xyz.
```

As shown here, generators are normally separated by commas; however, plus signs (+) are also allowed. For example, the following is valid:

```
MIM->fact a2b2c2d2
MIM->model abc+bcd
```

A.3.1   Deleting Edges

The command `DeleteEdge` removes edges from the current model. That is to say, it changes the current model to a new model, defined as the maximal submodel of the current model without the specified two-factor interactions. The edges to be removed are separated by commas. For example,

```
MIM->factor a2b2c2; cont xyz; model ab,bc/abx,cy,az,cz/yz,xy,bx
MIM->delete ab,bx
MIM->print
The current model is: bc,a/cz,cy,az,ax/yz,xy.
```

illustrates deletion of the edges [ab] and [bx] from the previous model.
Note that if larger variable subsets are specified, edges corresponding to all
variable pairs in the subsets are deleted. For example,

```
MIM->delete abc,de
```

removes the edges [ab], [bc], [ac], and [de].

Note also that if the current model is graphical, the model obtained by
deleting an edge will also be graphical.

A.3.2 Adding Edges

The command **AddEdge** adds edges to the current model. To be precise,
the new model contains the additional two-factor interactions together
with all the higher-order relatives for which the lower-order terms were
in the current model. The edges to be added are separated by commas. For
example,

```
MIM->fact a2b2c2; cont xyz; mod bc,a/cz,cy,az,ax/yz,xy
MIM->add aby,bx; pr
The current model is: bc,ab/cz,bcy,az,aby,abx/yz,bxy,aby.
```

illustrates addition of the edges [ab], [ay], [by], and [bx] to the previous
model.

Note that if the current model is heterogeneous graphical, then the model
obtained by adding an edge will also be heterogeneous graphical. However,
if an edge between a discrete and a continuous variable is added to a homogeneous graphical model, the resulting model will be heterogeneous: for
example,

```
MIM->mod a/ax,y/xy
MIM->add ay
MIM->pr
The current model is: a/ay,ax/xy,ay.
```

The command **HAddEdge** (*homogeneous* add edge) works like **AddEdge** except that it adds only the higher-order relatives that do not lead to variance
heterogeneity:

```
MIM->mod a/ax,y/xy
MIM->hadd ay
```

```
MIM->pr
The current model is: a/ay,ax/xy.
```

A.3.3   Other Model-Changing Commands

The command `SatModel` changes the current model to the saturated model
(or maximum model, if this has been set— see Section A.12.6). Similarly,
the `HomSatModel` command changes to the homogeneous saturated model,
and the `MainEffects` command changes the current model to the main
effects model. These commands take no operands. For example,

```
MIM->fact a2b2c2; cont wxyz
MIM->satmod; print
The current model is: abc/abcw,abcx,abcy,abcz/abcwxyz.
MIM->homsat; print
The current model is: abc/abcw,abcx,abcy,abcz/wxyz.
MIM->maineff; print
The current model is: a,b,c/w,x,y,z/z,y,x,w.
```

The command `BackToBase` (which takes no operands) changes the current
model to the base (alternative) model, if one such model has been set using
the `Base` command (Section A.7.1).

A.3.4   Model Properties

Some properties of the current model can be obtained by using the command `Show`. For mixed models (i.e., those involving both discrete and
continuous variables), the following information is given: whether the model
is collapsible onto the discrete variables, whether it is mean linear, whether
it is graphical or homogeneous graphical, and whether it is decomposable.
For example,

```
MIM->mod AB,AC,BC; show
The current model is AB,AC,BC.
It is not graphical.
It is not decomposable.

MIM->mod AB,BC,CD,AD; show
The current model is AB,BC,CD,AD.
It is graphical.
It is not decomposable.

MIM->mod //WX,XY,YZ,ZW; show
The current model is //YZ,XY,WZ,WX.
It is graphical.
```

```
It is not decomposable.

MIM->mod AB/ABX,BY/XY; show
The current model is AB/ABX,BY/XY.
It is collapsible onto the discrete variables.
It is not mean linear.
It is homogeneous graphical.
It is decomposable.

MIM->mod AB/AX,BX,BY/BXY; show
The current model is AB/AX,BX,BY/BXY.
It is collapsible onto the discrete variables.
It is not mean linear.
It is not graphical.
It is not decomposable.
```

The command Collapse, which has syntax

```
COLLAPSE varlist
```

determines whether the current model is collapsible onto the variable set specified. For example,

```
MIM->mod ab,bc,ac/ax,bx,cx/x; collapse abc
The model is collapsible onto abc.
```

A.4   Block-Recursive Models

Block-recursive (or chain graph) models can also be used. A current block-recursive model may be defined, which may coexist with a current undirected model.

The first step in using block-recursive models is to define the block structure.

A.4.1   Defining the Block Structure

This is done using the SetBlocks command, which takes the form:

```
SetBlocks v1 | v2 < | v3 < | v4>>
```

where v1, v2 etc are sets of variables. For example,

```
SetBlocks abcx | dy | ez
```

The variables in v1 are prior to those in v2, which are prior to those in v3, etc. The command should be used prior to working with block-recursive models. Note that changing the block structure (by repeating the command) destroys any current block-recursive information, so this should be done with caution. The current block structure is displayed using the command Show B.

Note also that when SetBlocks is invoked, block mode is turned on.

A.4.2   Block Mode

The BlockMode command switches between block mode and ordinary mode. In block mode, certain operations (see Table A.1) act upon the current block-recursive model, whereas in ordinary mode they act upon the current undirected model. It is useful to be able to switch back and forth between these modes. The syntax of the command is

```
BlockMode <+|->
```

where + turns block mode on, - turns it off, and blank shows the current mode.

In block mode, the graph window displays the current block-recursive model, rather than the current undirected model.

A.4.3   Defining Block-Recursive Models

Corresponding to the Model command, the BRModel command is used to define a block-recursive model. This must be consistent with the block structure set previously by SetBlocks. The syntax is

```
BRModel mf1 | mf2 < | mf3 < |mf4 >>
```

where mf1, mf2 etc are model formulae. For example,

```
MIM->Fact a2b2c2; Cont xyz
MIM->SetBlock ax|by|cz
Block structure set.
MIM->BRModel a/ax/x | ab/ax,by | ab,bc/x,y,bz/bxyz
MIM->pr
The current block-recursive model is:
  1  a/ax/x
  2  ab/ax,by/y,ax
  3  ab,bc/bz,abx,aby/bxyz,abxy
```

Each component model specifies the model for the variables in that block conditional on the variables in the prior blocks. Note that interaction terms

Command	Modified Action
HomSatModel SatModel MainEffects	set the current block-recursive model.
AddEdge DeleteEdge TestDelete	act on the appropriate component model.
Print	(without operands) displays the formula of the block-recursive model when in block mode. Note that Print m shows the current undirected model, and Print b the current block-recursive model, whatever the mode.
Fit	fits all component models by using Fit on each component undirected model in turn. Note that if some component models have discrete response(s) and continuous covariate(s), then the parameter estimates obtained will not maximize the conditional likelihood.
CGFit	fits all component models by using Fit on the first component and CGFit on the remaining components. (For large models this can be very time-consuming).
Stepwise	selects a block-recursive model by means of a stepwise selection in each block in turn.
DeleteLSEdge	deletes the least significant edge from the current block-recursive model after a one-step stepwise selection.
DeleteNSEdge	deletes the non-significant edges from the current block-recursive model after a one-step stepwise selection.

TABLE A.1. The modified actions of commands in block mode

between the conditioned variables are added to each component model (see Section 4.5).

A.4.4 Working with Component Models

This is done by means of the PutBlock and GetBlock commands. The PutBlock command stores the current (undirected) model as a component in the current (block-recursive) model. The reverse operation is performed by GetBlock. The syntax is

```
PutBlock k
GetBlock k
```

where k is a block number.

A validity check is performed before the current model is stored when PutBlock is used: to be valid, the model must contain all interactions between variables in the prior blocks. In addition to retrieving a model, GetBlock also redefines the set of fixed variables to be the variables in the prior blocks.

Note that the commands store and retrieve the fitted values, likelihood, etc. for the component models, when these are available.

A.5 Reading and Manipulating Data

Data can be read either (i) as raw (case-by-case) data using the Read command, or (ii) in summary form, i.e., as counts, means, and observed covariances using StatRead.

If data are read in case-by-case form, they can subsequently be transformed, new variables can be calculated, and observations can be restricted using the Restrict command (to be described later). Various commands also require raw data; for example, BoxCox, EMFit and CGFit.

If data are read in summary form, transformations and restrictions are not possible. (However, if purely discrete data are entered in contingency table form, they may be converted to case-by-case form using the command Generate).

Only one set of data can be handled at a time. Existing data are deleted when Read or StatRead is called. To add variables to a dataset, see Section B.5.

A.5.1 Reading Casewise Data

The command Read is used for reading casewise data, particularly data stored on files (for example, data transferred from other programs). (For interactive data entry, refer to the EnterData command in Section B.5).

The syntax of Read is

```
Read varlist
```

where varlist is a list of variables, i.e., single letters. The letters in the list can be separated by blanks; they must have been declared in advance as variable names. The data should be entered on the subsequent lines, each number being separated by one or more blanks, commas and/or tabs. The data for one case need not be on one line. The data should be terminated by an exclamation mark (!). For example,

```
MIM->fact a2b3; cont wxyz; read awbx
DATA->1 3.456 2 5.67654
DATA->2 3.656 3 2.53644 1,3.5354 1 2.4352 !
```

reads three cases. Missing values can be given as asterisks (*). Factors are entered as integers between 1 and k, where k is the number of levels.

The use of input files is described below (Section A.12.1).

A.5.2 Reading Counts, Means, and Covariances

Data can also be entered in the form of counts, means, and covariances. These are just the sufficient statistics for the full model. The command StatRead reads the statistics in standard cell order.

The syntax is:

```
StatRead varlist
```

where varlist is a variable list — first the discrete and then the continuous variables.

For example, for a purely discrete model, the sufficient statistics are just the counts of the contingency table. Thus, for a three-way table we write:

```
MIM->fact a2b2c2; statread abc
DATA->12 32 34 23 34 4 12 19 !
```

The order of the cells is as follows: (1,1,1), (1,1,2), (1,2,1), ..., (2,2,2). In other words, the last index changes fastest.

For purely continuous data (graphical Gaussian models), there is only one "cell" and the sufficient statistics are as follows: the number of observations (N), followed by the q empirical means, followed by the $q(q + 1)/2$ empirical covariances, where q is the number of variables. (The maximum likelihood estimates, i.e., with divisor N, not $N - 1$, should be used for the covariances.) For example, with $q = 2$, we might have

$$N = 47, \bar{y} = \begin{pmatrix} 2.36 \\ 9.20 \end{pmatrix}, S = \begin{pmatrix} 0.0735 & 0.1937 \\ 0.1937 & 1.8040 \end{pmatrix},$$

as the observed count, mean vector, and covariance matrix. These can be entered as follows:

```
MIM->Cont xy
MIM->StatRead xy
DATA->47 2.36 9.20 0.0735 0.1937 1.8040 !
```

After the count and the sample means, the elements of the lower triangle of the empirical covariance matrix are entered, row by row.

To check that the data have been entered correctly, it is wise to give the command `Print S` (i.e., print the sufficient statistics for the full model). In the current example, we would get:

```
Empirical counts, means and covariances
      X       0.074
      Y       0.194    1.804
  Means       2.360    9.200    47.000
              X        Y        Count
```

For mixed models, the sufficient statistics for the full model are the cell count, empirical cell means, and empirical cell covariances for each cell in the underlying table. The `StatRead` command expects these to be entered in standard cell order. For example, we might have

```
MIM->fact a 2; cont xy
MIM->statread axy
DATA->47 2.36 9.20 0.0735 0.1937 1.8040
DATA->54 2.87 9.41 0.0837 0.1822 1.9423 !
```

We mention in passing a trick that is useful in the analysis of mixed data. Often published sources do not report the sample cell covariance matrices

$$S_j = \sum_{k:i^{(k)}=j} (y^{(k)} - \bar{y}_j)(y^{(k)} - \bar{y}_j)'/n_j,$$

but instead just the overall covariance matrix

$$S = \sum_j n_j S_j / N.$$

Now we cannot enter $\{n_i, \bar{y}_i, S_i\}_{i \in \mathcal{I}}$ as we would like. However if we enter $\{n_i, \bar{y}_i, S\}_{i \in \mathcal{I}}$ and set the maximum model to the saturated homogeneous model (see subsection A.12.6), then as long as we fit only *homogeneous* models, the results are correct. This can be verified from the likelihood equations.

A.5.3   Transforming Data

The `Calculate` command is used to transform existing variables or to calculate new ones. The functions `SQRT`, `SQR`, `SIN`, `COS`, `ARCTAN`, `LN`, `EXP`, `FACT`, and the operators `+`, `-`, `*`, `/`, and `^` can be used. For example,

```
Calculate x=sin((x+y)^2)
```

The identifier on the left-hand side can be either a variate or factor. New variates need not be declared in advance. The right-hand side can contain variates, factors, or constants. Invalid operations, e.g., $\ln(-1)$, result in missing values. Similarly, if any value of the variables on the right-hand side is missing, the result is also missing.

Expressions can also include $<$, $>$, $=$, $<=$, and $>=$. These return 1 when the expression is true and 0 when it is false. For example,

```
MIM->calc x=y<z
```

calculates x as 1 for those cases where $y < z$; otherwise it is 0.

When the left-hand side is a factor, the right-hand side is rounded down to the nearest integer, and if this is not a valid factor value (1 to k, where k is the number of levels of the factor), then the result is set to missing. For example,

```
MIM->Fact f2
MIM->Calc f = 1 + (v>4.5)
```

discretizes v, i.e., calculates a factor f such that

$$f = \begin{cases} 1 & \text{if } v \le 4.5 \\ 2 & \text{otherwise.} \end{cases}$$

Recoding a factor is illustrated in the following fragment:

```
MIM->fact f3
<set values to f>
MIM->fact g2
MIM->calc g = 1*(f=1) + 2*(f=2) + 2*(f=3)
```

Five special functions are also available. OBS, UNIFORM, and NORMAL (which do not take operands) return the observation number, a uniform $(0,1)$ random variable, and a standard normal random variable, respectively. PNORMAL(x) returns the right tail probability in the standard normal distribution, and PNORMALINV(p) is its inverse, i.e., returns the x for which PNORMAL(x)=p.

A.5.4   Restricting Observations

The Restrict command is used to analyze data subgroups by restricting observations used subsequently. The syntax is:

```
Restrict expression
```

where expression follows the same rules as with `Calculate`. The observations are restricted to those for which the result equals 1. For example,

```
Restrict v<w
```

restricts to observations for which V is less than W. Similarly,

```
Restrict (v<w)*(w<z)
```

restricts to observations for which V is less than W and W is less than Z. To remove all restrictions, use

```
Restrict 1
```

A caution: when restricting factor levels it may be necessary to construct new factor variables. Consider, for example,

```
MIM->fact a3; cont x; read ax
   <data>
MIM->mod a/ax/x; fit
MIM->rest 1-(a=1); fit
```

A is a factor with three levels. `Restrict` here has the effect of omitting level one from the following analysis; however, as described in Section 5.2, MIM usually requires that no levels are empty when fitting models or calculating degrees of freedom. Thus, the above program fragment will result in an error. To avoid this, a new factor B with two levels could be created, as illustrated in the following fragment:

```
fact b2; calc b=a-1; model b/bx/x; fit
```

A.5.5 Generating Raw Data

The command `Generate` generates casewise data given data in contingency table form. For example,

```
MIM->Fact a2b3;
MIM->StatRead ab
DATA->1 2 3 2 1 2 !
Reading completed.
MIM->Generate
MIM->Print d
Obs  A  B
  1  1  1
  2  1  2
  3  1  2
  4  1  3
```

```
 5  1  3
 6  1  3
 7  2  1
 8  2  1
 9  2  2
10  2  3
11  2  3
```

The command takes no operands.

A.5.6 Deleting Variables

Variables in the data can be deleted using the command **Erase**. This has syntax

```
Erase vlist
```

where **vlist** is a list of variables. The variables are erased from the raw data and are undeclared.

A.6 Estimation

Three commands for parameter estimation are available. The **Fit** command fits the current undirected model by maximum likelihood, using all complete cases. The **EMFit** command uses the EM-algorithm to fit the current undirected model by maximum likelihood, including all incomplete cases. The **CGFit** command fits CG-regression models, using all complete cases.

A.6.1 Undirected Models (Complete Data)

The command **Fit** is used to fit the current model to the data. Consider, for example,

```
MIM->Cont xy
MIM->StatRead xy
DATA->47 2.36 9.20 0.0735 0.1937 1.8040 !
Reading completed.
MIM->Model //x,y
MIM->Fit
```

The **Fit** command gave the following response:

```
Deviance:      15.6338 DF: 1
```

Here, the deviance and the associated degrees of freedom are shown. The deviance is relative to the maximal model if one has been defined, otherwise to the saturated model. In the latter case, if the saturated model cannot be fitted, the log likelihood of the current model is displayed instead of the deviance.

Controlling the Fitting Algorithm

The iterative fitting process using the MIPS algorithm can be controlled if desired using three commands: MaxCycles, ConvCrit, and Method.

MaxCycles sets the maximum number of cycles that may be performed, and ConvCrit sets the convergence criterion (maximum difference between parameter estimates from one cycle to the next). Default values are 20 and 0.00001, respectively. These may be adjusted using the commands mentioned; for example,

```
MIM->maxcycle 50
MIM->convcrit 0.0001
```

The current values can be seen by giving the commands without operands.

Occasionally the Fit command may give a message like

```
No convergence after 20 cycles.
```

Before increasing MaxCycles or ConvCrit, it is advisable to investigate the convergence by using the command Summary. The syntax of this command is

```
Summary <+|->
```

where + turns the summary feature on, - turns it off, and blank shows the current mode. The following example illustrates the output:

```
MIM->mod ab/abx,aby/bxy
MIM->summary +
Fitting summary is now on.
MIM->fit
Calculating marginal statistics...
( 1,1,AB)  LL: 771.97828717 mdiff: 0.00000000 pdiff: 0.00000000
( 2,2,ABX) LL: 453.58828716 mdiff: 6.24968749 pdiff: 6.25000000
( 2,2,ABY) LL: 384.25495383 mdiff: 3.33316666 pdiff: 3.33333333
( 3,2,ABX) LL: 384.25495383 mdiff: 0.00000000 pdiff: 0.00000000
( 3,2,ABY) LL: 384.25495383 mdiff: 0.00000000 pdiff: 0.00000000
( 4,3,BXY) LL: 261.61509895 mdiff: 5.75913705 pdiff: 7.12489999
```

```
( 5,3,BXY) LL: 261.61509895 mdiff: 0.00000000 pdiff: 0.00000000
   Deviance:      15.7173 DF: 9
```

Each line reports a step in the fitting process. The parentheses enclose the cycle number, the part (discrete, linear, or quadratic), and the generator. LL denotes the minus twice the log likelihood attained after the step, mdiff denotes essentially the maximum increment in the moments parameters for the step, and pdiff denotes the maximum increment in an internal parametrisation (which one depends on the variant of the MIPS algorithm being used). More precisely, mdiff is calculated as

$$
\max_{i\in\mathcal{I},\gamma,\eta\in\Gamma}\left\{\frac{|\delta m_i|}{\sqrt{(m_i+1)}}, \frac{|\delta u_i^\gamma|}{\sqrt{\sigma_i^{\gamma\gamma}}}, \frac{|\delta\sigma_i^{\gamma\eta}|}{\sqrt{\sigma_i^{\gamma\gamma}\sigma_i^{\eta\eta}+(\sigma_i^{\gamma\eta})^2}}\right\}.
$$

For convergence, mdiff must be less than the value set by ConvCrit for all steps in the cycle.

By examining this information, it can be seen whether the algorithm is converging slowly or diverging. In the former case, the settings for ConvCrit or MaxCycles can be increased so that a further Fit command should result in convergence. Note that if a model has been fitted successfully (i.e., the current convergence criterion being satisfied) and the Fit command is used again, the algorithm will start from the fitted values that were previously obtained. So the convergence criterion can, if desired, be made successively more stringent between calls to Fit.

As mentioned above, several variants of the MIPS algorithm are implemented, as described in Appendix D. Per default, the most efficient variant is chosen— for example, for mean linear models, the mean linear variant is chosen. The Method command can be used to override this choice. The syntax is

    Method <letter>

where *letter* can be D (default), G (general) or S (step-halving). If the letter is omitted, the current variant is displayed.

A.6.2 Undirected Models (Missing Data)

The EMFit command fits the current model to the data, incorporating information from any incomplete observations. The syntax of EMFit is:

    EMFit <letter>

where letter can be R, S, F, or blank. This letter determines how the initial parameter estimates used by the algorithm are obtained, as described in detail below.

At each cycle, the expected likelihood and the change in this are printed out. For convergence, the change must be less than the current convergence criterion set by EMConvCrit. Often a less stringent criterion, say 0.001 or 0.01, will be appropriate.

The R option (default) causes the estimates to be based on substituting random values for the missing data before calculating sufficient statistics. The initial parameter estimates are then obtained by fitting the model in the standard way using these statistics.

The S option enables the user to supply start values for the missing data. As with the R option, these are used to calculate the sufficient statistics and thence the initial parameter estimates. The start values can be entered using EditData: first enter the desired value and then overwrite this with an asterisk (missing value). Afterwards, check using Print E that the values have been correctly entered.

Finally, the F option can be used when fitted values are already available. The initial parameter estimates are taken from the current fitted values. This is illustrated in the following fragment:

```
MIM->EMconvcrit 0.05
Convergence Criterion:    0.05000000
MIM->EMfit
 Cycle -2*Loglikelihood        Change
    1          215.6476
    2          203.7753   -11.872383
    3          201.2923    -2.482990
    4          200.8812    -0.411030
    5          200.8250    -0.056267
    6          200.8144    -0.010570
Successful convergence.
MIM->EMconvcrit 0.001
Convergence Criterion:    0.00100000
MIM->EMfit f
 Cycle -2*Loglikelihood        Change
    1          200.8108
    2          200.8091    -0.001664
    3          200.8083    -0.000818
Successful convergence.
```

The first call to EMFit used random starting values and a relatively lax convergence criterion (0.05). The criterion was then sharpened to 0.001 and the algorithm was restarted using the F option.

EMMaxCycles sets the maximum number of cycles that may be performed, and EMConvCrit sets the convergence criterion. The current values may be seen by giving the commands without operands.

After convergence, the fitted values are available in the usual way, using Print or Display. Similarly, the likelihood and the deviance can be written: note that these are calculated on the basis of the expected rather than the observed sufficient statistics.

Note also that Print E prints out the data, replacing any missing values with their imputed values. For discrete variables, these will be the most likely combination under the model, i.e., the cell with the largest probability given the remaining observed variables. For continuous variables, these will be the expected values given all the remaining variables (i.e., also any filled-in discrete variables). Print E does not actually replace the values in the data. This is done using the command Impute (which takes no operands).

The command EMFit requires raw (casewise) data. This can be inconvenient when analysing contingency table data in the form of cell counts. The command Generate can be used to create casewise data from a contingency table.

In some applications, it is of interest to see the quantities calculated in the E-step, as described in (D.21-D.22). In discrete latent variable models, for example, the probabilities of membership of the latent classes are important. When the Summary facility is turned on, estimates of these quantities are written out when they are calculated. This can be very voluminous, so it is sensible to do this only after convergence. A brief example follows:

```
MIM->fact a2; cont xy
MIM->read axy
DATA->1 2.3 4.5
DATA->2 3.4 5.6
DATA->1 3.5 4.3
DATA->*  *  4.7
DATA->1  *  *
DATA->2  2.5 *
DATA->2  2.1 3.2
DATA->1  2.1 3.4 !
Reading completed.
MIM->mod a/ax,ay/xy
MIM->emfit
EM algorithm: random start values.
  Cycle -2*Loglikelihood      Change
    1          44.2055
    2          35.3655     -8.840006
    3          31.6258     -3.739736
    4          30.5932     -1.032526
    5          30.4044     -0.188889
    6          30.3774     -0.026989
    7          30.3738     -0.003559
    8          30.3733     -0.000472
```

```
Successful convergence.
MIM->summary +
Fitting summary is now on.
MIM->emfit f
EM algorithm: previous fit as start values.
 E-step ...
Case:  Y=   4.700
   Pr( A= 1| Y=   4.700)= 0.525636
   E(X| A= 1 Y=   4.700)= (   2.959)
   Pr( A= 2| Y=   4.700)= 0.474364
   E(X| A= 2 Y=   4.700)= (   2.855)
Case:  A=1
   E(X,Y| A=1)= (   2.682,    4.161)
Case:  A=2 X=   2.500
   E(Y| A=2 X=   2.500)= (   4.204)
 M-step ...
 Cycle -2*Loglikelihood      Change
    1              30.3733
 E-step ...
Case:  Y=   4.700
   Pr( A= 1| Y=   4.700)= 0.525639
   E(X| A= 1 Y=   4.700)= (   2.959)
   Pr( A= 2| Y=   4.700)= 0.474361
   E(X| A= 2 Y=   4.700)= (   2.855)
Case:  A=1
   E(X,Y| A=1)= (   2.682,    4.161)
Case:  A=2 X=   2.500
   E(Y| A=2 X=   2.500)= (   4.204)
 M-step ...
    2              30.3733    -0.000000
Successful convergence.
```

Parameter estimates for continuous latent variables (i.e., variables for which all values are missing) are scaled so that their marginal means and variances are zero and one, respectively. In other words, so that $\sum_i \hat{m}_i \hat{\mu}_i = 0$ and $\sum_i \hat{m}_i \hat{\sigma}_i / N = 1$.

The EM algorithm is notoriously slow, and the present implementation in MIM is fairly simple. In many cases, therefore, it will converge slowly. It is recommended that only problems involving models of small to moderate dimension and small datasets be approached using the method.

The most important factors for computational complexity are the number of observations with missing values and, for each such observation, the number of cells in the underlying contingency table corresponding to the missing discrete variables. Thus, if there are many observations with many missing discrete values, extensive computation will be required at each cycle.

A.6.3 CG-Regression Models

The CGFit command fits a CG-regression model using ME-algorithm (see Appendix D). The syntax is

```
CgFit <options>
```

Variables in the model that are fixed (using the Fix command) are taken as covariates, while the remaining model variables are taken as responses. For example, in Hosmer and Lemeshow (1989) an example is given that studies the relationship between a binary response, coronary heart disease (c) and a covariate, age (a): The commands

```
MIM->Fix a; model c/ac/a; CGFit
```

give the output:

```
Convergence after 6 iterations.
-2*Cond. Loglikelihood:    107.187395
```

The parameter estimates may be shown by using the Display R, C command, where R is the set of responses and C the set of covariates. For example, continuing the above example, the command

```
MIM->Display c,a
```

gives the following output:

```
Linear predictors for c given a.
  c    Constant       a
  1       0.000    0.000
  2      -5.344    0.112
```

The estimated coefficients of the linear predictors for the discrete response variables (here, c) given the covariates (here, a) are shown.

Note the parameter estimates given by CGFit give the correct conditional distribution, but that the marginal distribution they describe will not, in general, be meaningful.

Two options for the CGFit command are V (verbose) and H (simple step-halving). The V option displays the progress of the algorithm, for example:

```
MIM->cgfit v
1   0  div: 0.052117 -2*CLL: 65.339445
2   0  div: 0.015578 -2*CLL: 65.328151
3   0  div: 0.005689 -2*CLL: 65.326994
4   0  div: 0.001792 -2*CLL: 65.326861
5   0  div: 0.000639 -2*CLL: 65.326845
6   0  div: 0.000210 -2*CLL: 65.326843
7   0  div: 0.000076 -2*CLL: 65.326843
8   0  div: 0.000026 -2*CLL: 65.326842
9   0  div: 0.000009 -2*CLL: 65.326842
Convergence after 9 iterations.
-2*Loglikelihood:    65.326842 DF:    9
```

Shown on each line are:

- the cycle number,
- the step-halving attempt number (0 means no step-halving was performed),
- `div`: the greatest divergence (prior to the current iteration) between the sides of the likelihood equations, and
- the conditional log likelihood.

Iteration stops when `div` is less than the convergence criterion as set by `MEConvCrit` or when the cycle number exceeds the number set by `MEMaxCycles`. The current values may be seen by giving the commands without operands.

The H-option specifies the simple step-halving method. This is as follows: each iteration starts with no step-halving (i.e., the step-halving constant κ is set to unity). If likelihood does not increase, then κ is halved. This continues until the likelihood increases: at most, however, 15 times. The default method (omitting the H-option) proceeds as follows: each iteration starts with $\kappa = min(1, 2\kappa_p)$, where κ_p is the value used in the previous iteration. As before, step-halving continues until the likelihood increases (again, at most 15 attempts).

The ME-algorithm involves iteratively fitting the current model: this fitting process is controlled in the same way as the `Fit` command, that is, using `MaxCycles`, `ConvCrit` and `Method`.

Like the EM-algorithm, the ME-algorithm can be very slow, and the present implementation in MIM is fairly unsophisticated. It is not recommended for high-dimensional models and large datasets.

A.7 Hypothesis Testing

This section describes various commands for performing significance tests: Test, BoxTest, FTest, TestDelete, SymmTest and RandomTest.

A.7.1 χ^2-Tests

The Test command performs an asymptotic likelihood ratio test of the current model against the base model. The test treats the deviance difference as asymptotically χ^2-distributed, with degrees of freedom given as the difference in the number of free parameters in the two models. Before using Test, the base (alternative) model must be set using the Base command. For example,

```
MIM->Fact a2b2c2; StatRead abc
DATA->32 86 11 35 61 73 41 70 !
Reading completed.
MIM->Model ab,bc,ac; Fit; Base
Calculating marginal statistics...
Deviance:      0.1493 DF: 1
MIM->Delete ab,ac; fit
Deviance:     24.4291 DF: 3
MIM->Test
Test of HO: BC,A
against H:  AB,BC,AC
LR:  24.2797    DF:   2    P: 0.0000
```

Normally the current model is a submodel of the base model, but if these roles are exchanged, it is detected by the program so that the correct test is performed.

The command requires that both models have been fit by the same command (i.e., by Fit, CGFit or EMFit). For example, the command can be used to compare CG-regression models:

```
MIM->model ij/ijx,ijy/xy; fix xy; cgfit
Fixed variables: xy
Convergence after 7 iterations.
-2*Conditional Loglikelihood:   64.893 DF:    9
MIM->base; del iy; cgfit; test
Convergence after 3 iterations.
-2*Conditional Loglikelihood:   72.584 DF:   11
Test of HO: ij/jy,ijx/xy
against H:  ij/ijx,ijy/xy
LR:   7.6909    DF:   2    P: 0.0214
```

It may not always be valid to compare models fit by `EMFit` in this way — this is entirely the user's responsibility.

Related commands include `BackToBase` (which changes the current model to the base model) and `SetBaseSat` (which sets the base model to be the saturated model or maximum model if this is defined). None of these commands (`Test`, `Base`, `BackToBase`, or `SetBaseSat`) require operands.

A.7.2   Test of Homogeneity

The `BoxTest` command computes Box's test of homogeneity (Section 5.13). The command is similar to the `Test` command except that the current model must be saturated homogeneous and the base model must be saturated heterogeneous. If some sample cell covariance matrices are singular, the expression for the degrees of freedom is corrected accordingly.

A.7.3   F-Tests

As with `Test`, the `FTest` command tests the current model against the base model. The output shows the test statistic, the degrees of freedom, and the p-value. For `FTest` to be appropriate, the following conditions must be satisfied:

1. There must be one continuous response variable. To specify this, all the variables in the models except one continuous variable must be fixed (using the `Fix` command).
2. There must be variance homogeneity in the conditional distribution of the response variable given the others under both the current and the base model. This just means that no quadratic generator containing the response variable may contain any discrete variables.
3. Both models must be collapsible over the response variable and the marginal models must be identical.

The conditions are checked when the command is given.

A.7.4   Edge Deletion Tests

The `TestDelete` command tests whether an edge can be deleted from the current model— in other words, whether the two variables are conditionally independent. The syntax is

```
TestDelete edge <options>
```

Option	Description	Other requirements
S	F-test	Homogeneous, one variable continuous
E	Exhaustive enumeration	Discrete separating set
M	Monte Carlo	Ditto
Q	Sequential Monte Carlo	Ditto
Z	Shows deviance decomposition	Ditto
D	Prints $R \times C \times L$ table	Ditto
C	Estimates size of reference set	Ditto
I	Use stratum-invariant scores	Ditto
L	Deviance test	Row variable discrete
P, F	Contingency table tests	Both variables discrete
W, X, Y, K, J	Rank tests	Row variable discrete

TABLE A.2. `TestDelete` options. All options require that the test corresponds to a decomposable edge deletion test.

If no options are specified, an asymptotic likelihood ratio test is performed (as with the `Test` command). This compares the deviance difference with a χ^2-distribution. When the test corresponds to a decomposable edge deletion test, the degrees of freedom are adjusted to account for parameter inestimability. Otherwise, the degrees of freedom are calculated in the same way as with the `Test` command, i.e., as the difference in the number of free parameters between the two models.

The options are summarized briefly in Table A.2.

What do we mean by this expression "corresponds to a decomposable edge deletion test"? We mean that either (i) both the initial and the resulting model are decomposable, or (ii) using collapsibility properties, the test is equivalent to a test of type (i). Figure A.1 illustrates the latter type.

A.7.5 Edge Deletion F-Tests

The S option causes an F-test to be performed instead of a χ^2 test. It requires that (i) both the initial and the resultant model are decomposable and variance homogeneous, and (ii) one or both of the vertices in the edge are continuous (see Section 5.3).

A.7.6 Exact Tests

The `TestDelete` command can also perform exact conditional tests. It can do so when

1. The test corresponds to a decomposable edge deletion test.

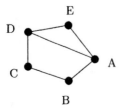

FIGURE A.1. The model is not decomposable, but is collapsible onto $\{A, D, E\}$. It follows that the test for the deletion of $[AE]$ corresponds to a decomposable edge deletion test.

2. The conditioning set — that is, the variables adjacent to both endpoints of the edge — is composed entirely of discrete variables.

Table A.3 shows which tests are available and which types of variables they can be used with. Nominal, ordinal, and binary types must be specified as factors. The rank tests and the randomised F-test are sensitive to the order of the variables specified: for example, `TestDelete AB J` will give different results to `TestDelete BA J`. The first variable specified is called the *row* variable, and the second is called the *column* variable, as described in Section 5.4. In all cases, the row variable must be discrete, while the column variable may be either discrete or continuous (depending on the test). Note that if the column variable is continuous, then raw data must be available.

To specify the method of computation, one of the following options should be given: `E` (exhaustive enumeration), `M` (Monte Carlo), `Q` (sequential Monte Carlo). If none are specified, only the asymptotic test is performed.

The maximum number of sampled tables in Monte Carlo sampling is set by the command `MaxSim`, the default being $1,000$. The prescribed maximum

Option	Test	Row variable	Column variable
L	LR test	Nominal	Nominal or variate
P	Pearson χ^2	Nominal	Nominal
F	Fisher's	Nominal	Nominal
W	Wilcoxon	Binary	Ordinal or variate
X	van Elteren (a)	Binary	Ordinal or variate
Y	van Elteren (b)	Binary	Ordinal or variate
K	Kruskal-Wallis	Nominal	Ordinal or variate
J	Jonckheere-Terpstra	Ordinal	Ordinal or variate
S	Randomised F-test	Nominal	Variate

TABLE A.3. `TestDelete` options for exact tests.

(a) Design-free stratum weights; (b) Locally most powerful stratum weights.

number of tables with $T(M_k) \geq t_{obs}$ in sequential Monte Carlo sampling is set by the command MaxExcess. Default is 20.

For example, the following fragment sets new values for the two parameters:

```
MIM->maxsim 5000
Max no of simulations:      5000
MIM->maxexcess 50
Max no of excesses:         50
```

Three further options are available. The option D prints out the underlying three-way contingency table, and the option C prints an estimate of the number of tables in the reference set Υ (see below) using the formula in Gail and Mantel (1977). Option I causes stratum-invariant scores to be used for the rank tests instead of the stratum-specific scores, which are default. A technical detail: when several tests are being computed and the sequential stopping rule is used, then it can only be applied to one test. Which test this is, is given as the first option appearing in the following list: LPFWXYKJS.

The exhaustive enumeration method can be used for problems for which $R \times C \times L < 1,900$, whereas there are no formal limits to the size of problems for which Monte Carlo sampling can be applied. In practice, however, time will usually be the limiting factor for large problems.

A.7.7   Symmetry Tests

The command SymmTest performs a test of multivariate symmetry (see Appendix C and Section 6.7). It is assumed that all variables in the current model are binary factors. The command takes no operands.

A.7.8   Randomisation Tests

The command RandomTest enables general randomisation tests to be performed— more general than those available using TestDelete. The syntax is

```
RandomTest Z <B> <-letter>
```

where Z is the randomisation factor, B is an optional blocking factor, and letter can be E (exhaustive enumeration), M (fixed sample Monte Carlo) or Q (sequential Monte Carlo). M is default.

The hypothesis test considered is identical to that defined in the Test command. That is, the current model is tested against the base model, using the deviance as test statistic.

A randomisation test is based on the randomisation distribution of the deviance, that is, its distribution over all possible randomisations. It is assumed that the total numbers receiving each level of Z are fixed. If a blocking factor B is supplied, then this is also true within each level of B.

The Monte Carlo process is controlled by parameters set by commands `MaxSim` and `MaxExcess` (for sequential Monte Carlo).

The command is inefficient compared with `TestDelete` and is only intended for hypothesis tests that are not available using `TestDelete`, that is, do not correspond to single edge removal tests.

A.8   Model Selection

A.8.1   Stepwise Selection

The syntax of the `Stepwise` command is

```
Stepwise options
```

where the available options are described below (see Table A.4 for a summary).

The default operation of `Stepwise` is backward selection using χ^2-tests based on the deviance difference between successive models. Unless specified otherwise, `Stepwise` runs in decomposable mode if the initial model is decomposable; otherwise, it runs in unrestricted mode. This may be overridden using the U option. The U option specifies underestricted mode as opposed to decomposable mode.

The F option results in forward selection so that instead of the least significant edges being successively removed, the most significant edges are added. In other words, at each step, the edge with the smallest p-value, as long as this is less than the critical level, is added to the current model (see, however, options O and H below).

Sparsity-corrected degrees of freedom calculations are used in decomposable mode (see Section 5.2).

The S option performs small-sample tests, namely F-tests, instead of χ^2 tests whenever appropriate. This option is only available in decomposable mode (see Section A.7.1).

In backward selection, coherence is default, whereas in forward selection, noncoherence is default. These settings can be overridden using the N and C options, standing for noncoherent, and coherent, respectively.

Option	Result
F	Forwards selection
U	Unrestricted, as opposed to decomposable, mode
S	Small-sample tests (F-tests)
C	Coherent mode
N	Noncoherent mode
O	One step only
H	Headlong
A	Reduce AIC
B	Reduce BIC
G	Use CG-regression models
I	Initial model as alternative
X	Maximum (saturated) model as alternative
V	Variance-heterogeneous edge addition
E	Exhaustive enumeration
M	Monte Carlo sampling
Q	Sequential Monte Carlo
L	Likelihood ratio test (G^2)
P	Pearson χ^2
W	Ordinal test

TABLE A.4. Options for stepwise model selection.

The O option causes one step only selection to be performed. After execution in backward selection, until the model changes, p-values can be written on the independence graph. Furthermore, the least significant edge can be deleted using the command DeleteLSEdge, and the nonsignificant edges using DeleteNSEdge.

The H option results in headlong model selection.

The I and X options specify the alternative hypotheses used in the significance tests carried out at each step. For example, in backward selection the tests performed will normally be of $\mathcal{M}_0$ versus $\mathcal{M}_1$, where $\mathcal{M}_1$ is the current model and $\mathcal{M}_0$ is obtained from $\mathcal{M}_1$ by removing an edge. The X option causes the test to be of $\mathcal{M}_0$ versus the saturated model, rather than versus $\mathcal{M}_1$. (If a maximum model is defined, that will be used instead of the saturated model.) The I option causes the test to be of $\mathcal{M}_0$ versus the initial model.

Normally, it will be preferable to use the tests between successive models (the default method). The I and M options have been included mainly to allow comparisons with other selection procedures.

Use of the I and X options will give rather silly results in conjunction with forward selection. They cannot be used with the A or S options.

The V option relates to a technical detail concerning edge addition and variance homogeneneity. Normally, in forward selection from an initial model that is variance homogeneous, all subsequent models considered will be variance homogeneous (corresponding to adding edges using HAddEdge instead of AddEdge). If the initial model is heterogeneous, then the usual operation (i.e., corresponding to AddEdge) is used. The V option forces the usual operation to be used when the initial model is variance homogeneous. For example,

    MIM->Model a,b/x,y,z/x,y,z; Stepwise f

steps through homogeneous graphical models, whereas

    MIM->Model a,b/x,y,z/x,y,z; Stepwise fv

steps through heterogeneous graphical models.

The A and B options allow stepwise selection based on the information criteria AIC and BIC (see Section 6.3). For each edge examined, the change in the criterion specified is calculated and displayed, and the edge corresponding to the greatest reduction in the criterion is chosen for addition or removal. This continues until no further reduction is possible. Options F, G, U, O, H, and V can also be used, but the remaining options have no effect.

The G option is used for stepwise selection among CG-regression models. The covariates are specified using the Fix command. The options A, B, F, U, C, N, O, H, and V can also be used, but the remaining options have no effect. The fitting process is controlled by the same parameters that control CGFit. For high-dimensional problems, this selection method may be very time-consuming.

Finally, the E, M, Q, L, and P options perform exact tests (see Section 5.4 for further details). These are useful for the analysis of sparse contingency tables and are only available in decomposable mode. The W option is useful when there are ordinal factors; the test chosen for the removal or addition of an edge between two factors is sensitive to ordinality. More precisely, the choice is shown in Table A.5.

The default critical level is 0.05. This can be changed using the command CritLevel, as in

	Nominal	Binary	Ordinal
Nominal	G^2	G^2	Kruskal-Wallis
Binary	G^2	G^2	Wilcoxon
Ordinal	Kruskal-Wallis	Wilcoxon	Jonckheere-Terpstra

TABLE A.5. Automatic choice of tests sensitive to ordinality.

```
MIM->CritLev
Critical level:  0.050
MIM->CritLev .001
Critical level:  0.001
```

If a new value is not given, the current value is written out.

A.8.2   The EH-Procedure

The EH-procedure is initialised using the command `InitSearch`. This command sets up internal lists for storing information on models that are fitted. Models that are rejected according to the criterion are stored in one list and models that are accepted (i.e., not rejected) are stored in another list.

Prior to initialisation, the model class should be set using the command `EHMode`, and if extensive output is to be written during the search process, the command `EHReport` should be used. In addition, the criterion for deciding the consistency of models with the data can be adjusted using the command `CritLevel`.

Search is started using the `StartSearch` command.

The command `Show S` shows the current status of the procedure, i.e., whether it has been initialised, and if it has, the current lists of accepted, rejected, minimal undetermined, and maximal undetermined models. The command `Clear S` clears the EH-procedure.

We now give a more detailed description of the individual commands mentioned above.

Choosing the Model Class

The `EHMode` command is used to specify the model class used by the EH-procedure. This can be either heterogeneous or homogeneous graphical models. Default is heterogeneous. The command has syntax

   `EHMode <letter>`

where *letter* can be X (heterogeneous) or H (homogeneous).

Setting the Reporting Level

The `EHReport` command controls the output from the EH-procedure. There are three levels of reporting detail: all, medium, and least. At the all level (option A), all four model lists (minimal accepted, maximal rejected, minimal undetermined, and maximal undetermined) are printed out at each

step, together with the results of fitting the individual models. At the
medium level (option M), lists of minimal and maximal undetermined models are printed out at each step. The default (corresponding to option L) is
to print out only the selected models. The syntax of EHReport is

```
EHReport <letter>
```

where *letter* is A, M, or L.

Initializing Search

The InitSearch command has the following syntax:

```
InitSearch <min_model> <-max_model>
```

The search is restricted to models that contain min_model as a submodel
and that are contained in max_model. For example,

```
InitSearch   A,B/AX,BY/AX,BY  - AB/ABX,ABY/ABX,ABY
```

Either or both of min_model and max_model can be omitted: by default,
the main effects model and the model corresponding to the complete graph
are used. The search algorithm is based on tests against max_model using
the χ^2-test on the deviance differences. The critical value is controlled by
the command CritLevel.

Note that the fitted models and min_model and max_model must be in the
appropriate model class.

Models fitted in the usual way (e.g., using Fit or Stepwise) after initialisation are added to the lists of accepted or rejected models as appropriate
(provided they are in the appropriate model class).

Starting the Search

The StartSearch command starts the EH-procedure after initialisation by
InitSearch. The syntax is:

```
StartSearch <MaxNoModels> <direction>
```

where MaxNoModels is a positive integer that controls the maximum number
of models to be fitted, and Direction is either D (downwards-only) or U
(upwards-only). The default value of MaxNoModels is unlimited, and the
default direction is bidirectional. For example,

```
StartSearch 128
```

This initiates the model search (bidirectional version), specifying that at most 128 models are to be fitted. Note that the upwards- and downwards-only versions are of primarily theoretical interest: the bidirectional version is much faster.

A.8.3 Selection Using Information Criteria

The `Select` command selects a model minimizing either Akaike's Information Criterion or Bayesian Information Criterion. The operation can be very time-consuming for large classes of models. The syntax is

```
Select  <letters>
```

where *letters* can be U, D, A, V, S, and M. The current model must be graphical. The default operation is to search among all graphical submodels of the current model, identify the model with the least BIC, and change to that model. No output is given.

If the U option is specified, then the search is upwards, that is, among all graphical models including the current model.

If the current model is homogeneous, then the search will be among homogeneous graphical models, otherwise it will be among heterogeneous graphical models. If the D option is specified, then the search will be restricted to decomposable models.

If the A option is specified, then the AIC criterion is used instead of BIC.

The V and S options control the amount of output written, V being the most verbose, and S giving some output.

The M option implements Monte Carlo search, i.e. search among a number of randomly chosen models in the search space. The number is specified using the `MaxSim` command.

The command is subject to the `Fix` command, in that edges fixed are not eligible for addition or removal. That is, all such edges that are present in the current model are present in all models searched, and all such edges absent from the current model are also absent from all models searched.

An example:

```
Model //vwxyz; Select
```

and

```
Model //v,w,x,y,z; Select u
```

have the same effect.

A.9   The Box-Cox Transformation

This command performs the power transformation proposed by Box and Cox (1964), i.e.,

$$f_\lambda(x) = \begin{cases} \dfrac{(x^\lambda - 1)}{\lambda} & \text{if } \lambda \neq 0 \\ \ln(x) & \text{if } \lambda = 0. \end{cases}$$

Note that the transformation requires x to be positive. The syntax of the BoxCox command is:

```
BoxCox var lowlim uplim no
```

where var is a variate in the data, lowlim and uplim are the lower and upper λ values, and no is the number of intervals between the λ values. For example,

```
BoxCox X -2 2 4
```

calculates the transformation for $\lambda = -2, -1, 0, 1, 2$. For each value, the following are calculated and displayed: minus twice the profile log likelihood of the full model, minus twice the profile log likelihood of the current model, and the deviance difference.

A.10   Residuals

The Residuals command calculates and saves residuals. Suppose the current model contains q continuous variables and that $q_2 < q$ of these have been fixed (using the Fix command), so that $q_1 = q - q_2$ remain unfixed. The syntax of the command is

```
Residuals <-> varlist
```

where varlist is a list of q_1 variable names. The residuals, i.e.,

$$y_1 - \hat{\mu}_{y_1 \mid i, y_2},$$

are calculated and saved in the variables. The variables need not be declared as variates in advance. If the minus sign is specified, the deletion residuals are calculated; otherwise, the (ordinary) residuals are calculated. Note that the deletion residuals require that the model be refitted for each observation: this may require substantial computation and is tolerable only in connection with decomposable models and datasets of moderate size.

An example using the lipids data from Section 4.1.10 is as follows:

```
MIM->fix UVW
Fixed variables: UVW
MIM->show v
```

Var	Label	Type	Levels	In Data	In Model	Fixed	Block
A	Treatment gp	disc	3	X	X	.	.
U	pre VLDL	cont	.	X	.	X	.
V	pre LDL	cont	.	X	X	X	.
W	pre HDL	cont	.	X	X	X	.
X	post VLDL	cont	.	X	X	.	.
Y	post LDL	cont	.	X	X	.	.
Z	post HDL	cont	.	X	X	.	.

```
MIM->resid LMN
MIM->show v
```

Var	Label	Type	Levels	In Data	In Model	Fixed	Block
A	Treatment gp	disc	3	X	X	.	.
U	pre VLDL	cont	.	X	.	X	.
V	pre LDL	cont	.	X	X	X	.
W	pre HDL	cont	.	X	X	X	.
X	post VLDL	cont	.	X	X	.	.
Y	post LDL	cont	.	X	X	.	.
Z	post HDL	cont	.	X	X	.	.
L	X residual	cont	.	X	.	.	.
M	Y residual	cont	.	X	.	.	.
N	Z residual	cont	.	X	.	.	.

Notice how labels are created indicating the corrrespondence with the response variables.

The Mahalanobis command calculates and saves Mahalanobis distances, together with the corresponding χ^2 quantiles. If some of the continuous variables in the model have been fixed (using the command Fix), they are treated as covariates in computation of the distances, e.g., for a case written as (i, y_1, y_2), with y_1 the covariates and y_2 the responses, the distance is calculated as

$$(y_2 - \hat{\mu}_{y_2|i,y_1})' \hat{\Omega}_{y_2|i,y_1} (y_2 - \hat{\mu}_{y_2|i,y_1}),$$

where $\hat{\mu}_{y_2|i,y_1}$ is the conditional mean of Y_2 given $I = i$ and $Y_1 = y_1$, and $\hat{\Omega}_{y_2|i,y_1}$ is the inverse of the conditional covariance of Y_2 given $I = i$ and $Y_1 = y_1$. The command has syntax

```
Mahalanobis <-> var1 <var2>
```

where `var1` and `var2` are variates (which need not be declared as such in advance). The Mahalanobis distances are stored in `var1` and the corresponding quantiles of the χ^2-distribution are stored in `var2`. If a minus sign is specified, the distance calculations utilize the deletion residuals and conditional covariances.

A.11   Discriminant Analysis

The `Classify` command is used in connection with discriminant analysis. The syntax is

```
Classify <-> G C <V1...Vk>
```

where `G` is a factor in the model. The command calculates a new factor `C` with the same number of levels as `G`. `C` contains the predicted classification using the maximum likelihood discriminant analysis method; `C` need not be declared in advance as a factor. Each observation is assigned to the level g with the largest $f(g, j, y)$.

The density f is estimated using the current model and, by default, all the available observations. By specifying the minus $(-)$, the leave-one-out method is used, i.e., the density for each observation is estimated using all available observations except the one in question. This option is computationally intensive.

`C` is computed for all unrestricted observations for which the measurement variables (j and y) are not missing.

For example,

```
MIM->Model G/GX,GY,GZ/XY,XZ
MIM->Classify GQ
MIM->Classify -GR
MIM->Model GQR
MIM->Print s
```

calculates and writes out information showing both the apparent and the leave-one-out error rates.

The values of the log densities can be stored, if so desired, in the variates `V1` to `Vk`, where k is the number of levels of `G`. This makes it possible, for example, to compute discriminant functions using arbitrary prior probabilities (sample proportions are implicitly used in the above method).

A.12   Utilities

In this section, we describe diverse general purpose utilities.

A.12.1   File Input

Input to MIM is normally from the keyboard. However, commands and data can be read from a file by using the `Input` command. The syntax is

```
Input filename
```

where filename is the usual filename (possibly including a path). For example,

```
MIM->input \data\iris
```

Subsequent input to MIM is from the file until the end of the file is reached, at which point it reverts to the keyboard.

All commands can be used in the file. Normally the command lines are not echoed to the screen, but the command `Echo` can change this. The syntax is

```
Echo <+|->
```

where + turns echoing on, – turns it off, and blank shows the current mode.

The use of comments is described in Section A.1.

Two commands, `Suspend` and `Revert`, are used in connection with file input, mainly for constructing interactive examples for teaching purposes. `Suspend` can be used on input files only, where it has the effect of allowing the user to enter commands from the keyboard. When the command `Revert` is entered, input is again taken from the input file, starting at the line following the `Suspend` command.

Input files can be nested, with up to nine levels.

A.12.2   The Workspace

The `Save` command saves the whole workspace on a file. The syntax is

```
Save filename
```

The whole of the workspace is saved on the file, with two minor exceptions: model search results and information about input files are not saved. This is a convenient way of storing data, labels, fitted models, the current graph, and other settings on one file.

The `Retrieve` command retrieves a workspace from a file created using `Save`. The syntax is

```
Retrieve filename
```

Workspace files have a special format that only MIM can read. It is inadvisable to use these files for long-term storage of data, since the format is in general version-specific. That is, the format is not backwards compatible.

The `Show` command displays the current model, the variable declarations, or information about the workspace. The syntax is:

```
Show letters
```

where letters can be:

P (or blank): properties of the current model (see Section A.3.4).

V: the variable declarations, labels, etc. (see the example in Section A.10).

W: the current state of the workspace.

D: comment lines (see Section A.1).

S: the current status of the model search.

The `Clear` command clears the contents of the workspace or the current results of the model search procedure. The syntax is:

```
Clear letters
```

where letters can be:

A (or blank): clears the whole workspace and sets defaults.

S: clears the model search procedure.

A.12.3   Printing Information

This subsection describes the use of the `Print` command, which is used for printing information about the data or the fitted model. The syntax is `Print letters`, where `letters` can be blank or can be one or more letters from the following list: S,T,U,V,F,G,H,I,M,Z. These have the following effect:

M (or blank): prints the current model formula.

The letters F, G, H, and I cause information about the fitted model to be printed. More precisely:

F: prints the fitted count, variate means, and covariance matrix for each cell, i.e., $\{m_i, \mu_i, \Sigma_i\}$ are printed for each i.

G: prints the corresponding canonical parameters, i.e., $\{\alpha_i, \beta_i, \Omega_i\}$ for each cell i.

H: prints the fitted count, means, and correlation matrix for each cell.

I: prints the estimated discrete and linear canonical parameters and the matrix of estimated partial correlations for each cell. The relation between the matrix of partial correlations, $\Upsilon_i = \{v_i^{\gamma\varsigma}\}_{\gamma,\varsigma \in \Gamma}$, say, and the inverse covariance matrix $\Omega_i = \{\omega_i^{\gamma\varsigma}\}_{\gamma,\varsigma \in \Gamma}$ is almost the same as the relation between the correlation and covariance matrix: for $\gamma \neq \varsigma$, $v_i^{\gamma\varsigma} = -\omega_i^{\gamma\varsigma} / \{\omega_i^{\gamma\gamma} \omega_i^{\varsigma\varsigma}\}^{\frac{1}{2}}$.

The letters S, T, U, and V cause corresponding sample quantities to be printed. More precisely

S: prints the observed count, variate means, and covariance matrix for each cell.

T: prints the corresponding discrete, linear, and quadratic canonical parameters for each cell.

U: prints the count, means, and correlation matrix for each cell.

V: prints the discrete and linear parameters and the matrix of partial correlations for each cell.

Note that the variable set for which the empirical statistics are calculated is taken from the current model (provided the model variables are present in the data). This is illustrated in the following fragment:

```
MIM->fact a2; cont x
MIM->sread ax
DATA->12 4.5 2.1 24 6.5 3.2 !
Reading completed.
MIM->print s
Calculating marginal statistics...
Empirical counts, means and covariances.
A
1     X     2.100
      Means  4.500   12.000
             X      Count

2     X     3.200
      Means  6.500   24.000
             X      Count
```

```
MIM->model //x; print s
Calculating marginal statistics...
Empirical counts, means and covariances.
      X     3.722
    Means   5.833   36.000
              X      Count
```

Similarly, the G, H, and I options give the corresponding canonical parameters, correlations, partial correlations, etc.

The letters D and E cause the raw (case-by-case) data to be printed out when such data is available.

D: prints the raw data. Missing values are printed as *'s. Only data for the unrestricted observations are printed.

E: is the same as D, except that if the results of EMFit are available, then instead of the missing values, their imputed values are printed out. See EMFit. Note that data values are stored as an ordered pair (missing flag, value). PRINT D prints out values for which the missing flag is false, whereas PRINT E prints all values.

Finally, the letters X, Y, and Z cause information about the current and the saturated (full) model to be printed out. More precisely,

X: prints, when this is available, minus twice the log likelihood of the saturated model or maximum model if such has been defined,

Y: prints, when this is available, minus twice the log likelihood of the current model, and

Z: prints, when this is available, the deviance and degrees of freedom of the current model.

The format used in Print and other commands can be controlled by the PrintFormat command. The syntax of this command is

```
PrintFormat fw d
```

where fw and d are integers specifying the field width and number of decimals, respectively. For example:

```
MIM->print u
Empirical counts, means and correlations
      W    1.000
      X    0.553   1.000
      Y    0.547   0.610   1.000
      Z    0.409   0.485   0.711   1.000
    Means 38.955  50.591  50.602  46.682  88.000
            W       X       Y       Z      Count
```

```
MIM->printformat 8 4
Printing format: 8,4
MIM->print u
Empirical counts, means and correlations
        W    1.0000
        X    0.5534  1.0000
        Y    0.5468  0.6096  1.0000
        Z    0.4094  0.4851  0.7108  1.0000
    Means 38.9545 50.5909 50.6023 46.6818  88.0000
             W       X       Y       Z      Count
```

A.12.4   Displaying Parameter Estimates

The `Display` command prints out parameter estimates, after the current model has been fitted using `Fit`, `CGFit` or `EMFit`. The syntax is:

```
Display Rset<, Cset <, options>>
```

where `Rset` is a set of response variables; `Cset` is a set of covariates; and `options` consists of zero, one, or two of letters `C` and `S` (these stand for canonical and standardized, respectively). For example, the following fragment

```
MIM->Model AB/AX,BX,ABY/XY
MIM->Fit
MIM->Display AB,XY
```

displays parameter estimates for the conditional distribution of A and B, given X and Y. Similarly,

```
MIM->Display AB,X
```

would display parameter estimates for the conditional distribution of A and B, given X.

Note that the variable sets `Rset`, `Cset`, and `options` are each written as generators, i.e., without commas. Both `Cset` and `options` may be omitted. If `Cset` is omitted, the marginal distribution of `Rset` is shown.

When `Rset` consists of continuous variables only, the parameters printed out are determined by `options`. Per default, if `options` is blank, the moments parameters are printed out: If `options=C`, then the canonical parameters are shown. If `options=S`, then counts, means, and correlations are displayed, and if `options=SC`, then partial correlations are shown. For example,

```
MIM->model //VWX,XYZ; fit;
Deviance:        0.8957 DF: 4
MIM->Display VWYZ,X
Fitted conditional means and covariances.
     V    -6.579    0.900   211.927
     W    12.418    0.754    50.020 107.368
     Y    -3.574    0.993     0.000   0.000 107.795
     Z   -12.323    1.080     0.000   0.000  34.107 164.297
                       X        V       W       Y       Z
```

The output shows the conditional means and conditional covariance of
V, W, Y, and Z given X. The last four columns show the conditional
covariance matrix, while the first two show the conditional means (constant
term and coefficient of X). For example, we see that

$$E(V \mid X = x) = -6.579 + 0.900x.$$

When Rset consists of discrete variables only, and Cset contains continuous
variables, then the linear predictor for each combination of levels for the
variables in Rset is shown. Here options has no effect. It should be noted
that only the linear predictors are shown.

When Rset and Cset both contain discrete variables only, then either the
conditional probabilities of Rset given Cset, or the corresponding linear
predictors, are shown, depending on whether or not option C is specified.

When Rset consists of both discrete and continuous variables, the parameters for the discrete responses given Cset are shown first, followed by
the parameters for the continuous responses given Cset and the discrete
responses.

A.12.5   Displaying Summary Statistics

The Describe command displays univariate statistics for a variable. The
syntax is

        Describe letter

where letter is a variable. The raw data must be available. For a discrete variable, the marginal counts are displayed, and for a continuous
variable, the maximum, minimum, 95%, 75%, 50%, 25%, and 5% fractiles
are displayed.

The DisplayData command displays sample statistics; the syntax and
operation is the same as Display.

A.12.6   Setting the Maximum Model

The deviance of a model is normally defined as $2(\hat{\ell}_f - \hat{\ell}_m)$ where $\hat{\ell}_m$ is the log likelihood of the current model and $\hat{\ell}_f$ is the log likelihood of the saturated (unrestricted) model. However, if there are insufficient data to ensure existence of the maximum likelihood estimate under the saturated model, $\hat{\ell}_f$ and hence the deviance are undefined.

The command MaxModel sets the current model to be the maximum model so that the deviance for subsequent models is defined as $2(\hat{\ell}_x - \hat{\ell}_m)$, where $\hat{\ell}_x$ is the log likelihood of the maximum model. Similarly, the degrees of freedom are calculated in respect to the maximum model. For example, in the two-way ANOVA setup with one observation per cell, the models $AB/ABX/ABX$ and $AB/ABX/X$ do not have MLEs. The maximum model can be set to $AB/AX, BX/X$. This makes the deviance of submodels, for example, AB/AX/X, well-defined.

This is illustrated in the following program fragment:

```
MIM->fact A2B2; cont X; read ABX
DATA->1 1 4.3 1 2 5.6 2 1 3.7 2 2 3.6 !
Reading completed.
MIM->mod AB/AX,BX/X; fit
Calculating marginal statistics...
Warning: the deviance is undefined, probably because there are
insufficient data to fit the full model. Set a maximum model using
the MaxModel command.
Likelihood:      14.0433 DF: 4
MIM->maxmodel; fit
MIM->fit
Deviance:        0.0000 DF: 0
MIM->del AX; fit
Deviance:        5.9707 DF: 1
MIM->fix AB; ftest
Fixed variables: AB
Test of H0: AB/BX/X
against H:  AB/AX,BX/X
F:     3.4490     DF:   1,   1    P: 0.3145
```

Notice that, initially, the deviance of $AB/AX, BX/X$ was undefined since MLEs for the full model do not exist. In this case, Fit writes out minus twice the log likelihood instead of the deviance.

A.12.7   Fixing Variables

The Fix command is used in various contexts to mark variables as fixed. The syntax is

```
Fix vlist
```

where `vlist` is a list of variables. This has an effect on stepwise model selection, CG-estimation, calculating residuals and Mahalanobis distances, and computing F-tests. To remove all fixing, use `Fix` without operands.

A.12.8 Macros

A simple macro facility is supported. This is useful when sequences of commands are repeated often, for example, in simulation. The facility performs simple text substitution into the command line before parsing. Macros are called as follows:

```
@filename(parameterlist)
```

where `filename` is the name of the file containing the commands and the item `parameterlist` is a list of parameters, separated by commas. There may be up to nine parameters. A parameter is a sequence of characters, not including comma "," or right parenthesis ")". The file should contain ordinary commands and data, but may also include the symbols &1, ..., &9. When the macro is executed, these are replaced by the corresponding parameters. If no corresponding parameter has been given, a blank is substituted. For example, if there is a file `tt` in the current directory, containing the line

```
model &1&2,&1&3,&2&3; fit; test
```

then it may be called using

```
@tt(A,B,C)
```

After substitution, the line becomes

```
model AB,AC,BC; fit; test
```

which is then processed in the usual way.

While developing macros, it can be useful to switch echoing on using the `Echo` command. This echoes the command lines after parameter substitution, before they are parsed. Note also that to pass model formulae as parameters, it is necessary to use "+" signs to separate the generators, instead of commas.

Macros can be nested, with up to nine levels.

Appendix B

Implementation Specifics of MIM

This appendix describes various implementation-specific aspects of MIM 3.1, in particular the user interface. For a variety of reasons, it makes sense to separate the command-based numerical engine from the interface. For example, the latter is highly dependent on the operating system, which makes it subject to rapid change. So by the time you read these lines, the user interface may well have changed substantially.

B.1 Calling MIM

MIM is invoked by clicking the MIM icon, and responds by displaying the main window, which looks something like Figure B.1.

Several features of the main window are worthy of note. At the top is the main menu, for selecting commands and other operations. Sometimes items may be grayed out, if the features are not currently available. For example, if no variables have been defined, a model cannot be specified.

Below the main menu is the work area where output from MIM is shown, and the prompt `MIM->` where the user is prompted for input. At the bottom of the window there is a status bar. This is used for displaying status messages about calculations that may take some time. At the bottom right there is a status indicator (coloured circle). This functions as an "interrupt" button; click it to abort any lengthy calculations. To the left of this is a small rectangle containing a block mode indicator: a plus sign is shown here in block mode. When appropriate, a scroll bar appears, with which the work area may be scrolled up and down.

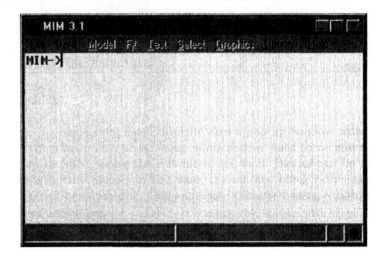

FIGURE B.1. The MIM interface

The key feature of the interface is that it is both menu-driven and command-driven. Many menu items supply easy access to the command language, in the form of pop-up dialogues eliciting options, variables, or other information from the user, and then calling the appropriate command. With experience, some may prefer to enter commands directly at the prompt. Most of the pop-up dialogues are self-explanatory and are not described in detail here; see the online help for further information.

When invoked, MIM searches for a configuration file called `mim31.msf` in the working directory. This contains information about the interface settings (screen fonts, colours, and the like). If the file is not found or has an invalid format, default values are used. If the interface settings are changed in a session, on exiting the user will be asked whether the new settings should be changed. If the answer is affirmative, the settings will be stored in the `mim31.msf` file.

It is possible to specify a workspace file (see Section A.12.2) when invoking MIM. The normal way to do this is to use a certain extension for workspace files, for example `.mim`, and use View|Options|FileTypes in Windows Explorer to associate this extension with MIM. Then clicking such a file in Windows Explorer, will cause MIM to be opened and the workspace retrieved.

B.2   The Main Menu

Here a short overview of the menu is given.

- File
  - New: this clears the workspace (corresponding to the `Clear` command).
  - Retrieve; Save; SaveAs: these call standard Windows dialogues to retrieve and save workspace files (see Appendix A.12.2).
  - Editor: see Section B.4 below.
  - Input: calls a standard Windows dialogue to identify a file, and then calls the `Input` command for that file (see Section A.12.1). Used to input text files with commands and (often) data.
  - Printer Setup: calls a standard Windows printer setup dialogue.
  - Print: prints the work area.
  - Save Output: calls a standard Windows dialogue to identify a file, and then saves the work area to that file.
  - Clear Output: clears the work area.
  - Exit: closes MIM.
- Data
  - Declare Variate: elicits user input to declare a variate.
  - Declare Factor: elicits user input to declare a factor.
  - Show Variables: calls the `Show W` command.
  - Enter Data: see below.
  - Edit Data: see below.
  - Access Database: see below.
  - List Data: calls the `Print D` command.
  - Univariate Statistics: elicits user input to choose a variable, then calls the `Describe` command.
  - Show Summary Statistics: calls the `Print` command to show empirical statistics.
  - Erase Variables: elicits user input to choose some variables, then calls the `Erase` command.
  - Calculate Variable: elicits user input for a calculate expression, then calls the `Calculate` command.
  - Restrict Observations: elicits user input for a restrict expression, then calls the `Restrict` command.

- Model
  - Saturated Model: sets the current model to the saturated model.
  - Homogeneous Saturated Model: sets the current model to the homogeneous saturated model.
  - Main Effects Model: sets the current model to the main effects model.
  - Model Formula: elicits user input for a model formula, then calls the `Model` command.
  - Show Properties: calls the `Show p` command.
  - Delete Edge: elicits user input to choose one or more edges, then calls the `DeleteEdge` command.
  - Add Edge: elicits user input to choose one or more edges, then calls the `AddEdge` command.
- Fit
  - Fit: fits the current model to the data, using the `Fit` command.
  - Show Estimates: elicits user input to choose the parameters to be shown, then calls the `Display` command.
  - Set Maximum Model: sets the maximum model as the current model.
  - Residuals: elicits user input to calculate residuals (using the `Residuals` command).
  - Mahalanobis: elicits user input to calculate Mahalanobis distances (using the `Mahalanobis` command).
  - EMFit: calls the `EMFit` command.
  - CGFit: calls the `CGFit` command.
- Test
  - Set Base: calls the `Base` command.
  - Test: calls the `Test` command.
  - Delete edge: elicits user input to perform an edge deletion test using the `TestDelete` command.
- Select
  - Stepwise: elicits user input to specify a stepwise selection using the `Stepwise` command.
  - EH-procedure|Initialize: initializes the EH-procedure.
  - EH-procedure|Start: starts the EH-procedure.
  - EH-procedure|Show Status: shows the current status of the EH-procedure.
  - EH-procedure|Clear: clears the EH-procedure.

  – Select|Global: elicits user input to specify model selection by minimum AIC or BIC.

- Graphics

  – Independence Graph: shows the independence graph of the current model.

  – Scatterplot: elicits user input to specify a simple scatter plot.

  – Histogram: elicits user input to specify a simple histogram.

  – BoxPlot: elicits user input to specify a simple box plot.

- Options

  – Iterative Fitting: allows modifications of parameters controlling the MIPS algorithm.

  – Font: allows change of screen font.

  – Colours: allows change of screen colours.

  – Show Settings: shows various current settings.

  – Command Recall: determines whether up and down arrows scroll through the work area or recall previously entered command lines (see below).

- Help

  – Help: calls MIM's hypertext help system.

  – About: shows information about the current version, user license, and the like.

B.3   Entering Commands and Navigating the Work Area

When typing a line into MIM, it may of course happen that one makes a mistake. Before ENTER is pressed, an error may be corrected by using the backspace key on the keyboard and retyping. Various other editing keys can also be used: left and right arrows, <Ins>, <Del>, and <End>.

To browse around the work area, the <PgUp> and <PgDn> keys can be used, or the scroll bar can be dragged up or down. Normally, the up and down arrows can also be used to browse through the output (one line at a time). However, in command recall mode, the up and down arrow keys can be used to recall previously entered lines. Command recall mode is obtained using the menu item Options|Command Recall.

B.4   The Built-In Editor

Selecting the menu item File|Editor invokes a pop-up window with a simple built-in editor. This allows commands and/or data to be entered and submitted to MIM, using the Submit menu item. Blocks can be marked as usual using the mouse; in this case, the marked block rather than the whole file will be submitted. Cut-and-paste (accessed using a right mouse click) is also available.

The menu items File|GetCommands, File|GetOutput and File|GetBoth read input, output, or both input and output, respectively, from the work area.

B.5   Interactive Data Entry

The menu item Data|Enter Data is used for entering data interactively using a simple full-screen spreadsheet-like editor. It calls the **EnterData** command, whose syntax is

```
EnterData no varlist
```

where **no** is the number of observations and **varlist** is a list of variables. The letters in the list can be separated by blanks; they must have been declared in advance as variable names. Discrete variables take the values $1, \ldots, k$, where k is the number of levels declared for that variable. Missing values can be entered as asterisks (*). For example,

```
fact a2; cont xyz; enter 20 axyz
```

enables data entry of 20 observations for the variables a, x, y, and z. The entry screen is shown in Figure B.2.

The cursor is moved around using the arrow keys, as well as <Home> (move to the left end of the line), <End> (move to the right end), <PgUp>, and <PgDn> keys. Values are entered into the fields in the ordinary way. Only valid factor values are accepted — others are registered as missing. The order of the variables and observations can /be changed by dragging and dropping the row and column headings with the mouse.

Selecting the menu item Labels causes variable labels rather than names to be displayed. After the data have been entered they are saved into MIM using the menu item File|Save.

The menu item Data|Edit Data allows data editing, using precisely the same full-screen spreadsheet method as is used for data entry. It can also

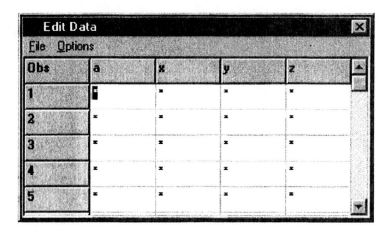

FIGURE B.2. Spreadsheet-style data entry.

be invoked using the `EditData` command (which takes no operands). The feature is useful for both editing and browsing through data.

The following example shows how to add a variable to an existing dataset:

```
MIM->cont x; read x
DATA->2 5 8 0 3 5 7 9 1 3 6 7 !
MIM->cont y; calc y=sqrt(-1)
MIM->editdata
```

First, 12 values of a variable X are entered using `Read` in the usual way. To add a continuous variable Y to this dataset, the `Calculate` command is used to initialize the Y values to missing (see next subsection). Then `EditData` can be used to enter the Y values.

B.6 Independence Graphs

Selecting the menu item Graphics|IndependenceGraph causes a pop-up window to appear, displaying the independence graph of the current model. An example is shown in Figure B.3. The default form of the graph is as a circle, as shown. Discrete nodes are shown as filled circles, and continuous nodes as hollow circles. The variable letters are shown in the circles.

The window stays open until explicitly closed (using File|Close), displaying the current model even when this changes. So the progress of, for example, a stepwise selection can be observed in the graph window.

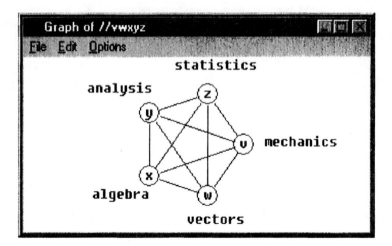

FIGURE B.3. An independence graph

The graph can be adjusted in various ways. Firstly, labels can be displayed by selecting the Options|Labels item from the menu on the graph window. Secondly, the graph can be rearranged by dragging and dropping the nodes with the mouse. The variable labels follow along, retaining their relative position to the nodes. They can also be moved, using the mouse.

The menu item Options|Rearrange examines permutations of the node positions so as to minimize the number of crossovers.

Two features are available after a one-step stepwise selection (see Appendix A.8.1). Any nonsignificant edges are shown with dotted lines, and the menu item Options|PValues causes the p-values associated with the edges to be shown on the graph.

Colours and text font can be controlled using the menu items Options|Font and Options|Colours. Options|Grid and Options|SnapToGrid cause, respectively, a grid to appear and the nodes to be placed on the nearest grid point. These are useful in connection with LaTeX graphics (see Appendix B.8).

In block mode, the current block-recursive model is drawn with arrows, and Options|Draw Blocks draws blocks on the graph, as shown in Figure B.4. The blocks can be resized and moved using the mouse.

B.7   Simple Data Graphics

Three simple forms of data graphics are available: scatter plots, histograms and box plots. Each can be one- or two-dimensional.

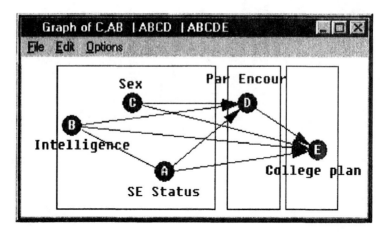

FIGURE B.4. A chain graph

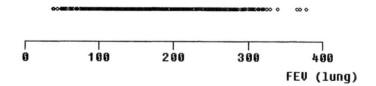

FIGURE B.5. A univariate scatter plot

B.7.1 Scatter Plots

When the menu item Graphics|Scatterplot is selected, a dialog box appears, which allows one or two variables (either variates or factors) to be specified. For example, specifying an x-variable only gives a one-dimensional plot shown in Figure B.5. Such plots can be useful for examining marginal distributions.

Bivariate plots are illustrated in Section 6.6. Clicking on a data point gives information about that point. Various parts of the plots (for example, the axis endpoints) can be changed using the menu item Options|Specifications.

The colours used in the plot, and the font can also be modified using the menu. The plot can be copied to the clipboard for transfer to other programs, and it can be saved in various formats, as described in Section B.8.

B.7.2   Histograms

When the menu item Graphics|Histogram is selected, a dialog box appears, allowing the selection of either one variate (giving an ordinary univariate histogram, either horizontal or vertical), or one variate and one factor (giving a grouped histogram).

Various parts of the plot (for example, the axis endpoints) can be changed using the menu item Options|Specifications.

The colours used in the plot, and the font can also be modified using the menu. The plot can be copied to the clipboard for transfer to other programs, and it can be saved in various formats, as described in Section B.8.

B.7.3   Box Plots

A box plot is a simple representation of a univariate distribution. It consists of a box from the 25% to the 75% fractile, a vertical bar showing the median, and lines from the box to the 5% and 95% fractiles. Values outside the 5% and 95% fractiles are shown as points.

When the menu item Graphics|Boxplot is selected, a dialog box appears, allowing the selection of either one variate (giving an ordinary univariate boxplot, either horizontal or vertical), or one variate and one factor (giving a grouped boxplot). Figure B.6 shows a grouped boxplot.

Various parts of the plot (for example, the axis endpoints) can be changed using the menu item Options|Specifications.

The colours used in the plot, and the font can also be modified using the menu. The plot can be copied to the clipboard for transfer to other programs, and it can be saved in various formats, as described in Section B.8.

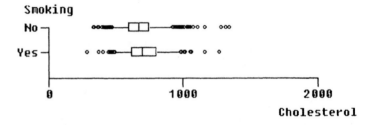

FIGURE B.6. A grouped boxplot

B.8   Graphics Export Formats

Graphics windows (i.e., independence graphs, scatter plots, histograms, and box plots) may be exported as a bitmap (to a file), in Windows enhanced metafile format (to a file or to the clipboard), in JPeg format (to a file), or in LaTeX format (to a file).

Bitmaps are used by a wide variety of Windows programs. For example, most paint programs allow input of bitmap files. An exact copy of the window is exported. Bitmap files have the extension .bmp.

Windows enhanced metafile format is useful in connection with other Windows-based word processing or graphics software — for example, MSWord, CorelDraw and others. What are essentially instructions to Windows on how to draw the graphic are stored on the file. The advantage compared with bitmaps is that the graphics may be rescaled without loss of resolution, and the individual elements in the graphics — for example, the text, or line segments — may be individually selected and edited with various software packages. Windows metafile format files have the extension .wmf.

The JPeg format is a compressed format widely used in HTML documents; the standard extension is .jpg.

LaTeX format is useful for inclusion in documents prepared using LaTeX. LaTeX code generating a near-copy of the contents of the window is saved on file. Due to limitations in LaTeX, some features cannot be transferred (for example, font and colour). Moreover, some versions of LaTeX restrict the line gradients that are permitted, so that if invalid gradients are used, the LaTeX code will not compile. This occurs particularly with independence graphs: the user must check manually before generating the LaTeX code that the gradients of the edges are of the form i:j, where i and j are integers between −6 and +6. (For directed edges (arrows) i and j must lie between −4 and 4). It is straightforward to do this using the grid feature. If there are invalid gradients, the graph must be rearranged before generating the LaTeX code. LaTeX code files have the extension .tex.

B.9   Direct Database Access

It can be time-consuming to prepare data for entry using the Read or StatRead commands. If the data are stored in a database, it is generally far easier to import them directly from the database. Some versions of MIM can access data in desktop databases such as Access, Paradox and dBase; from spreadsheet programs such as Excel; from statistical programs such as SAS and SPlus; as well as from other databases for which ODBC drivers are available. Moreover, remote (client/server) databases for which ODBC drivers are available can also be accessed. For further information, consult the online help system.

B.10   Program Intercommunication

MIM provides a COM automation server. This enables other software to access MIM programatically. Lines with commands or data may be passed to MIM from a COM client, processed, and the output lines transferred to the client. For further information, consult the online help system.

Appendix C

On Multivariate Symmetry

Suppose $X = (X_1, \ldots, X_q)'$ is a random vector in $\mathcal{R}^{|\Gamma|}$ following the $\mathcal{N}(\mu, \Sigma)$ distribution. Suppose further that the components of X are dichotomized at their means, generating a 2^q table $\{p_i\}$ of orthant probabilities. For a cell i in this table, define i^c to be its mirror image, i.e., $i^c = (j^\delta)$, where

$$j^\delta = \begin{cases} 1 & \text{if } i^\delta = 2 \\ 2 & \text{otherwise} \end{cases}.$$

We are interested in the property of multivariate symmetry, namely that $p_i = p_{i^c}$ for all $i \in \mathcal{I}$.

Lemma 1. *The orthant probabilities satisfy multivariate symmetry.*

Proof: Follows from $X \sim \mathcal{N}(\mu, \Sigma) \Leftrightarrow -X \sim \mathcal{N}(-\mu, \Sigma)$.

Lemma 2. *The orthant probabilities $\{p_i\}$ satisfy a nonhierarchical loglinear model in which all interactions involving an odd number of factors vanish.*

Proof: We construct a loglinear expansion with the given property. We write the expansion as

$$\ln(p_i) = \sum_{a \subseteq \Gamma} u_i^a, \tag{C.1}$$

where, as usual, it is understood that the term u_i^a depends on i only through i_a. Each interaction term has one degree of freedom, so we can write

$$u_i^a = \begin{cases} +u^a & \text{if } \#2(i_a) \text{ is even} \\ -u^a & \text{otherwise} \end{cases},$$

where $\#2(i_a)$ is the number of 2's in i_a. We rewrite (C.1) as

$$\ln(p_i) = \sum_{a \subseteq \Gamma} (-1)^{\#2(i_a)} u^a.$$

The relation $p_i = p_{i^c}$ implies that

$$\sum_{a \subseteq \Gamma} (-1)^{\#2(i_a)} u^a = \sum_{a \subseteq \Gamma} (-1)^{\#2(i_a^c)} u^a,$$

giving

$$\sum_{a \subseteq \Gamma, \#a \text{ odd}} (-1)^{\#2(i_a)} u^a = \sum_{a \subseteq \Gamma, \#a \text{ odd}} (-1)^{\#2(i_a^c)} u^a,$$

where $\#a$ is the number of elements of a. It follows that

$$\sum_{a \subseteq \Gamma, \#a \text{ odd}} (-1)^{\#2(i_a)} u^a = 0. \tag{C.2}$$

Now take some $d \subseteq \Gamma$, let $e = \Gamma \setminus d$, and sum (C.2) over all i for which $i_d = (1, 1, \ldots, 1)$:

$$0 = \sum_{i: i_d = (1,1,\ldots,1)} \sum_{a \subseteq \Gamma, \#a \text{ odd}} (-1)^{\#2(i_a)} u^a$$

$$= \sum_{i: i_d = (1,1,\ldots,1)} \sum_{a \subseteq \Gamma, \#a \text{ odd}} (-1)^{\#2(i_{a \cap e})} u^a$$

$$= \sum_{a \subseteq d, \#a \text{ odd}} u^a, \tag{C.3}$$

since if $a \not\subseteq d$, the number of cells with $\#2(i_{a \cap e})$ even is equal to the number of cells with $\#2(i_{a \cap e})$ odd. Taking d in (C.3) successively as all singletons, all triplets, etc. implies that $u^b = 0$ whenever $\#b$ is odd, thereby proving Lemma 2.

Notice that the only constraint on the $\{p_i\}$ used in Lemma 2 was that of multivariate symmetry, i.e., that $p_i = p_{i^c}$ for all $i \in \mathcal{I}$. So the result is also valid for more general symmetric distributions.

The nonhierarchical loglinear model described in Lemma 2 has explicit MLEs. To see this, partition $\mathcal{I}$ into $\mathcal{I}_1 = \{i \in \mathcal{I} : i_1 = 1\}$ and $\mathcal{I}_2 = \{i \in \mathcal{I} : i_1 = 2\}$, say. The likelihood is proportional to

$$\prod_{i \in \mathcal{I}} p_i^{n_i} = \prod_{i \in \mathcal{I}_1} p_i^{n_i} p_{i^c}^{n_{i^c}} = \prod_{i \in \mathcal{I}_1} p_i^{n_i + n_{i^c}}.$$

We can regard this as a model for an incomplete table with cells $\mathcal{I}_1$ and cell counts $n_i + n_{i^c}$, subject only to the constraint $\sum_{i \in \mathcal{I}_1} p_i = \frac{1}{2}$. It follows that the MLEs are

$$\hat{p}_i = \frac{n_i + n_{i^c}}{2N}.$$

A simple test of multivariate symmetry is the deviance

$$2\sum_{i \in \mathcal{I}} n_i \ln\left(\frac{2n_i}{n_i + n_{i^c}}\right),$$

whose asymptotic null distribution is χ^2 with 2^{q-1} degrees of freedom.

Alternatively, a conditional test can be constructed. For each pair of cells i and i^c, this conditions on the sum $n_i + n_{i^c}$ and tests whether

$$\Pr(I = i \mid I \in \{i, i^c\}) = \frac{1}{2}.$$

In other words, it is equivalent to 2^{q-1} binomial tests.

Appendix D

On the Estimation Algorithms

This appendix explains the computational details of the estimation algorithms used in MIM. It is primarily intended for those who wish to implement the algorithms in other systems.

D.1 The MIPS Algorithm

We describe an algorithm for finding the maximum likelihood estimates of the parameters of a general hierarchical interaction model with formula

$$d_1, \ldots, d_r / l_1, \ldots l_s / q_1, \ldots q_t. \tag{D.1}$$

This method is an implementation of the MIPS algorithm (Frydenberg and Edwards, 1989), a general procedure for finding maximum likelihood estimates. See also Lauritzen (1996, section 6.4.3).

D.1.1 Notation

As in Section 4.1.5, let $(n_j, T_j, \bar{y}_j, SS_j, S_j)_{j \in \mathcal{I}}$ be the observed counts, variate totals, variate means, uncorrected sums of squares and products, and sample cell variances for cell j. To remove any possible ambiguity we can give the formulae for these:

$$n_j = \#\{k : i^{(k)} == j\}$$
$$T_j = \sum_{k:i^{(k)}=j} y^{(k)}$$

$$\bar{y}_j \quad = T_j/n_j$$

$$SS_j = \sum_{k:i^{(k)}=j} y^{(k)} y^{(k)'}$$

$$S_j \quad = \sum_{k:i^{(k)}=j} (y^{(k)} - \bar{y}_j)(y^{(k)'} - \bar{y}_j)'/n_j$$

$$= SS_j/n_j - \bar{y}_j \bar{y}_j'.$$

For $a \subseteq \Delta$, we write the marginal cell corresponding to i as i_a; likewise, for $d \subseteq \Gamma$, we write the subvector of y as y^d.

Similarly, we write the marginal cell counts as $\{n_{j_a}\}_{j_a \in \mathcal{I}_a}$, marginal variate totals as $\{T_{j_a}^d\}_{j_a \in \mathcal{I}_a}$, marginal variate means as $\{\bar{y}_{j_a}^d\}_{j_a \in \mathcal{I}_a}$, marginal uncorrected sums of squares and products as $\{SS_{j_a}^d\}_{j_a \in \mathcal{I}_a}$, and marginal corrected sums of squares and products as $\{S_{j_a}^d\}_{j_a \in \mathcal{I}_a}$. In formulae,

$$n_{j_a} \quad = \sum_{k:k_a=j_a} n_k$$

$$T_{j_a}^d \quad = \sum_{k:k_a=j_a} T_k^d$$

$$= \sum_{k:k_a=j_a} n_k \bar{y}_k^d$$

$$\bar{y}_{j_a}^d \quad = T_{j_a}^d/n_{j_a}$$

$$SS_{j_a}^d = \sum_{k:k_a=j_a} SS_k^d$$

$$S_{j_a}^d \quad = SS_{j_a}^d/n_{j_a} - \bar{y}_{j_a}^d (\bar{y}_{j_a}^d)',$$

where T_k^d and SS_k^d are the subvectors and -matrices of T_k and SS_k corresponding to d.

We also need some corresponding "fitted" quantities. That is, for some $\theta = \{m_j, \mu_j, \Sigma_j\}_{j \in \mathcal{I}}$, we define

$$m_{j_a} \quad = \sum_{k:k_a=j_a} m_k$$

$$ET_{j_a}^d \quad = \sum_{k:k_a=j_a} m_k \mu_k^d$$

$$E\bar{y}_{j_a}^d \quad = ET_{j_a}^d/m_{j_a}$$

$$ESS_{j_a}^d = \sum_{k:k_a=j_a} m_k \{\Sigma_k^d + \mu_k^d (\mu_k^d)'\}$$

$$ES_{j_a}^d \quad = ESS_{j_a}^d/m_{j_a} - E\bar{y}_{j_a}^d (E\bar{y}_{j_a}^d)'.$$

Note that the observed and fitted marginal quantities can be derived using similiar computations based on $\{n_j, T_j, SS_j\}_{j \in \mathcal{I}}$ for the observed quantities and $\{m_j, m_j \mu_j, m_j(\Sigma_j + \mu_j \mu_j')\}_{j \in \mathcal{I}}$ for the fitted quantities. We use the E-notation to limit the number of symbols, and to emphasise this parallelism; it is *not* used here as an expectation operator.

D.1.2   The Likelihood Equations

As also described in Section 4.1.5, the likelihood equations are constructed by equating the minimal canonical statistics, U, with their expectations under the model. That is, if we denote the model parameters by θ and the parameter space of the model by Θ_m, then the maximum likelihood estimate, when it exists, is the unique $\theta \in \Theta_m$ for which $E_\theta(U) = u$, where u is the realisation of U. The minimal canonical statistics are in the present case a set of marginal counts corresponding to the discrete generators, a set of marginal variate totals corresponding to the linear generators, and a set of variate sums of squares corresponding to the quadratic generators.

Thus the likelihood equations take the following form. For each discrete generator a, we require that for all $i_a \in \mathcal{I}_a$,

$$m_{i_a} = n_{i_a}. \tag{D.2}$$

Similarly, for each linear generator l, setting $a = l \cap \Delta$ and $\gamma = l \cap \Gamma$, we require that for all $i_a \in \mathcal{I}_a$,

$$ET^\gamma_{i_a} = T^\gamma_{i_a}. \tag{D.3}$$

Finally, for each quadratic generator q, setting $a = q \cap \Delta$ and $d = q \cap \Gamma$, we require that for all $i_a \in \mathcal{I}_a$,

$$ESS^d_{i_a} = SS^d_{i_a}. \tag{D.4}$$

It is convenient to rewrite these equations in a different form. Firstly, exploiting the first syntax rule described in Section 4.1.3, we can rewrite equation (D.3) in terms of means as

$$E\bar{y}^d_{i_a} = \bar{y}^d_{i_a}. \tag{D.5}$$

This holds for those set-pairs (a, d) for which, for each $\gamma \in d$, there exists a linear generator containing $a \cup \gamma$.

Similarly, again exploiting the syntax rules in Section 4.1.3, we can rewrite equation (D.4) in terms of sample covariances as

$$ES^d_{i_a} = S^d_{i_a}. \tag{D.6}$$

For certain sets a and d, as explained shortly, we have similar results for homogeneous sample covariances, of the form

$$ES_a^d = S_a^d. \tag{D.7}$$

In this equation, S_a^d is the homogeneous $\bar{y}_{i_a}^d$-corrected covariance estimate, i.e., $\sum_{i_a \in \mathcal{I}_a} n_{i_a} S_{i_a}^d / N$, and ES_a^d is the corresponding fitted quantity $\sum_{i_a \in \mathcal{I}_a} m_{i_a} ES_{i_a}^d / N$. We use the subscript a in (D.7) to indicate that these quantities depend on which set a is used.

We thus rearrange the likelihood equations as follows:

Q A list of set-pairs (a_i^1, d_i^1), for $i = 1, \ldots, k_1$, say, corresponding to the heterogeneous quadratic generators q_k, i.e., those for which $q_k \cap \Delta \neq \emptyset$. For these the equations (D.2), (D.5), and (D.6) hold.

H A list of set-pairs (a_i^2, d_i^2), for $i = 1, \ldots, k_2$, say, defined as the maximal set-pairs (a, d) such that

   a. $\emptyset \subset d$,

   b. $d \subseteq q_k$ for some quadratic generator q_k, and

   c. for each $\gamma \in d$, $a \cup \gamma \subseteq l_k$ for some linear generator l_k.

   For these sets (a_i^2, d_i^2) the equations (D.2), (D.5), and (D.7) hold.

D A list of sets a_i^3, for $i = 1, \ldots, k_3$, say corresponding to the discrete generators. For these, the equations (D.2) hold.

The lists are then pruned in the sense that any set-pair in list **H** contained in a set-pair in list **Q** is omitted from **H**. Similarly, any set a_i^3 contained in some a_j^1 or a_j^2 is omitted from list **D**. For homogeneous models, list **Q** will be empty, and for heterogeneous graphical models, list **H** will be empty.

D.1.3   The General Algorithm

This consists of a number of *cycles*, each cycle consisting of a series of steps, where each step corresponds to a set-pair in one of the three lists. It requires a starting value for $\theta \in \Theta_m$, but appears very insensitive to this, so a convenient choice is to set $m_i = 1$, $\mu_i^\gamma = 0$, and $\Sigma_i = I$ for all $i \in \mathcal{I}$ and $\gamma \in \Gamma$.

Write the current parameter estimates as $\{m_i, \mu_i, \Sigma_i\}_{i \in \mathcal{I}}$ in moment form, or equivalently as $\{\alpha_i, \beta_i, \Omega_i\}_{i \in \mathcal{I}}$ in canonical form. We now describe how these current parameter estimates are updated during the steps.

For the first type of update, corresponding to a set-pair (a, d) in list **Q**, we compute the sample statistics from the right-hand sides of (D.2), (D.5) and (D.6), namely

$$\{n_{i_a}, \bar{y}_{i_a}^d, S_{i_a}^d\}_{i_a \in \mathcal{I}_a}.$$

We also compute the corresponding right-hand sides of the equations, which
we write as

$$\{m_{i_a}, E\bar{y}_{i_a}^d, ES_{i_a}^d\}_{i_a \in \mathcal{I}_a}.$$

Note that each consists of an array of elements of the form (scalar, $|d|$-
vector, $|d| \times |d|$ symmetric matrix). Using equations (4.3—4.5), we then
transform these to canonical form, to, say,

$$\{\tilde{\alpha}_{i_a}, \tilde{\beta}_{i_a}^d, \tilde{\Omega}_{i_a}^d\}_{i_a \in \mathcal{I}_a}$$

and

$$\{\breve{\alpha}_{i_a}, \breve{\beta}_{i_a}^d, \breve{\Omega}_{i_a}^d\}_{i_a \in \mathcal{I}_a},$$

respectively. These are then used to update the canonical parameters: for
each $i \in \mathcal{I}$, and $\gamma, \eta \in d$,

$$\alpha_i \ := \alpha_i + \tilde{\alpha}_{i_a} - \breve{\alpha}_{i_a} \tag{D.8}$$

$$\beta_i^\gamma \ := \beta_i^\gamma + \tilde{\beta}_{i_a}^\gamma - \breve{\beta}_{i_a}^\gamma \tag{D.9}$$

$$\omega_i^{\gamma\eta} := \omega_i^{\gamma\eta} + \tilde{\omega}_{i_a}^{\gamma\eta} - \breve{\omega}_{i_a}^{\gamma\eta}, \tag{D.10}$$

The second type of update, corresponding to a set-pair (a, d) in list $\mathbf{H}$, is
very similar, the quantities in equations (D.7) replacing those from (D.6).
That is to say, we calculate the sample statistics from the right-hand sides
of (D.2), (D.5), and (D.7), namely,

$$\{n_{i_a}, \bar{y}_{i_a}^d, S_a^d\}_{i_a \in \mathcal{I}_a},$$

and also the corresponding fitted quantities from the left-hand sides,
namely,

$$\{m_{i_a}, E\bar{y}_{i_a}^d, ES_a^d\}_{i_a \in \mathcal{I}_a}.$$

As before, we transform each to canonical form, to, say,

$$\{\dot{\alpha}_{i_a}, \dot{\beta}_{i_a}^d, \dot{\Omega}^d\}_{i_a \in \mathcal{I}_a}$$

and

$$\{\ddot{\alpha}_{i_a}, \ddot{\beta}_{i_a}^d, \ddot{\Omega}^d\}_{i_a \in \mathcal{I}_a},$$

respectively. These are then used to update the canonical parameters: for
each $i \in \mathcal{I}$, and $\gamma, \eta \in d$,

$$\alpha_i \ := \alpha_i + \dot{\alpha}_{i_a} - \ddot{\alpha}_{i_a} \tag{D.11}$$

$$\beta_i^\gamma \ := \beta_i^\gamma + \dot{\beta}_{i_a}^\gamma - \ddot{\beta}_{i_a}^\gamma \tag{D.12}$$

$$\omega_i^{\gamma\eta} := \omega_i^{\gamma\eta} + \dot{\omega}^{\gamma\eta} - \ddot{\omega}^{\gamma\eta}. \tag{D.13}$$

The third type of update, corresponding to a set a in list $\mathbf{D}$, consists simply
of updating the discrete canonical parameters: for each $i \in \mathcal{I}$,

$$\alpha_i := \alpha_i + \ln(n_{i_a}) - \ln(m_{i_a}). \tag{D.14}$$

We do not specify any particular order of steps in each cycle, just that
an update is performed for every element of the lists $\mathbf{Q}$, $\mathbf{H}$, and $\mathbf{D}$. The
algorithm continues until a convergence criterion is satisfied. This is based
the changes in moments parameters in an updating step, written δm_i, $\delta \mu_i^\gamma$,
and $\delta \sigma_i^{\gamma\eta}$. More precisely, the criterion states that

$$\texttt{mdiff} = \max_{i \in \mathcal{I}, \gamma, \eta \in \Gamma} \left\{ \frac{|\delta m_i|}{\sqrt{(m_i + 1)}}, \frac{|\delta \mu_i^\gamma|}{\sqrt{\sigma_i^{\gamma\gamma}}}, \frac{|\delta \sigma_i^{\gamma\eta}|}{\sqrt{\sigma_i^{\gamma\gamma} \sigma_i^{\eta\eta} + (\sigma_i^{\gamma\eta})^2}} \right\}$$

should be less than some predetermined (small) value for all steps in a
cycle.

When $q = 0$ this algorithm reduces to the IPS algorithm (see Section 2.2.1),
and when $p = 0$ to Speed and Kiiveri's algorithm (see Section 3.1.2).
For decomposable models, the algorithm will converge after the first iteration provided the update steps are performed in a particular sequence
(see Section 4.4).

Note that the algorithm requires much conversion between the natural
(moments) form and the canonical form, for both the model parameters
and the updating quantities. For some models these computations can be
reduced substantially, as we now show.

D.1.4   The Δ-Collapsible Variant

This applies to models that are collapsible onto the set of discrete vari-
ables Δ. For such models, the logarithms to the cell probabilities $\{p_i\}_{i \in \mathcal{I}}$
obey the same factorial constraints as the discrete canonical parameters.
In other words, the cell counts follow the loglinear model that is specified
by the discrete generators. They can therefore be fitted using the standard
IPS algorithm, and the fitted values held constant thereafter. The advan-
tage of this is that in the updates corresponding to the lists $\mathbf{Q}$ and $\mathbf{H}$, the
discrete canonical parameters need not be updated. By using the mixed
parametrisation $\{p_i, \beta_i, \Omega_i\}_{i \in \mathcal{I}}$ rather than the full canonical parametrisation $\{\alpha_i, \beta_i, \Omega_i\}_{i \in \mathcal{I}}$, we avoid having to recalculate p_i and α_i whenever we
transform from moments to canonical form and vice versa. This applies
both for the model parameters and for the updating quantities. So, for example, an update corresponding to a set-pair (a, d) in list $\mathbf{Q}$ would involve
calculating

$$\{\bar{y}_{i_a}^d, S_{i_a}^d\}_{i_a \in \mathcal{I}_a}$$

and the corresponding fitted quantities

$$\{E\bar{y}_{i_a}^d, ES_{i_a}^d\}_{i_a \in \mathcal{I}_a},$$

deriving from these the linear and quadratic canonical parameters, and then using these in updates (D.9) and (D.10).

We can summarize this variant as follows:

1. Fit the cell counts by use of the standard IPS algorithm, using the discrete generators.
2. Repeat until convergence:
   a. Perform the updates (D.9) and (D.10) for the set-pairs in list **Q**.
   b. Perform the updates (D.12) and (D.13) for the set-pairs in list **H**.

D.1.5   The Mean Linear Variant

Further computational savings can be made for models that are both collapsible onto Δ and mean linear (see Section 4.3). For such models, the cell means $\{\mu_i\}_{i\in\mathcal{I}}$ obey the same factorial constraints as the linear canonical parameters $\{\beta_i\}_{i\in\mathcal{I}}$. Here we can use the mixed parametrisation $\{p_i, \mu_i, \Omega_i\}_{i\in\mathcal{I}}$ to allow replacement of the updating step (D.9) by

$$\mu_i^\gamma := \mu_i^\gamma + \bar{y}_{i_a}^\gamma - E\bar{y}_{i_a}^\gamma \tag{D.15}$$

for all $i \in \mathcal{I}$ and $\gamma \in d$. This process, in which the fitted means are incremented by the difference between the observed and fitted marginal means, is the basis of the sweep algorithm (Wilkinson, 1970) for ANOVA models. For balanced designs it converges after one cycle.

To fit the covariance matrices, we utilize (D.4) and calculate

$$\tilde{S}_{i_a}^d = SS_{i_a}^d / m_{i_a} - \sum_{k:k_a=i_a} m_k \mu_k^d (\mu_k^d)' / m_{i_a}$$

$$\tilde{ES}_{i_a}^d = \sum_{k:k_a=i_a} m_k \Sigma_k^d / m_{i_a},$$

and their inverses, say, $\tilde{\Omega}_{i_a}^d = (\tilde{S}_{i_a}^d)^{-1}$ and $\check{\Omega}_{i_a}^d = (\tilde{ES}_{i_a}^d)^{-1}$, and perform the update

$$\omega_i^{\gamma\eta} := \omega_i^{\gamma\eta} + \bar{\omega}_{i_a}^{\gamma\eta} - \check{\omega}_{i_a}^{\gamma\eta}. \tag{D.16}$$

for all $i \in \mathcal{I}$ and $\gamma, \eta \in d$. We can summarize this variant as follows:

1. Fit the cell counts by use of the standard IPS algorithm, using the discrete generators.
2. Fit the cell means by use of the sweep algorithm (D.15), using the linear generators. For balanced designs this converges after one cycle.
3. Fit the covariance matrices by means of update (D.16), using the quadratic generators. Repeat this until convergence.

D.1.6   The Q-Equivalent Variant

Still further computational savings are available for models that, in addition to being mean linear and collapsible onto Δ, constrain the cell covariances in the same way they constrain the cell precision matrices. A condition for the stated property is that the quadratic generators induce a partition of Γ, i.e., for all pairs of quadratic generators (q_k, q_l) it holds that $q_k \cap q_l \subseteq \Delta$. For such models we have that

$$\Sigma_i^d = \tilde{S}_{i_a}^d \qquad\qquad\qquad (D.17)$$

and so we can summarize the variant as follows:

1. Fit the cell counts by use of the standard IPS algorithm, using the discrete generators.

2. Fit the cell means by use of the sweep algorithm, using the linear generators. For balanced designs, this converges after one cycle.

3. Fit the covariance matrices by applying (D.17), using the quadratic generators (no iteration is necessary).

D.1.7   The Step-Halving Variant

Finally, we describe a variant which is not designed to improve computational efficiency, but rather to ensure convergence in all cases. For pure models, the algorithm is known to converge to the maximum likelihood estimate whenever this exists (Ireland and Kullback, 1968; Speed and Kiiveri, 1986), but for mixed models this does not necessarily hold. Occasionally the algorithm does not succeed, either because it leads to covariance matrices that are not positive definite (preventing matrix inversion), or because the likelihood does not increase at each step.

One recalcitrant example is given in the following fragment, due to M. Frydenberg (personal comm.):

```
fact i2j3; cont xy; statread ijxy
10 1   2   1.0 0.9 1
20 3   4   2.0 1.4 1
 5 5   6   1.0 0.0 1
15 7   8   1.0 2.0 7
 5 9  10   2.0 2.0 3
45 11 12   4.5 5.0 6 !
model i,j/ix,iy,jx,jy/ixy,jxy
```

Frydenberg and Edwards describe a modification to the general algorithm involving a step-halving constant κ. The increment to the model parameters are multiplied by this constant, so that for example, (D.8)-(D.10) become

$$\alpha_i := \alpha_i + \kappa(\tilde{\alpha}_{i_a} - \check{\alpha}_{i_a}) \tag{D.18}$$

$$\beta_i^\gamma := \beta_i^\gamma + \kappa(\tilde{\beta}_{i_a}^\gamma - \check{\beta}_{i_a}^\gamma) \tag{D.19}$$

$$\omega_i^{\gamma\eta} := \omega_i^{\gamma\eta} + \kappa(\tilde{\omega}_{i_a}^{\gamma\eta} - \check{\omega}_{i_a}^{\gamma\eta}). \tag{D.20}$$

Prior to the updating step, κ is set to unity. The update is attempted, but two things are checked: that the likelihood increases, and that the resulting covariance matrices are positive definite. If either condition does not hold, κ is halved and the update attempted again. This modification is believed to ensure convergence to the maximum likelihood estimate whenever this exists. At any rate, no counterexamples have as yet been found.

D.2 The EM-Algorithm

We here describe the computations required to implement the EM-algorithm together with the MIPS algorithm. The incomplete likelihood (4.34) is complex and would be difficult to maximize directly; however, the power of the EM-algorithm (Dempster et al., 1977) lies in the way that estimation algorithms for the complete likelihood (4.33) can be used to maximize the incomplete likelihood. Each cycle in the algorithm consists of two steps: an E (expectation) step and an M (maximisation) step. In the E-step, expected values of the sufficient statistics given the current parameter estimates and the observed data are calculated. In the M-step, new parameter estimates are calculated on the basis of the expected sufficient statistics using the ordinary algorithms.

We now explain the calculations behind the E-step in connection with a hierarchical interaction model. Suppose the current parameter estimates are $\{p_i, \mu_i, \Sigma_i\}_{i \in \mathcal{I}}$ and that the corresponding canonical parameters are $\{\alpha_i, \beta_i, \Omega_i\}_{i \in \mathcal{I}}$. Write the cell counts, variate totals, and variate sums of squares and products as $\{n_i, T_i, SS_i\}_{i \in \mathcal{I}}$. The minimal sufficient statistics for a model are sums of these quantities over margins corresponding to the generators of the model formula (see Section 4.1.5).

The complete observations (if there are any) will contribute to these sufficient statistics. We now consider the expected contribution of the incomplete cases, given the nonmissing values and the current parameter estimates.

Consider then an incomplete case of the form $(i_1, *, y_1, *)$, where the p_1 nonmissing discrete variables correspond to the subset $a \subseteq \Delta$ and the q_1 nonmissing continuous variables correspond to the subset $b \subseteq \Gamma$. The marginal distribution of (I, Y_1) is given by $\{p_i, \mu_i^1, \Sigma_i^{11}\}_{i \in \mathcal{I}}$ with corresponding canonical parameters, say, $\{\tilde{\alpha}_i, \tilde{\beta}_i, \tilde{\Omega}_i\}_{i \in \mathcal{I}}$.

For each i_2, writing $i = (i_1, i_2)$, we calculate

$$p(i_2|i_1, y_1) = \frac{\exp(\tilde{\alpha}_i + \tilde{\beta}'_i y_1 - \frac{1}{2} y'_1 \tilde{\Omega}_i y_1)}{\sum_{j:j_a = i_1} \exp(\tilde{\alpha}_j + \tilde{\beta}'_j y_1 - \frac{1}{2} y'_1 \tilde{\Omega}_j y_1)} \tag{D.21}$$

and similarly

$$\mu_i^{2 \cdot 1} = E(y_2|i, y_1) = (\Omega_i^{22})^{-1}(\beta_i^2 - \Omega_i^{21} y_1) \tag{D.22}$$

and

$$E(y_2 y'_2|i, y_1) = (\Omega_i^{22})^{-1} + \mu_i^{2 \cdot 1}(\mu_i^{2 \cdot 1})'.$$

The sufficient statistics are incremented for the case at hand as follows:

$$n_i := n_i + p(i_2|i_1, y_1),$$

$$T_i := T_i + \left(\begin{array}{c} y_1 \\ \mu_i^{2 \cdot 1} \end{array} \right) p(i_2|i_1, y_1),$$

and

$$SS_i := SS_i + \left[\left(\begin{array}{c} y_1 \\ \mu_i^{2 \cdot 1} \end{array} \right) \left(\begin{array}{c} y_1 \\ \mu_i^{2 \cdot 1} \end{array} \right)' + \left(\begin{array}{cc} 0 & 0 \\ 0 & (\Omega_i^{22})^{-1} \end{array} \right) \right] p(i_2|i_1, y_1),$$

where $i = (i_1, i_2)$ for all i_2. This process is repeated for all incomplete cases. For efficiency, all cases with identical nonmissing values are processed in the same step.

In the M-step, new parameter estimates are calculated on the basis of the expected sufficient statistics in the usual way: that is to say, using these expected sufficient statistics in place of the observed ones. For decomposable models, the explicit formulae of Section 4.4 are used. For nondecomposable models, rather than iteration until convergence with the MIPS algorithm, only one iteration may be performed at each M-step. This is the so-called GEM- (generalized EM) algorithm, which is generally much more efficient than iterating until convergence at each step.

Since likelihood functions of the form (4.34) typically are not convex, the algorithm may converge to a saddle point or to a local maximum. The algorithm can be started from a random point, so this may be used to find the global maximum.

Little and Schluchter (1985) describe application of the EM-algorithm in a context closely related to the present one. Lauritzen (1995), Geng (2000), and Didelez and Pigeot (1998) study methods for efficient calculation of the E-step.

D.3   The ME-Algorithm

This section describes the estimation algorithm used for CG-regression models, the so-called ME-algorithm introduced by Edwards and Lauritzen (1999). For a given hierarchical interaction model $\mathcal{M}$, write the minimal canonical statistics as $T = (T_1, \ldots, T_K)$. As described in Section 4.1.5 and Appendix D.1, the likelihood equations are constructed by equating the expectation of T under the model to its observed value, t. That is, if we denote the model parameters by θ and the model parameter space by Θ, then the maximum likelihood estimate, when it exists, is the unique solution of $E_\theta(T) = t$ such that $\theta \in \Theta$.

Consider now the CG-regression model $\mathcal{M}_{b|a}$. Write $T = (U, V)$, where V only involves variables in a and so is fixed under the conditional model. Then the maximum likelihood estimate under the conditional model, when it exists, is the unique solution of $E_\theta(U|a) = u$ such that $\theta \in \Theta_{b|a}$. Here $\Theta_{b|a}$ is the parameter space for the conditional model and $E_\theta(U|a)$ is the conditional expectation of U given the covariates a. The computation of these conditional expectations was described in Appendix D.2.

Let $\hat{\theta}(u, v)$ denote the estimate found by the MIPS algorithm as applied to observed statistics (u, v). Then the ME-algorithm is simply described as follows: set $u_0 = u$ and $\theta_0 = \hat{\theta}(u, v)$, then repeat

$$u_{n+1} = u_n + u - E_{\theta_n}(U|a); \quad \theta_{n+1} = \hat{\theta}(u_{n+1}, v)$$

until convergence. In words, the algorithm is based on a running set of adjusted statistics u_n. These are incremented at each step by the quantity $u - E_{\theta_n}(U|a)$, that is, the difference between the observed statistics and their current conditional expectation. Not uncommonly the algorithm as just described diverges; however, a simple modification forces convergence. This uses the increment $\kappa(u - E_{\theta_n}(U|a))$, where κ is a step-halving constant. That is to say, initially at each step the unmodified update is tried (corresponding to $\kappa = 1$). If the conditional likelihood does not increase in this step, κ is halved and the update is attempted. This step-halving process is repeated until a κ is found for which the conditional likelihood increases.

As mentioned, the ME-algorithm resembles the EM-algorithm closely but maximizes the conditional rather than the marginal likelihood. In that and other senses it can be regarded as dual to the EM-algorithm. It is of wide applicability (Edwards and Lauritzen, 1999).

References

[1] Aalen, O.O. (1987). Dynamic modelling and causality. *Scand. Actuar. J.*, 177–190.

[2] Agresti, A. (1984). *Analysis of Ordinal Categorical Data*. Wiley, New York.

[3] Agresti, A. (1990). *Categorical Data Analysis*. Wiley, New York.

[4] Akaike, H. (1974). A new look at the statistical model identification. *IEEE Transactions in Automatic Control* 19: 716–23.

[5] Altman, D.G. (1991). *Practical Statistics for Medical Research*. Chapman and Hall, London.

[6] Andersen, T.W. (1971). *An Introduction to Multivariate Statistical Analysis*. Wiley, New York.

[7] Andersson, S.A., Madigan, D. and Perlman, M.D. (1997). On the Markov equivalence of chain graphs, undirected graphs, and acyclic digraphs. *Scand. J. Statist.* 24: 81–102.

[8] Andersson, S.A., Madigan, D. and Perlman, M.D. (2000). Alternative Markov properties of chain graphs. *Scand. J. Statist.*, to appear.

[9] Anscombe, G.E.M. (1971). *Causation and Determination*, Cambridge University Press. Reprinted in Sosa, E. and Tooley, M. (eds.), 1993. *Causation*. Oxford University Press, Oxford, 88–104.

[10] Asmussen, S. and Edwards, D. (1983). Collapsibility and response variables in contingency tables. *Biometrika* 70, 3: 566–78.

[11] Badsberg, J.H. (1991). A Guide to CoCo. Research Report R 91-43, Institute for Electronic Systems, University of Aalborg.

[12] Besag, J. and Clifford, P. (1991). Sequential Monte Carlo p-values. *Biometrika* 78, 2: 301–4.

[13] Bishop, Y.M., Fienberg, S., and Holland, P. (1975). *Discrete Multivariate Analysis*. M.I.T. Press, Cambridge, Mass.

[14] Box, G.E.P. (1949). A general distribution theory for a class of likelihood criteria. *Biometrika* 36: 317–346.

[15] Box, G.E.P. and Cox, D.R. (1964). An analysis of transformations (with discussion). *J. R. Stat.. Soc. B* 26: 211–50.

[16] Brillinger, D.R. (1996). Remarks concerning graphical models for time series and point processes. *Revista de Econometria* 16: 1-23.

[17] Bronars, S. G. and Grogger, J. (1994). The economic consequences of unwed motherhood: using twin births as a natural experiment. *Amer. Econ. Review* 84: 1141–1156.

[18] Buhl, S.L. (1993). On the existence of maximum likelihood estimators for graphical Gaussian models. *Scand. J. Statist.* 20: 263–270.

[19] Caputo, A., Heinicke, A., and Pigeot, I. (1999). A graphical chain model derived from a model selection strategy for the sociologists graduates study. *Biom. Jour.* 41, 2: 217–234.

[20] Chatfield, C. (1995). Model uncertainty, data mining and statistical inference (with discussion). *J. R. Stat. Soc. A* 158, 3: 419–466.

[21] Christensen, R. (1990). *Log-linear Models.* Springer-Verlag, New York.

[22] Christiansen, S. K. and Giese, H. (1991). Genetic analysis of the obligate barley powdery mildew fungus based on RFLP and virulence loci. *Theor. Appl. Genet.* 79: 705–712.

[23] Cochran, W. G. (1965). The planning of observational studies of human populations (with discussion). *J. R. Stat. Soc. B* 53: 79–109.

[24] Cornfield, J., Haenszel, W., Hammond, C. Lilienfeld, A.M., Shimkin, M.B. and Wynder, E.L. (1959). Smoking and lung cancer: recent evidence and a discussion of some questions. *J. of Nat. Cancer Inst.* 22, 1: 173–203.

[25] Cowell, R.G., Dawid, A.P., Lauritzen, S.L., and Spiegelhalter, D.J. (1999). *Probabilistic Networks and Expert Systems.* Springer-Verlag, New York.

[26] Cox, D.R. and Wermuth, N. (1990). An approximation to maximum likelihood estimates in reduced models. *Biometrika* 77: 747–761.

[27] Cox, D.R. and Wermuth, N. (1992) Response models for mixed binary and quantitative variables. *Biometrika,* 79: 441–61.

[28] Cox, D.R. (1992). Causality: Some statistical aspects. *J. R. Stat. Soc. A.* 155: 291–301.

[29] Cox, D.R. (1993). Causality and graphical models. *Bull. Int. Stat. Inst.* Proceedings 49th Session 1: 365–372.

[30] Cox, D.R. and Wermuth, N. (1993). Linear dependencies represented by chain graphs (with discussion). *Statist. Sci.* 8: 204–218, 247–277.

[31] Cox, D.R. and Wermuth, N. (1994). Tests of linearity, multivariate normality and the adequacy of linear scores. *Appl. Statist.* 43: 347–355.

[32] Cox, D.R. and Wermuth, N. (1996). *Multivariate Dependencies: Models, Analysis and Interpretation.* Chapman and Hall, London.

[33] D'Agostino, R.B. (1998). Propensity score methods for bias reduction in the comparison of a treatment to a non-randomized control group. *Statistics in Medicine* 17: 2265–2281.

[34] Dahlhaus, R. (2000). Graphical interaction models for multivariate time series. *Metrika,* to appear.

[35] Dahlhaus, R., Eichler, M. and Sandkuhler, J. (1997). Identification of synaptic connections in neural ensembles by graphical models. *J. Neuroscience Methods* 77:93-107.

[36] Darroch, J.N., Lauritzen, S.L., and Speed, T.P. (1980). Markov fields and log-linear interaction models for contingency tables. *Ann. Stat.* 8: 522–539.

[37] Dawid, A.P. (1979). Conditional independence in statistical theory (with discussion). *J. R. Stat. Soc. B* 41: 1–31.

[38] Dawid, A.P. and Lauritzen, S.L. (1993). Hyper Markov laws in the statistical analysis of decomposable graphical models. *Ann. Stat.* 21: 1272–1317.

[39] Day, N.E. (1969). Estimating the components of a mixture of normal distributions. *Biometrika* 56: 463–474.

[40] Deming, W.E. and Stephan, F.F. (1940). On a least squares adjustment of a sampled frequency table when the expected marginal totals are known. *Ann. Math. Statist.* 11: 427–444.

[41] Dempster, A.P. (1971). *Elements of Continuous Multivariate Analysis*. Addison Wesley, Reading, Mass.

[42] Dempster, A.P. (1972). Covariance selection. *Biometrics* 28: 157-75.

[43] Dempster, A.P., Laird, N.M. and Rubin, D.B. (1977). Maximum likelihood from incomplete data via the EM algorithm (with discussion). *J. R. Stat. Soc. B* 39: 1–38.

[44] Didelez, V. (1999). Local Independence Graphs for Composable Markov Processes. Research Report no. 158, Sonderforschungsbereich 386, Ludwigs-Maximilians-Universitet, Munich.

[45] Edwards, D. and Kreiner, S. (1983). The analysis of contingency tables by graphical models. *Biometrika* 70: 553-65.

[46] Edwards, D. (1984). A computer intensive approach to the analysis of sparse multidimensional contingency tables. *COMPSTAT '84: Proceedings in Computational Statistics* (Havránek, T., Sidak, Z., and Novak, M., eds.). Physica-Verlag, Vienna, 355–360.

[47] Edwards, D. (1990). Hierarchical interaction models (with discussion). *J. R. Stat. Soc. B* 52: 3–20.

[48] Edwards, D. (1992). Linkage analysis using loglinear models. *Comp. Stat. and Data Anal.*, 13: 281–290.

[49] Edwards, D. (1993). Some computational aspects of graphical model selection. In Antoch, J. (ed.), *Computational Aspects of Model Choice*, Physica-Verlag, Heidelberg.

[50] Edwards, D. (1995). Graphical modelling. In Krzanowski, W.J. (ed.) *Recent Advances in Descriptive Multivariate Analysis*, Oxford University Press, Oxford, 127–148.

[51] Edwards, D. (1999). On model prespecification in confirmatory randomized studies. *Statist. Med.* 18: 771–785.

[52] Edwards, D. and Havránek, T. (1985). A fast procedure for model search in multidimensional contingency tables. *Biometrika* 72: 339–351.

[53] Edwards, D. and Havránek, T. (1987). A fast model selection procedure for large families of models. *J. Amer. Stat. Assoc.* 82: 205–213.

[54] Edwards, D. and Lauritzen, S. L. (1999). The ME Algorithm for Maximizing a Conditional Likelihood Function. Research Report R-99-2015, Univ. of Aalborg.

[55] Eells, E. (1991). *Probabilistic Causality*. Cambridge University Press.

[56] Eerola, M. (1994). *Probabilistic Causality in Longitudinal Studies*. Springer-Verlag, New York.

[57] Eriksen, P.S. (1996). Tests in covariance selection models. *Scand. J. Statist.* 23: 275–284.

[58] Everitt, B.S. (1977). *The Analysis of Contingency Tables*. Chapman and Hall, London.

[59] Farewell, V.T. (1998). Hidden Markov Models. In Armitage, P.and Colton, T. (1998). *Encyclopedia of Biostatistics* 3: 1908–1916.

[60] Fienberg, S.E. (1980). *The Analysis of Cross-Classified Categorical Data*. MIT Press.

[61] Freedman, D., Pisani, R., and Purves, R. (1978). *Statistics*. Norton, New York.

[62] Freeman, G. H. and Halton, J. H. (1951). Note on an exact treatment of contingency, goodness of fit and other problems of significance. *Biometrika* 38: 141–149.

[63] Frydenberg, M. and Edwards, D. (1989). A modified iterative scaling algorithm for estimation in regular exponential families. *Comp. Stat. and Data Anal.* 8: 142–153.

[64] Frydenberg, M. and Lauritzen, S.L. (1989). Decomposition of maximum likelihood in mixed graphical interaction models. *Biometrika* 76: 539–55.

[65] Frydenberg, M. (1989). The chain graph Markov property. *Scand. J. Statist.* 17: 333–353.

[66] Frydenberg, M. (1990). Marginalization and collapsibility in graphical interaction models. *Annals of Stat.* 18: 790–805.

[67] Gabriel, K.R. (1969). Simultaneous test procedures: Some theory of multiple comparisons. *Ann. Math. Stat.* 40: 224–250.

[68] Gail, M. and Mantel, N. (1977). Counting the number of $r \times c$ contingency tables with fixed margins. *J. Amer. Stat. Assoc.* 72: 859–862.

[69] Galles, D. and Pearl, J. (1997). An Axiomatic Characterisation of Causal Counterfactuals. Technical Report, Cognitive Systems Laboratory, U.C.L.A.

[70] Geiger, D. and Pearl, J. (1988). On the logic of influence diagrams. In *Proceedings of 4th workshop on uncertainty in artificial intelligence*, Minneapolis, Minnesota, 136–47.

[71] Geiger, D. and Pearl, J. (1993). Logical and algorithmic properties of conditional independence and graphical models. *Ann. Stat.* 21: 2001–2021.

[72] Ghosh, J.K and Sen, P.K. (1985). On the asymptotic performance of the log likelihood ratio statistic for the mixture model and related results. In *Proceedings of the Berkeley Conference in honour of Jerzy Neyman and Jack Kiefer, Vol. II*, Wadsworth.

[73] Gibbs, W. (1902). *Elementary Principles of Statistical Mechanics*. Yale University Press.

[74] Glonek, G.F.V., Darroch, J.N., and Speed, T.P. (1988). On the existence of maximum likelihood estimators for hierarchical loglinear models. *Scand. J. Statist.* 15: 187–193.

[75] Good, I.J. (1961). A Causal Calculus, I-II. *British Journal for the Philosophy of Science*, 11: 305–318, 12: 43–51.

[76] Goodman, L.A. (1970). The multivariate analysis of qualitative data: Interactions among multiple classifications. *J. Amer. Stat. Assoc.* 65: 226–56.

[77] Goodman, L.A. (1973a). Causal analysis of data from panel studies and other kinds of surveys. *Amer. J. Sociol.* 78: 1135-1191.

[78] Goodman, L.A. (1973b). The analysis of multidimensional contingency tables when some variables are posterior to others: A modified path analysis approach. *Biometrika* 60: 179–92.

[79] Goodman, L.A. (1981). Association models and canonical correlation in the analysis of cross-classifications having ordered categories. *J. Amer. Stat. Assoc.* 76: 320–334.

[80] Greenland, S. (1998). Basic methods for sensitivity analysis and external adjustment. Ch. 19: Rothman, K.J., Greenland, S. eds. *Modern Epidemiology.* Lippincott-Raven, Philadelphia, 343–358.

[81] Greenland, S., Pearl, J. and Robins, J.M. (1999). Causal diagrams for epidemiological research. *Epidemiology* 10, 1: 37–48.

[82] Halkin, H., Sheiner, L.B., Peck, C.C. and Melmon, K.L. (1975). Determinants of the renal clearance of digoxin. *Clin. Pharmocol. Theor.* 17: 385–394.

[83] Hand, D. (1992). On comparing two treatments. *The American Statistician* 46: 190–192.

[84] Havránek, T. (1987). Model search in large model families. *Proceedings of the First World Congress of the Bernoulli Society* (Prokhoruv, Yu. A. and Sazonov, V.V., eds.), vol. 2, VNU Sci. Press, Utrecht, 767–778.

[85] Holland, P.W. (1986). Statistics and causal inference (with discussion). *J. Amer. Stat. Assoc.* 81, 396: 945–970.

[86] Horáková, M. (1991). Implementation of fast model selection for graphical mixed interaction models. *Comp. Stat. Quar.* 6: 99–111.

[87] Hotelling, H. (1931). The generalization of Student's ratio. *Ann. Math. Stat.* 2: 360–378.

[88] Hume, D. (1748). *An Enquiry concerning Human Understanding.*

[89] Hosmer, D.W. and Lemeshow, S. (1989). *Applied Logistic Regression,* Wiley, New York.

[90] Højsgaard, S. (1998). Split Models for Contingency Tables, PhD dissert., Biometry Research Unit, Danish Institute of Agricultural Sciences.

[91] Ireland, G.T and Kullback, S. (1968). Contingency tables with given margins. *Biometrika* 55: 179–188.

[92] Jöreskog, K.G. and Sörbom, D. (1989). *LISREL 7. A Guide to the Program and Applications* 2nd ed. SPSS Inc, Chicago.

[93] Kalviainen, R., Brodie, M.J., Duncan, J., Chadwick, D., Edwards, D., and Lyby, K. (1998). A double-blind, placebo-controlled trial of tiagabine given 3-

times daily as add-on therapy for refractory partial seizures. *Epilepsy Research*, 30, 1: 31–40.

[94] Kauermann, G. (1996). On a Dualization of Graphical Gaussian Models. *Scand. J. Statist.* 23: 105–116.

[95] Kiiveri, H. and Speed, T.P. (1982). In S. Leinhardt (ed.), Structural analysis of multivariate data: A review. *Sociological Methodology*. Jossey-Bass, San Francisco.

[96] Kiiveri, H., Speed, T.P., and Carlin, J.B. (1984). Recursive causal models. *Journ. Austr. Math.* 36: 30–52.

[97] Kim, J. (1971). Causes and events: Mackie on causation. *Journal of Philosophy* 60: 426-41. Reprinted in Sosa, E. and Tooley, M. (eds.), 1993. *Causation*. Oxford University Press, Oxford, 60–74.

[98] Koch, G.G., Amara, J., Atkinson, S., and Stanish, W. (1983). Overview of categorical analysis methods. *SAS-SUGI* 8: 785–794.

[99] Koster, J.T.A. (1996). Markov properties of nonrecursive causal models, *Ann. Stat.*, 24, 5: 2148–77.

[100] Koster, J.T.A. (1996). On the validity of the Markov interpretation of path diagrams of Gaussian structural equations systems with correlated errors, *Scand. J. Statist.* 26: 413–31.

[101] Kreiner, S. (1987). Analysis of multidimensional contingency tables by exact conditional tests: Techniques and strategies. *Scand. J. Statist.* 14: 97–112.

[102] Kreiner, S. (1989). Graphical modelling using DIGRAM. Research report 89/11, Statistical Research Unit, Univ. of Copenhagen.

[103] Krusinka, E.M. (1992). Discriminant analysis in graphical and hierarchical interaction models. In *Statistical Modelling*. van der Heijden, P.G.M., Francis, B., and Seeber, E.U.H. (eds.) North-Holland, Amsterdam, 175–184.

[104] Krzanowski, W.J. (1975). Discrimination and classification using both binary and continuous variables. *J. Amer. Stat. Assoc.* 70: 782-90.

[105] Krzanowski, W.J. (1980). Mixtures of continuous and categorical variables in discriminant analysis. *Biometrics* 36: 493–9.

[106] Krzanowski, W.J. (1988). *Principles of Multivariate Analysis*. Clarendon Press, Oxford.

[107] Lamport, L. (1986). *LaTeX, A Document Preparation System*. Addison-Wesley, Massachusetts.

[108] Lauritzen, S.L. (1989a). Lectures on Contingency Tables (3rd ed.). Technical Report R-89-29, Institute for Electronic Systems, Aalborg Univ.

[109] Lauritzen, S.L. (1989b). Mixed graphical association models (with discussion). *Scand. J. Statist.* 16: 273–306.

[110] Lauritzen, S.L. (1992). Graphical association models. Draft Version. Research Report, Institute for Electronic Systems, University of Aalborg.

[111] Lauritzen, S.L. (1996). *Graphical Models*. Clarendon Press, Oxford.

[112] Lauritzen, S.L. (1999). Causal inference from graphical models. Research Report R-99-2021, Institute for Electronic Systems, Aalborg Univ.

[113] Lauritzen, S.L., Dawid, A.P., Larsen, B.N., and Leimer, H.-G. (1990). Independence properties of directed Markov fields. *Networks* 20: 491–505.

[114] Lauritzen, S.L. and Spiegelhalter, D.J. (1988). Local computations with probabilities on graphical structures and their application to expert systems (with discussion). *J. R. Stat. Soc. B* 50: 157–224.

[115] Lauritzen, S.L. and Wermuth, N. (1989). Graphical models for associations between variables, some of which are qualitative and some quantitative. *Ann. Stat.* 17: 31–57.

[116] Lehmann, E.L. (1975). *Non-parametrics: Statistical Methods Based on Ranks.* Holden-Day, San Francisco.

[117] Leimer, H.-G. (1989). Triangulated graphs with marked vertices. In *Graph Theory in Memory of G. A. Dirac*, (ed. L.D. Andersen, C. Thomassen, B. Toft, and P.D. Vestergaard), pp. 311–24. Elsevier Science Publishers B.V. (North-Holland), Amsterdam. *Annals of Discrete Mathematics, Volume 41.*

[118] Lewis, D. (1973). Causation. *Journal of Philosophy* 70: 556–67.

[119] Lingjærde, O., Ahlfors, U.G., Bech, P., Densker, S.J., and Elgen, K. (1987). The UKU side effect rating scale. *Acta. Psych. Scand.* 334 supplementum.

[120] Lynggaard, H. and Walther, K.H. (1993). Dynamic Modelling with Mixed Graphical Association Models. Research Report, Institute for Electronic Systems, University of Aalborg.

[121] Little, R.J.A. and Schluchter, M.D. (1985). Maximum likelihood estimation for mixed continuous and categorical data with missing values. *Biometrika* 72: 497–512.

[122] Mackie, J.L. (1965). Causes and conditions. *American Philosophical Quarterly* 2/4: 245–55 and 261–4. Reprinted in Sosa, E. and Tooley, M. (eds.), 1993. *Causation.* Oxford University Press, Oxford, 33–55.

[123] Marcus, R., Peritz, E. and Gabriel, K.R. (1976). On closed testing procedures with special reference to ordered analysis of variance. *Biometrika* 63: 655–660.

[124] Mardia, K.V., Kent, J.T, and Bibby, J.M. (1979). *Multivariate Analysis.* Academic Press.

[125] McCullagh, P. and Nelder, J.A. (1989). *Generalized Linear Models*, 2nd ed. Chapman and Hall, London.

[126] McFadden, J.A. (1955). Urn models of correlation. *Ann. Math. Statist.* 26: 478–489.

[127] McLachlan, G.F. and Basford, K.E. (1987). *Mixture Models.* Marcel Dekker, New York.

[128] Mehta, C. and Patel, N. (1983). A network algorithm for performing Fisher's exact test in $r \times c$ contingency tables. *J. Amer. Stat. Assoc.* 78: 427–434.

[129] Mehta, C. and Patel, N. (1991). *StatXact User Manual*, Version 2. Cytel Software Corporation, Cambridge MA.

[130] Mohamed, W.N., Diamond, I.D. and Smith, P.W.F. (1998). The determinants of infant mortality in Malaysia: A graphical chain modelling approach. *J. R. Stat. Soc. A* 161: 349-366.

[131] Morrison, D.F. (1976). *Multivariate Statistical Methods.* McGraw-Hill.

[132] Muirhead, R.J. (1982). *Aspects of Multivariate Statistical Theory.* Wiley.

[133] Munk-Jensen, N., Ulrich, L.G., Obel, E.B., Pors Nielsen, S., Edwards, D., and Meinertz, H. (1994). Continuous combined and sequential estradiol-norethindrone acetate treatment of postmenopausal women: Effect on plasma lipoproteins in a two-year placebo-controlled trial. *Am. J. Obstet. Gyn.* 171: 132–138.

[134] Murphy, B.P., Rohl, J.S., and Cribb, R.L. (1986). Recursive techniques in statistical programming. *COMPSTAT 1986 Proceedings*, 338–344.

[135] Murphy, B.P. (1981). *The LOLITA Program Manual.* Raine Medical Statistics Unit, University of Western Australia.

[136] Neil-Dwyer, G., Lang, D.A., Smith, P.W.F. and Iannotti, F. (1998). Outcome after aneurysmal subarachnoid haemorrhage: The use of a graphical model in the assessment of risk factors. *Acta Neurochirurgica* 140: 1019–1027.

[137] Neyman, J. (1923). On the application of probability theory to agricultural experiments. Essay on Principles, Section 9. Transl. (1990) in *Statist. Sci.* 5: 465–80.

[138] Norušis, M.J. (1988). *SPSS^X Advanced Statistics Guide*, 2nd. ed. McGraw-Hill, New York.

[139] Patefield, W.M. (1981). Algorithm AS 159: An efficient method of generating random $r \times c$ tables with given row and column totals. *App. Stat.* 30: 91–7.

[140] Pearl, J. (1986a). A constraint propagation approach to probablistic reasoning. In *Uncertainty in Artificial Intelligence*, (ed. L.M. Kanal and J. Lemmer), pp. 357–70. North-Holland, Amsterdam.

[141] Pearl, J. (19986b). Fusion, propagation and structuring in belief networks. *Artificial Intelligence*, 29: 241–88.

[142] Pearl, J. and Paz, A. (1987). Graphoids. A graph-based logic for reasoning about relevancy relations. In *Advances in Artificial Intelligence-II* (ed. B.D. Boulay, D. Hogg and L. Steel), 357–63. North-Holland, Amsterdam.

[143] Pearl, J. and Verma, T.S. (1987). The logic of representing dependencies by directed graphs. In *Proceedings of 6th Conference of the American Association of Artificial Intelligence*, 374–9. American Association of Artificial Intelligence.

[144] Pearl, J. (1988). *Probabilistic Inference in Intelligent Systems.* Morgan Kaufman, San Mateo.

[145] Pearl, J. (1993). Aspects of graphical models connected with causality. *Bull. Int. Stat. Inst., Proceedings 49th Session* 1: 391–403.

[146] Pearl, J. (1995a). Causal Diagrams for Empirical Research (with discussion). *Biometrika* 82, 4: 669–710.

[147] Pearl, J. (1995b). Causal Inference from Indirect Experiments. *Artificial Intelligence in Medicine* 7: 561–582.

[148] Pearl, J. and Verma, T. (1991). A theory of inferred causation. In J.A. Allen, R. Fikes and E. Sandewell, (ed.) *Principles of Knowledge Representation and Reasoning: Proc. 2nd Int. Conf.*, 441–452, San Mateo, CA. Morgan Kaufmann.

[149] Pearl, J. (1996). Structural and Probabilistic Causality. *Psychology of Learning and Motivation* 34: 393–435.

[150] Pearl, J. (2000). *Causality.* Cambridge Univ. Press, to appear.

[151] Pirie, W. (1983). Jonckheere tests for ordered alternatives. In *Encyclopaedia of Statistical Sciences* (Kotz, S. and Johnson, N.L., eds) 4: 315–318, Wiley, New York.

[152] Porteous, B.T. (1989). Stochastic inequalities relating a class of log likelihood ratio statistics to their asymptotic χ^2 distribution. *Ann. Stat.* 17: 1723–1734.

[153] Radelet, M. (1981). Racial characteristics and the imposition of the death penalty. *Amer Sociol. Rev.* 46: 918–927.

[154] Reinis, Z., et al. (1981). Prognostic significance of the risk profile in the prevention of coronary heart disease. (in Czech). *Bratis. Lek. Listy* 76: 137–150.

[155] Robertson, T. and Wright, F.T. (1981). Likelihood ratio tests for and against a stochastic ordering between multinomial populations. *Ann. Stat.*, 9: 1248–1257.

[156] Robins, J. M. (1986). A new approach to causal inference in mortality studies with a sustained exposure period — applications to control of the healthy worker effect. *Math. Model.* 7: 1393–512.

[157] Rosenbaum, P.R. and Rubin, D.B. (1983). The central role of the propensity score in observational studies for causal effects. *Biometrika* 70, 1: 41–55.

[158] Rosenbaum, P.R. (1984). From association to causation in observational studies: The role of tests of strongly ignorable treatment assignment. *J. Amer. Stat. Assoc.* 79, 385: 41–48.

[159] Rosenbaum, P.R. and Rubin, D.B. (1984). Reducing bias in observational studies using subclassification on the propensity score. *J. Amer. Stat. Assoc.* 79, 387: 516–524.

[160] Rosenbaum, P.R. (1995). *Observational Studies.* Springer-Verlag, New York.

[161] Rosenbaum, P.R. (1999). Choice as an alternative to control in observational studies (with discussion). *Statistical Science* 14, 3: 259–304.

[162] Roverato, A. and Whittaker, J. (1993). Standard Errors for Parameters of Graphical Gaussian Models. Technical Report, Univ. of Lancaster.

[163] Rubin, D.B. (1974). Estimating causal effects of treatments in randomized and non-randomized studies. *J. Educ. Psychol.* 66: 688–701.

[164] Rubin, D.B. (1976). Inference and missing data (with discussion). *Biometrika* 63: 581–92.

[165] Rubin, D.B. (1978). Bayesian inference for causal effects: the role of randomisation. *Ann. Statist.* 7: 34–58.

[166] Ruggeri, M., Biggeri, A., Rucci, P., and Tansella, M. (1998). Multivariate analysis of outcome of mental health care using graphical chain models. *Psychol. Med.* 28, 6: 1421–31.

[167] Salmon, W. (1980). Causality: Production and propagation. *Proceedings of the 1980 Biennial Meeting of the Philosophy of Science* vol. 2, ed. Asquith, P.D. and Giere, R.N., 46–69. Reprinted in Sosa, E. and Tooley, M (eds). *Causation.* Oxford University Press, Oxford, 154–171.

[168] Schoener, T.W. (1968). The *anolis* lizards of Bimini: Resource partitioning in a complex fauna. *Ecology* 49: 704–726.

[169] Schultz-Larsen, K., Avlund, K. and Kreiner, S. (1992). Functional ability of community dwelling elderly: Criterion-related validity of a new measure of functional ability. *J. Clin. Epidem.* 45: 1315–1326.

[170] Schwarz, G. (1978). Estimating the dimension of a model. *Ann. Statist.* 6: 461–4.

[171] Schweder, T. (1970). Composable Markov Processes. *J. Appl. Prob.* 7: 400–10.

[172] Seber, G.A.F. (1984). *Multivariate Observations.* Wiley, New York.

[173] Shapiro, A. (1986). Asymptotic theory of overparameterized structural models, *J. Amer. Stat. Assoc.* 81, 393: 142–149.

[174] Silvey, S.D. (1970). *Statistical Inference.* Chapman and Hall, London.

[175] Simpson, C.H. (1951). The interpretation of interaction in contingency tables. *J. R. Stat. Soc. B* 13: 238–41.

[176] Smith, J. Q. (1989). Influence diagrams for statistical modelling. *Ann. Stat.* 17: 654–72.

[177] Smith, P.W.F. (1992). Assessing the power of model selection procedures used when graphical modelling. In Dodge, Y. and Whittaker, J. (eds.) *Computational Statistics, Proceedings, Vol I*: 275–280. Physica-Verlag, Heidelberg.

[178] Sobel, M.E. (1996). An Introduction to Causal Inference. *Soc. Meth. and Res.* 24, 3: 353–379.

[179] Sosa, E. and Tooley, M. (eds). *Causation.* Oxford University Press, Oxford.

[180] Speed, T.P and Kiiveri, H.T. (1986). Gaussian Markov distributions over finite graphs. *Ann. Stat.* 14: 138–150.

[181] Spiegelhalter, D.J., Dawid, A.P., Lauritzen, S.L., and Cowell, R.G. (1993). Bayesian analysis in expert systems (with discussion). *Statist. Sci.* 8: 219–283.

[182] Spielberger, C.D., Gorsuch, R.L., and Luschene, R.E. (1970). *Manual for the State-Trait Anxiety Inventory.* Consulting Psychologists Press, Palo Alto, CA.

[183] Spielberger, C.D., Russell, S., and Crane, R. (1983). Assessment of anger. In J.N. Butcher and C.D. Spielberger (eds.), *Advances in Personality Assessment* 2: 159–187. Erlbaum, Hillsdale.

[184] Spirtes, P., Glymour, C. and Scheines, R. (1993). *Causation, Prediction and Search.* Springer-Verlag, New-York.

[185] Spirtes, P., Richardson, T., Meek, C., Scheines, R. and Glymour, C. (1998). Using path diagrams as a structural equations tool. *Sociological Methods and Research*, 27, 182–225.

[186] Stanghellini, E. (1987). Identification of a single-factor model using graphical Gaussian rules. *Biometrika* 84: 241–4.

[187] Stanghellini, E., McConway, K.J. and Hand, D.J. (1999). A discrete variable chain graph for applicants for credit. *Appl. Stat.* 48: 239–251.

[188] Stone, R. (1993). The assumptions on which causal inferences rest. *J. R. Stat. Soc. B* 55: 455–466.

[189] Strotz, R.H. and Wold, H.O.A. (1960). Recursive versus nonrecursive systems: An attempt at synthesis. *Econometrika* 28: 417–27.

[190] Suppes, P. (1970). *A Probabilistic Theory of Causation*. North-Holland, Amsterdam.

[191] Suzuki, D.T., Griffiths, A.J.F., Miller, J.H., and Lewontin, R.C. (1989). *An Introduction to Genetic Analysis*. W.H. Freeman, New York.

[192] van Elteren, P.H. (1960). On the combination of independent two-sample tests of Wilcoxon. *Bull. Inst. Intern. Statist.* 37: 351–361.

[193] Verma, T. and Pearl, J. (1990a). Causal Networks: Semantics and expressiveness. In *Uncertainty in artificial intelligence IV*, (ed. R.D. Scachter, T.S. Levitt, L.N. Kanal, and J.F. Lemmer), pp. 69–76. North-Holland, Amsterdam.

[194] Verma, T. and Pearl, J. (1990b). Equivalence and synthesis of causal models. In *Proceedings of the 6th conference on uncertainty in artificial intelligence*, (ed. P. Bonissone, M. Henrion, L.N. Kanal and J.F. Lemmer), pp. 255–70. North-Holland, Amsterdam.

[195] Vicard, P. (2000). On the identification of a single-factor model with correlated residuals. *Biometrika*, to appear.

[196] Wermuth, N. (1976). Model search among multiplicative models. *Biometrics* 32: 253–263.

[197] Wermuth, N. (1987). Parametric collapsibility and the lack of moderating effects in contingency tables with a dichotomous response variable. *J. R. Stat. Soc. B.* 49: 353–364.

[198] Wermuth, N. (1989). Moderating effects in multivariate normal distributions. *Methodika* III: 74–93.

[199] Wermuth, N. (1990). Association models with few variables: characteristics and examples. In: K. Dean (ed.), *Population Health Research*, 181–203. Sage, London.

[200] Wermuth, N. (1992). On block-recursive linear regression equations (with discussion). *Brazilian J. of Prob. and Stat. (REBRAPE)* 6: 1–56.

[201] Wermuth, N. and Lauritzen, S.L. (1983). Graphical and recursive models for contingency tables. *Biometrika* 70: 537–552.

[202] Wermuth, N. and Lauritzen, S.L. (1990). On substantive research hypotheses, conditional independence graphs and graphical chain models (with discussion). *J. R. Stat. Soc. B* 52: 21–72.

[203] Wermuth, N., Cox, D.R., and Pearl, J. (1994). Explanations for multivariate structures derived from univariate recursive structures. Technical Report, Johannes Gutenberg University, Mainz.

[204] Wermuth, N., Wehner, T., and Gönner, H. (1976). Finding condensed descriptions for multi-dimensional data, *Computer Programs in Biomedicine* 6: 23–38.

[205] Whittaker, J. (1990). *Graphical Models in Applied Multivariate Statistics*. Wiley.

[206] Whittemore, A.S. (1978). Collapsibility of multidimensional contingency tables. *J. R. Stat. Soc. B* 40: 328–40.

[207] Wilkinson, G. (1970). A general recursive procedure for the analysis of variance. *Biometrika*, 57, 19–46.

[208] Wright, S. (1921). Correlation and causation. *J. Agric. Res.* 20: 557–585.

Index

Springer Texts in Statistics *(continued from page ii)*